Culture, Space and Climate Change

Ways of handling climate change vary worldwide. Differences can be observed in the perception of potential threats and opportunities as well as in the appraisal of adequate coping strategies. Collective efforts often fail not because of technical restrictions, but as a result of social and cultural differences between the actors involved. Consequently, there is a need to explore in greater depth those zones of cultural friction which emerge when actors deal with climate change.

This book examines how cultural differences in the handling of climate change can be described and explained. The work develops the concept of culture as relational space, elaborates explanatory approaches, and investigates them by surveying more than 800 actors responsible for spatial development of the European coastal regions in the Netherlands, Germany, Denmark, and Poland. In doing so, this book engages with debates on cultural globalisation, in which the attachment of culture to place is increasingly being questioned. Adopting the approach of culture as relational space allows possible cultural formations to be examined across diverse fields of application from the local to the global scale. In addition, the book investigates how far different value orientations, beliefs, and identities can explain diverse perceptions of problems and opportunities right up to preferences for climate-mitigation and adaptation measures.

Providing comprehensive insights into the diverse zones of cultural friction which scholars and practitioners face when handling climate change locally and globally, this book will be of great interest to those studying climate change, environmental sociology, and sustainable planning.

Thorsten Heimann is a research associate at the Leibniz Institute for Research on Society and Space in Erkner/Berlin. He holds a PhD in sociology from the Freie Universität Berlin, Germany.

Routledge Advances in Climate Change Research

The Global Climate Regime and Transitional Justice
Sonja Klinsky and Jasmina Brankovic

Climate Justice and the Economy
Social Mobilization, Knowledge and the Political
Edited by Stefan Gaarsmand Jacobsen

A Critical Approach to Climate Change Adaptation
Discourses, Policies and Practices
Edited by Silja Klepp and Libertad Chavez-Rodriguez

Contemplating Climate Change
Mental Models and Human Reasoning
Stephen M. Dark

Climate Change, Moral Panics and Civilization
Amanda Rohloff
Edited by André Saramago

Climate Change and Social Inequality
The Health and Social Costs of Global Warming
Merrill Singer

Cities Leading Climate Action
Urban Policy and Planning
Sabrina Dekker

Culture, Space and Climate Change
Vulnerability and Resilience in European Coastal Areas
Thorsten Heimann

For more information about this series, please visit:
www.routledge.com/Routledge-Advances-in-Climate-Change-Research/book-series/RACCR

Culture, Space and Climate Change

Vulnerability and Resilience in European Coastal Areas

Thorsten Heimann

Translated by Jonathan Styles

Routledge
Taylor & Francis Group
LONDON AND NEW YORK

earthscan
from Routledge

First published 2019
by Routledge
2 Park Square, Milton Park, Abingdon, Oxon OX14 4RN

and by Routledge
52 Vanderbilt Avenue, New York, NY 10017

First issued in paperback 2020

Routledge is an imprint of the Taylor & Francis Group, an informa business

Translation from the German language edition:
Klimakulturen und Raum. Umgangsweisen mit Klimawandel an europäischen Küsten by Thorsten Heimann Copyright © Springer VS 2017. Springer Fachmedien Wiesbaden GmbH is part of Springer Nature. All Rights Reserved.

British Library Cataloguing-in-Publication Data
A catalogue record for this book is available from the British Library

Library of Congress Cataloging-in-Publication Data
Names: Heimann, Thorsten, author.
Title: Culture, space and climate change : vulnerability and resilience in European coastal areas / Thorsten Heimann ; translated by Jonathan Styles.
Other titles: Klimakulturen und Raum. English
Description: Abingdon, Oxon ; New York, NY : Routledge, 2019. | Series: Routledge advances in climate change research | Includes bibliographical references and index.
Identifiers: LCCN 2018039400 (print) | LCCN 2018049726 (ebook) | ISBN 9780429436659 (Master) | ISBN 9780429791611 (Adobe Reader) | ISBN 9780429791604 (ePub) | ISBN 9780429791598 (Mobipocket unencrypted) | ISBN 9781138348455 (hardback) | ISBN 9780429436659 (ebook)
Subjects: LCSH: Climatic changes–Social aspects–Europe. | Climate change mitigation–Social aspects–Europe. | Integrated coastal zone management–Europe. | Coastal ecosystem health–Europe.
Classification: LCC QC903.2.E85 (ebook) | LCC QC903.2.E85 H4513 2019 (print) | DDC 363.738/747094–dc23
LC record available at https://lccn.loc.gov/2018039400

ISBN 13: 978-0-367-58281-4 (pbk)
ISBN 13: 978-1-138-34845-5 (hbk)

Typeset in Goudy
by Wearset Ltd, Boldon, Tyne and Wear

Contents

Figures

Tables

Note on translation

Translated by Jonathan Styles

Funded by

Federal Ministry of Education and Research

IRS Leibniz Institute for Research on Society and Space

PROGRESS
Potsdam Research Cluster for Georisk Analysis, Environmental Change and Sustainability

Leibniz Institute for Research on Society and Space (IRS) part funded the translation of this work.

Acknowledgements

This study explores cultural spaces as relational spaces, an approach that might also offer something in the attempt to describe contemporary academic communities. On the way to finalising this work, many people around the world accompanied and supported me. I would like to offer my sincere thanks to all those colleagues and friends, locally and globally, who accompanied and supported me in completing this work.

Special thanks go to the Leibniz Institute for Research on Society and Space for providing a grant in aid of the translation of this work. The German Federal Ministry of Education and Research supported the 'Potsdam Research Cluster for Georisk Analysis, Environmental Change, and Sustainability' (PROGRESS) which enabled my collection of data.

For the translation of this study from the German original into English I am greatly thankful to Jonathan Styles. He achieved this with high competence, patience, and reliability. Theresa Kim helped me to edit the final version. Many thanks also to the unknown reviewers who supported me with their helpful hints.

I would also like to thank the many politicians, administrators, economic representatives, non-governmental organisations, and academics who donated their precious time to answer my questions. They became the protagonists of this project and made its success a possibility.

Finally I would like to express my deep gratitude to my parents, family and friends, from whom I received unconditional support at all stages of my work, for their good advice and words of encouragement, as well as for their supply of marzipan and green tea. Finally, I would like to thank Daniela.

Abbreviations

ARK	Nationaal programma Adaptatie ruimte en klimaat (Dutch National Programme for Spatial Adaptation to Climate Change)
ARL	Akademie für Raumforschung und Landesplanung (Academy for Spatial Research and Planning)
BauGB	Baugesetzbuch der Bundesrepublik Deutschland (Building Code of the Federal Republic of Germany)
BMBF	Bundesministerium für Bildung und Forschung (Germany Federal Ministry of Education and Research)
BMUB	Bundesministerium für Umwelt, Naturschutz, Bau und Reaktorsicherheit (German Federal Ministry for the Environment, Nature Conservation, Building, and Nuclear Safety)
CCS	Carbon Capture and Storage
CFC	chlorofluorocarbon
CHP	combined heat and power
COP	Conference of the Parties
DAS	Deutsche Anpassungsstrategie an den Klimawandel (German Strategy for Adaptation to Climate Change)
EC	European Commission
ECCP	European Climate Change Programme
EEA	European Environment Agency
EEG	Erneuerbare Energien Gesetz (German Renewable Energy Sources Act)
EU	European Union
EU ETS	European Union Emissions Trading System
Fr.	French
GARP	Global Atmospheric Research Programme
GDP	gross domestic product
GEMS	Global Environmental Monitoring System
GHG	greenhouse gas
HEP	Human Exceptionalism Paradigm
IMWG	Interministerial Working Group

IOŚ-PIB	Instytut Ochrony Środowiska – Państwowy Instytut Badawczy (Polish Institute for Environmental Protection – National Research Institute)
IPCC	Intergovernmental Panel on Climate Change
KWI	Kulturwissenschaftliches Institut Essen (Cultural Sciences Institute, Essen)
Land	One of the 16 federal states of Germany (plural: Länder)
LPT	local public transport
LTA	long-term agreement
LULUCF	Land use, land-use change, and forestry
MKRO	Ministerkonferenz für Raumordnung (Ministerial Conference for Spatial Development)
NAS	Nationale Adaptatiestrategie (Dutch National Adaptation Strategy)
NEP	New Environmental Paradigm
NEPP	National Environmental Policy Plan of the Netherlands
NGO	non-governmental organisation
PLN	Polish złoty
pos.	position
RGW	Rat für gegenseitige Wirtschaftshilfe (Council for Mutual Economic Assistance)
ROG	Raumordnungsgesetz der Bundesrepublik Deutschland (Spatial Development Law of the Federal Republic of Germany)
SD	standard deviation
SPA 2020	Strategiczny plan adaptacji 2020 (Polish Adaptation Strategy 2020)
TEU	Treaty on the European Union
TFEU	Treaty on the Functioning of the European Union
UN	United Nations
UNEP	United Nations Environment Programme
UNFCCC	United Nations Framework Convention on Climate Change
VROM	Ministerie van Volkshuisvesting, Ruimtelijke, Ordening en Mileubeheer (Dutch Ministry for Housing, Spatial Planning and Environment)
WBGU	Wissenschaftlicher Beirat der Bundesregierung Globale Umweltveränderungen (German Advisory Council on Global Change)
WCP	World Climate Programme
WMO	World Meteorological Organisation
WG I, II, III	Working Groups of the Intergovernmental Panel on Climate Change

1 Introduction

Culture, space, and differences in handling climate change

This study explores the ways of handling climate change on the North and Baltic Sea coasts, namely those of Germany, the Netherlands, Denmark, and Poland. Global discourse on human-made climate change was established, at the latest, at the implementation of the Intergovernmental Panel on Climate Change (IPCC) by the United Nations in 1988. While in its early phase discourse was initially restricted to various scientific actors, it manifested diverse resonances in such other fields as politics and administration, media, economics, and civil society (see Weingart *et al.* 2002). At the present point in time, it is not only similarities but also significant variations that are observable, for example with regard to the specific focus of attention being paid by various actors, confidence in the findings of scientific research, or the assessment of opportunities and threats. These phenomena are also observable on the coasts of Europe.

Since 2000, in the aftermath of global efforts at the United Nations, the European Union has been developing its own programme for the handling of climate change: the European Climate Change Programme (ECCP). First, within this framework, the EU member state actors involved drafted common measures for the reduction of greenhouse gas emissions.[1] Since 2005 they have focused, in addition, on adaptation to the possible impacts of climate change. Among other things, the member states have been earmarking funding for handling sea-level rise[2] along European coastlines (see Wilson and Piper 2010: 52ff. and Chapter 6).

By 2010, in the German federal state of Schleswig-Holstein, much construction work was already being carried out along the dykes of the North and Baltic Sea coasts, explicitly in the context of sea-level rise related to climate change. At the same time, a small step across the border into the neighbouring Land of Mecklenburg-Vorpommern seemed to mark a step into a world of different problems. Here, the responsible coastal protection agency presented any possible threat in starkly different terms, despite the similar physical constitution of the location. The 'Regulations for Coastal Protection of Mecklenburg-Vorpommern' posited that there was no acute requirement to act in relation to a possible sea-level rise related to climate change (MLUV 2009: 99).

Differences between actors have also been observed across national borders. Referring to Poland, Raabe and Härtel observed in direct connection with the publication of the forth IPCC report in 2007: 'Climate change is discussed in Polish politics and media only rarely, and is not perceived as an acute threat' (Raabe and Härtel 2007: 1). At the same time in the Netherlands, by contrast, climate change found its way explicitly into many action strategies. Thus in 2007 a new Delta Commission was mandated with the task of developing new measures against the impacts of climate change, in particular against intensified sea-level rise (see also Section 6.3). In September 2008, under the guiding principle 'Working together with water' (original: 'Samen werken met water') (Deltacommissie 2008), the Commission focused on a paradigm shift:[3] since coastal protection in the Netherlands had proceeded for centuries by using technological measures against water, more room should now be created for water through modified spatial structures. Initial compulsory purchase and resettlement plans were executed, so that embankments could be renaturalised and new flood plains could be created (Asendorpf 2009).

These examples hint at the diversity of possible knowledge constructions of vulnerability and resilience to climate change. (Knowledge) constructions of vulnerability should here be understood in terms of perceived problems and opportunities connected to climate change. Resilience constructions represent measures that are conceived, discussed, or implemented to reduce vulnerability and thereby increase resilience (see also Section 2.1.4). The vulnerability and resilience constructions shared by actors will, throughout this work, be termed '*climate cultures*' (Welzer *et al.* 2010: 13; Leggewie *et al.* 2009: 176; see also Section 2.1), and the opening examples imply that these cultures might constellate at various scales of (socio-)spatial variations among the actors concerned. The following central questions arise:

- Which forms of *climate cultures* can be observed along selected European coasts?
  - What kinds of knowledge constructions of threats and opportunities do actors share? (Vulnerability constructions)
  - What kinds of preferences for practical measures do actors share? (Resilience constructions)
- How can *differences in climate culture* be explained, or *what are the discriminating factors* of climate cultures?

Various *climate cultures* lead to a diverse range of challenges. Collective endeavours between actors can only be successful if a certain consensus about a joint problem definition or the practical measures to be taken exists (see also Heimann 2013: 22). Collective efforts often fail not because of technical restrictions or available expertise, but owing to social and cultural differences between the actors involved (Heimann 2015: 441, see also Christmann *et al.* 2012, Heidbrink *et al.* 2010, Goethe-Institut 2009, and Leggewie and Welzer 2009). Thus in 2010, Heidbrink, Leggewie, and Welzer ask: 'Where is the contribution of cultural studies to the climate debate to be found?' (2010: 429). They too invite us to take a closer look at the differences between actors' shared knowledge, perception of problems, and preferences; that is, their '*zones of cultural friction*' (Heimann 2015: 441; Crate 2011: 188; see also Heimann 2013: 16; Christmann *et al.* 2012: 23).

This appeal by the authors suggests that there has been little contribution from German social or cultural studies to climate-change research. Look a little closer, though, and there most certainly is a related body of work to be found. Most apparent are the many qualitatively oriented studies that, moreover, often engage ethnographically with understanding the logics of climate-change interpretation and action. They frequently orient themselves to local communities or other small social groups (see for more detail Section 2.1 and Chapter 5). The advantage of this work is that actors' climate-related knowledge can be traced back in detail to local constellations or conflicts. In the German-speaking sphere one also finds quantitative work concerned in particular with explanatory factors for the acceptance or rejection of measures related to climate mitigation. It is only since the 2010s there have been a number of studies which focus specifically on adaptation practices (ibid.).

Social-science related lines of research into climate change are thus much in evidence, and can be drawn upon in the present work. In particular, the last-mentioned, quantitatively oriented studies draw their mostly action-theoretical explanatory approach from the long-standing *debates within Anglo-American environmental sociology*, as well as the adjacent disciplines of anthropology, psychology, and geography, some of which reach back to the 1970s. They have their origins in the *issue of explanation* for 'environmental concern' (Dunlap and Catton 1979: 249) and 'environmental behaviour' (Dunlap and van Liere 1977: 202).

With the appearance of the discourse around explanation for 'climate change [...] behavior' (Scannell and Gifford 2013: 66), it becomes apparent that many things still remain unclear, and to date such differing forms of background knowledge as values, beliefs, and self-identification have been discussed as explanatory factors. Frequently, though, these appear in the literature mostly as isolated fragments. There is hardly any work that attempts a synthesis or even an integrated explanatory model for the differences in ways of handling climate change. This is true not only for explanatory designs, but also for the previously mentioned practices, as often only individual strategies and preferences for climate mitigation or adaptation are considered. Thus the diversity in the spectrum of possible measures related to climate change usually remains unclear.

In addition, a work on the subject of culture and climate change must confront a wider debate, namely the *discourse within social science on cultural globalisation* of the last decades. The exact (socio-)spatial constellation of any '*zone of cultural friction*' in the handling of climate change is only sketchily understood. Climate change has been recorded as a global phenomenon by many scientific actors (see Chapter 3). In connection with this, several authors from the social sciences see in the global handling of climate change a paradigmatic example of the emergence of a global society (see Erhardt-Martinez *et al.* 2015: 218f.;

Weiss and Burke 2011: 1057). Global definitions of problems and the necessity of finding common solutions should accordingly contribute to the *process of cultural homogenisation*, in which, for example, global institutions are established and common stocks of knowledge and practice are increasingly shared across borders (see also Chapters 3, 4, and 6). (Climate-)cultural boundaries would then have to arise in the present in completely new social formations, while traditional (for instance, national boundaries should lose their differentiating impact. Conversely, there are authors who continue above all to represent locally and regionally oriented *differentiation theories*. They continue to emphasise the high relevance of local particularities in climate-related thought and practice (see for example Adger *et al.* 2013; Christmann *et al.* 2012; Crate 2011; Biesbroek *et al.* 2010; O'Brien 2009; Brooks *et al.* 2005; Bulkeley and Kern 2004).

Why do these authors come to such different conclusions? The above-mentioned repertoire of qualitative analyses often appears too slender to be able to make assertions about larger 'zones of cultural friction'. But the quantitative research has also reached its limits. Authors such as Beck (2005) or Wimmer and Glick-Schiller (2002) point out that in recent decades researchers have often continued to cling to the same traditional units of comparison, in particular by comparing nation states. In this connection, Beck has criticised 'sociology's blind spot for Europe' (Beck 2005: 3, in detail Section 2.2.1). New socio-spatial formations could thus easily be overlooked or not taken into account at all. At the same time, other authors such as Wimmer and Glick-Schiller (2002: 301) emphasise that the possibly persistent effects of historic borders should still be investigated as well, as they could indeed be of continued relevance.

It can be shown that various implicit spatial concepts in the discourse on the global cultural constellation in the past decades have found application (see Section 2.2). Consequently, the understanding of (cultural) space should be made clear and a procedure to explore climate cultures needs to be elaborated (see Section 2.3). In the present work a *relational concept of cultural space* is developed, making possible an exploration of actors' cultural affiliation exclusively by their socially shared knowledge. Relevant factors contributing to the differentiation of cultures can subsequently be determined, and thus cultural variations explained. In the course of this:

- existing '*orders of knowledge*' (Reckwitz 2000: 30) are revealed, i.e. regular patterns of socially shared climate knowledge and their systematic interactions with other forms of knowledge held by actors. Here in particular the question arises, to what extent shared *values, beliefs, and self-identification* can explain differences in climate knowledge.
- In connection with the socio-spatial discourse, the question of the relevance of other spatial constructions as explanatory factors also arises. To what extent are climate-change knowledge or other existing 'orders of knowledge' simultaneously separated by historical boundaries, in particular

the *national affiliation* of actors? Or are the *alternative (climate-)cultural formations* postulated by the authors mentioned much more observable?

This work should provide a contribution both to the *discourse within environmental sociology* on the explanation of climate interpretation and action, and to the *discourse within sociology of space* on cultural globalisation. The work can be divided broadly into five parts: these range through the socio-theoretical development of cultural and spatial understanding (Chapter 3); the description of relevant climate-related discourse in respect to debated problems and opportunities, and potential spatial planning measures (Chapters 3 and 4); the question of explanatory or, respectively, differentiating factors for varying climate cultures (Chapters 5 and 6); the empirical procedure (Chapter 7); and the results of the empirical study (Chapters 8 to 10).

In Chapter 2 the social-theoretical principles for the subsequent empirical investigation of climate cultures along European coasts are laid out. The current state of research in the cultural and social science on climate will be introduced in order to contextualise the present work and its approach. In the process, it will be argued that the concept of culture needs to be more specifically detailed. To this end, a knowledge-oriented understanding of culture will be drawn on. A social-constructivist perspective will be chosen that allows for the exploration of actors' existing knowledge within its social conditionality. It is for this reason that a positivistic perspective on climate change is avoided: my object of research is not climate change itself, but rather actors' knowledge about it. Having delimited the concept of climate cultures, section 2.2 then delves further into the discourse within sociology of space on (cultural) globalisation. In connection with this, my own understanding of culture as relational space outlined above will be further developed, allowing for a precise, empirical description of climate cultures.

In both the following chapters (Chapters 3 and 4), the existing discourse on possible problems, opportunities, and climate-change measures will be introduced. Chapter 3 highlights the scientific construction of knowledge about climate change and at the same time describes the genesis of key institutions whose global activity increased over the previous centuries. Chapter 4 then introduces the discussion about possible measures. These are conceivable in practically all social areas of activity. The present work will concentrate on the area of spatial development, which offers the advantage that both climate-mitigation and adaptation measures are open for consideration.

The discourses mentioned already take actors' possible shared knowledge constructions into account. In the empirical section they will be more closely investigated. Before this, there is the question about the differentiating factors that explain differences in actors' stocks of knowledge. Similar approaches regularly appear to explain both variant constructions of problems and opportunities, as well as climate-mitigation and adaptation preferences. These originate in the already-mentioned research traditions, and will be comprehensively treated in Chapter 5. As mentioned, there are hardly any approaches to be found that consolidate the central concepts discussed into a unified explanatory model. This integration will also be carried out in Chapter 5. In addition, central positions in relation to other social factors for the explanation of differences in climate culture will be considered. I shall discuss in particular the relevance of national affiliation as well as the affiliation to functionally differentiated fields of administration, politics, economics, and civil society. Through this, an action-theoretical explanatory model will be developed whose capability to explain differences in climate culture can be tested in the empirical section of this work.

Connected to whether regional or national background, or other alternative formations, emerge as differentiating factors for climate cultures is the question of overriding, and specifically local, political and administrative conditions. In Chapter 6 I shall investigate their historical developments, focusing on the main global and European developments. In addition, I shall cast a glance at developments within the individual states.

These are the premises of the empirical study. Chapter 7 describes its methodology and the technical details of the survey design more closely. The study uses a sample of actors ($n = 830$) in the field of spatial development responsible for the coastal municipalities of all the named regions. Subsequently, the constructions and indicators developed for the questionnaire will be introduced.

In the final step, the findings of the empirical study will be presented (Chapters 8 to 10). Chapter 8 will investigate which conditions are in place that allow climate knowledge to be shared across the study area by briefly examining existing communications about climate. In Chapter 9, actors' vulnerability constructions relating to climate change will be considered. Chapter 10 follows with a comprehensive investigation of discernible preferences for particular measures over others. The analysis proceeds along a uniform pattern, investigating first of all which common or different vulnerability and resilience constructions are shared by actors. Once the climate cultures have been determined, it will then be tested which factors are able to account for the differences between cultures. In conclusion, the study's main findings will be summarised, and their relation to the study's central questions discussed.

Notes

1 On the difference between climate mitigation and adaptation, see Chapter 4.
2 The IPCC has projected a possible sea-level rise of between 26 cm and 98 cm during the twenty-first century (IPCC 2013: 25, Section 3.3).
3 This paradigm shift had already become apparent in the Netherlands in the 1990s (see Chapter 6).

References

Adger, W. Neil, Barnett, Jon, Brown, Katrina, Marshall, Nadine, and O'Brien, Karen (2013) 'Cultural Dimensions of Climate Change Impacts and Adaptation'. *Nature Climate Change* 3 (2): 112–117.

Asendorpf, Dirk (2009) 'Platz für Wasser. Für die Niederländer ist der steigende Meeresspiegel eine Herausforderung, die sie gelassen meistern können'. *Die Zeit*, 20 August 2009 No. 35. Available at: www.zeit.de/2009/35/U-Niederlande-Polder (accessed 20 October 2010).

Beck, Ulrich (2005) 'Europäisierung. Soziologie für das 21. Jahrhundert'. *Aus Politik und Zeitgeschichte* (APuZ) 34–35: 3–11.

Biesbroek, G., Robbert, Rob, J., Swart, Carter, Timothy R., Cowan, Caroline, Heinrichs, Thomas, Mela, Hanna, Morecroft, Michael D., and Rey, Daniela (2010) 'Europe Adapts to Climate Change: Comparing National Adaptation Strategies'. *Global Environmental Change* 20: 440–450.

Brooks, N., Adger, W.N., and Kelly, P.M. (2005) 'The Determinants of Vulnerability and Adaptive Capacity at the National Level and the Implications for Adaptation'. *Global Environmental Change* 15 (2): 151–163.

Bulkeley, Harriet and Kern, Kristine (2004) 'Local Climate Change Policy in the United Kingdom and Germany'. Discussion Papers No. SP IV 2004–103. Abteilung Zivilgesellschaft, Konflikte und Demokratie des Wissenschaftszentrums Berlin für Sozialforschung.

Christmann, G.B., Heimann, T., Mahlkow, N., and Balgar, K. (2012) 'Klimawandel als soziale Konstruktion?' *Zeitschrift für Zukunftsforschung* 1 (urn:nbn:de:0009–32–34126). Available at: www.zeitschrift-zukunftsforschung.de/ausgaben/2012/1/3412/Christmann_et_al (accessed 28 October 2013).

Crate, Susan A. (2011) 'Climate and Culture: Anthropology in the Era of Contemporary Climate Change'. *The Annual Review of Anthropology 2011*. 40: 175–194. DOI: 0.1146/annurev.anthro.012809.104925 (accessed 29 October 2013).

Deltacommissie (2008) *Working Together with Water: A Living Land Builds for its Future. Findings of the Deltacommissie 2008* [*Samen werken met water. Een land dat leeft, bouwt aan zijn toekomst. Bevindingen van de Deltacommissie 2008*]. Alkmaar: Hollandia Printing. Available at: www.deltacommissie.com/doc/deltareport_full.pdf (accessed 15 May 2012).

Dunlap, Riley E. and Catton William R. (1979) 'Environmental Sociology'. *Annual Review of Sociology* 5: 243–273.

Dunlap, Riley E. and Van Liere, Kent D. (1977) 'Land Ethic or Golden Rule: Comment on "Land Ethic Realized" by Thomas A. Heberlein'. *Journal of Social Issues* 33 (3): 200–2007.

Erhardt-Martinez, Karen, Rudel, Thomas K., Norgaard, Kari Marie, and Broadbent, Jeffrey (2015) 'Mitigating Climate Change'. In: Dunlap, Riley E. and Brulle, Robert J. (ed.): *Climate Change and Society. Sociological Perspectives*. New York: Oxford University Press, pp. 199–234.

Goethe-Institut e.V. (ed.) (2009) 'Es gibt einen Globus, aber mehrere kulturelle Klimawelten'. In: *goethe.de/klima. Kultur und Klimawandel.* Available at: www.goethe.de/ges/umw/prj/kuk/the/kul/de4139355.htm (accessed 19 August 2015).

Heidbrink, Ludger, Leggewie, Claus, and Welzer, Harald (2010) 'Von der Natur- zur sozialen Katastrophe: Wo bleibt der Beitrag der Kulturwissenschaften zur Klima-Debatte? Ein Aufruf '. In: Bieber, Christoph, Drechsel, Benjamin, and Lang, Anne-Katrin (ed.): *Kultur im Konflikt. Claus Leggewie Revisited.* Bielefeld: Transcript Verlag, pp. 429–432.

Heimann, Thorsten (2013) 'Konsens oder Dissens. Herausforderungen des Klimawandels für die deutschen Küstenstädte und –gemeinden aus Expertensicht'. In: Heimann, Thorsten and Christmann, Gabriela B.: *Klimawandel in den deutschen Küstenstädten und -gemeinden. Befunde und Handlungsempfehlungen für Praktiker*, pp. 22–72. Available at: www.klimabuendnis-kieler-bucht.de/wp-content/uploads/2014/12/progress-abschlussbericht.pdf.

Heimann, Thorsten (2015) 'Review of Schmitt, Lea: *Klima, Raum und Zeit im Wandel. Eine ethnographische Untersuchung von Adaptionen und Konflikten auf der westfriesischen Insel Ameland*'. *Raumforschung und Raumordnung* 73 (6): 441–443.

IPCC Intergovernmental Panel on Climate Change (ed.) (2013) *Climate Change 2013: The Physical Science Basis*. Contribution of Working Group I to the Fifth Assessment Report of the Intergovernmental Panel on Climate Change. Cambridge/New York: Cambridge University Press. Available at: www.climatechange2013.org/images/report/WG1AR5_ALL_FINAL.pdf (accessed 27 April 2014).

Leggewie, Claus and Welzer, Harald (2009) *Das Ende der Welt, wie wir sie kannten: Klima, Zukunft und die Chancen der Demokratie*. Frankfurt am Main: Fischer Verlag.

Leggewie, Claus, Welzer, Harald, and Heidbrink, Ludger (2009) 'Klimakultur: ein transdisziplinärer Projektverbund'. In: Lutz, Petra and Macho, Thomas (ed.): *Zwei Grad. Das Wetter, der Mensch und sein Klima*. Göttingen: Wallstein Verlag, pp. 176–182.

MLUV Ministerium für Landwirtschaft, Umwelt und Verbraucherschutz Mecklenburg-Vorpommern (ed.) (2009) *Regelwerk Küstenschutz Mecklenburg-Vorpommern. Übersichtsheft. Grundlagen, Grundsätze, Standortbestimmung und Ausblick*. Rostock: MLUV.

O'Brien, Karen (2009) 'Do Values Subjectively Define the Limits to Climate Change Adaptation?'. In: Adger, Neil, Lorenzioni, Irene, and O'Brien, Karen: *Adapting to Climate Change. Thresholds, Values, Governance*, pp. 164–181.

Raabe, Stephan and Härtel, Janina (2007) *Der Diskurs über den Klimawandel in Polen*. Warschau: Andere Prioritäten. Available at: www.kas.de/polen/de/publications/11117/ (accessed 15 October 2010).

Reckwitz, Andreas (2000) *Die Transformation der Kulturtheorien*. Weilerswist: Velbrück.

Scannell, Leila and Gifford, Robert (2013) 'Personally Relevant Climate Change: The Role of Place Attachment and Local Versus Global Message Framing in Engagement'. *Environment and Behavior* 45 (1): 60–85.

Weingart, Peter, Engels, Anita and Pansegrau, Petra (2002) *Von der Hypothese zur Katastrophe. Der anthropogene Klimawandel im Diskurs zwischen Wissenschaft, Politik und Massenmedien*. Opladen: Leske & Budrich.

Weiss, Thomas G. and Burke, Martin J. (2011) 'Legitimacy, Identity and Climate Change: Moving from International to World Society?' *Third World Quarterly* 32 (6): 1057–1072.

Welzer, Harald, Soeffner, Hans-Georg, and Giesecke, Dana (2010) 'KlimaKulturen'. In: Welzer, Harald, Soeffner, Hans-Georg, and Giesecke, Dana (ed.): *Klimakulturen. Soziale Wirklichkeiten im Klimawandel*. Frankfurt/New York: Campus Verlag, pp. 7–19.

Wilson, Elizabeth and Piper, Jake (2010) *Spatial Planning and Climate Change*. New York: Routledge.

Wimmer, Andreas and Glick-Schiller, Nina (2002) 'Methodological Nationalism and beyond: Nation–State Building, Migration and the Social Sciences. *Global Networks 2* (4): 301–334. Available at: www.princeton.edu/~awimmer/B52.pdf (accessed 5 August 2015).

2 Climate cultures and space

Theoretical foundations

In this chapter I shall describe what the term 'climate cultures' refers to in this study, why it appears necessary to consider it in connection to the concept of relational space, and what impacts result from this for the analysis of ways of dealing with climate change in Europe.[1]

First, the concept of culture will be defined more precisely (Section 2.1). In doing so, an insight into its application in social science climate-change-related research will be provided. The more than 2000-year-long evolution of the concept of culture means that the term manifests diverse semantic associations. In the social sciences, however, a primarily knowledge-oriented concept of culture has recently established itself. It is this concept which the present study takes up. Subsequently, the concept of knowledge must also be more closely considered. Building on this, the concept of 'climate cultures' can finally be more precisely defined.

It will become apparent that culture can hardly be considered apart from spatial categories. Until now, however, authors concerning themselves with cultures have rarely reflected upon their own often implicitly applied (cultural) concept of space. This becomes all the more problematic with respect to a precise analysis when one takes into account that in earlier discourse on globalisation, often new and differing relationships between territory, culture, and society were postulated, to which could also be regularly ascribed various implicit concepts of space (Section 2.1). To be able to better answer the question about the precise relationship between climate culture and space, implicitly applied historical concepts of space should first be explicitly reflected upon (Section 2.2). Building upon this, a relational understanding of cultural spaces will be developed that allows for a separate analysis of place/region and culture (Section 2.3).

2.1 Culture

2.1.1 Culture in social-science climate research

In the following, important core developments of social-science climate-change-related lines of research will be discussed, in particular studies devoted to *culture-related questions*. A comprehensive discussion of all social-science climate-change-related lines of research cannot be made here, particularly as, in the meantime, one can refer to existing secondary literature (see Hulme 2015; Grothmann *et al.* 2011; Renn *et al.* 2011; Biesbroek *et al.* 2010; Voss 2010).

Social-science climate-change-related research can be seen to have begun as an offshoot of the natural sciences. Following Biesbroek *et al.* (2010: 444), the development can be sub-divided into (1) climate-system research, (2) climate-mitigation and research on impacts of climate change, and (3) vulnerability and adaptation research. Thus (1) scientific discourse around *climate-system research* initially dominated, in which an understanding of climate change, its causes, and further connections was negotiated (see also Chapter 3). In the course of the 1990s, with the discussion of *anthropogenic climate change* as a problem for global society, there arose an interdisciplinarily oriented (2) *climate-mitigation research*, which in particular discussed technological, political, and economic solutions to reduce greenhouse gas emissions (see also Chapter 4). In Europe at the start of the new millennium a growing number of research projects were initiated to look at climate-change impacts on society and the environment (see Biesbroek *et al.* 2010: 444). Building upon this, in the 2000s the idea developed that mitigation measures alone would not be sufficient to cope with climate impacts. Thus projects came into being that dealt with the vulnerabilities of various areas of society such as coastal protection and spatial development, etc. Closely tied to this, since the second half of the 2000s (3) *climate-adaptation research* has developed to search for ways to adapt to such impacts of climatic changes (ibid., and see Chapter 4).

If one were to picture the general development of anthropogenic climate-change research from the scientific definition of the problem to its increasingly interdisciplinary research orientations, it would hardly be a surprise that sociologically oriented research has for a long time also been characterised, above all, by application-oriented projects that assumed from the outset a high level of vulnerability from climate change. Only recently have numerous authors appeared to indicate the blind spots of these earlier approaches (Schmitt 2015: 24; Christmann and Ibert 2012; Christmann *et al.* 2012; Krauss *et al.* 2012: 121; Conrad 2010: 101ff.). For example, Christmann and Ibert (2012) state from a social-constructivist perspective that actors often ascribe vulnerabilities anticipatorily (and thus, a priori, unfalsifiably!), without taking into account that these vulnerabilities exist at this time only as a human interpretation of threat. This shows that in the majority of investigations the aspect of social and cultural 'constructedness' has long been left comparatively unheeded. Fundamental social-science research seeking to investigate the variety of logics of interpretation and action used by participating actors, and wishing to do so from the perspective of a second-order observer, could hardly continue for long in this manner.

As this work has been translated in order to contribute views from German sociology on the subject to the English-speaking social sciences, I shall first start our journey in the *German-speaking sphere*. Here, the term *Klimakulturen* (climate cultures) has been coined by academics at the Kulturwissenschaftliches

Institut (KWI) in Essen, where research under the same name has been led by Claus Leggewie. Under the term 'climate cultures' (Welzer *et al.* 2010: 13), they seek to investigate a variety of aspects:

> This expertise concerns the historic stock of experience regarding anticipated, perceived, and lived catastrophes just as much as the interpretative frameworks that belong to them. They relate as much to the cultural practices and interpretative frameworks that have led to the cause of anthropogenic climate change as to the broad field of its social, psychological, and legal treatment. Not least, it challenges the human potential for interpretation and giving meaning: the philosophical treatment of aspects of justice and responsibility, as well as philological and literary language criticism and sociology of knowledge's analysis of collective interpretative frameworks.

This *broad concept of climate culture* serves to outline for the authors an expanded field of research, namely that of the social and cultural sciences. However, while it is capable of elucidating some rather broad research fields, for empirical studies, as a rule, it needs to be fleshed out further.

Nico Stehr can be named as one among the first sociologists to introduce the concept of culture into German climate research. Located in the sociology of knowledge tradition of Karl Mannheim, he points to the social conditionality of worldview and concepts of nature in the political discourse about climate change (Grundmann and Stehr 2003: 265ff.). He also makes the relation between physical space and culture a subject of discussion, and in doing so warns (for example in his collaborations with Hans von Storch) of climate-deterministic simplifications that assume direct causal relationships between local climatic conditions and cultural patterns of behaviour. He does not exclude the possibility of physical and cultural interactions, however (Goethe-Institut 2011; Stehr and von Storch 1999: 45ff.; Stehr 1996).

The developed fund of German *anthropological research* provides a further exception to the repertoire of essentialist studies in social-science climate research (Schmitt 2015; Greschke 2010: 4; Krauss 2007). The majority of these studies are devoted to the question of how climatic changes are processed and construed by actors in specific lifeworlds. The studies often focus on local 'community research' or are conducted within other smaller groups. A pioneer in the German-speaking sphere is anthropologist Werner Krauss. In the North- and Baltic-Sea regions he ethnographically explores various actor groups as 'cultural producers' (Krauss 2007: 8) of competing knowledge systems and perceptions of climate change, pointing to the variety of possible interpretive patterns in relation to material environments: 'The coast is a concept that should always be thought of in the plural' (ibid.). Comparing the ways of handling climate change in Germany and the USA, writing with Hans von Storch he comes to the conclusion: 'Culture contributes to perceptions of climate change' (von Storch and Krauss 2005). At this point, a concept of culture appears that seems to orient itself at presumed national 'container spaces' (see Section 2.2).

In summary, the concept of culture in current German ethnographically influenced studies encompasses various forms of knowledge and interpretive patterns for climate change in lifeworld contexts.

Within *the German-speaking sphere* it can be ascertained that empirical studies on the charged relationship between culture and climate change have until now been influenced above all by qualitative methods, and are usually rather idiographically oriented. In addition, many of these studies focus on somewhat smaller local groups or communities as the subject of research. Theory-testing, culture-comparative quantitative work on the handling of climate change could remedy omissions in the research to date. In addition, research from the *English-speaking sphere* can be tied in.

Next up are studies that share a *holistic understanding* of the term 'culture' (see Section 2.1.2, Hulme 2015). Bryson's (1994) objective, for example, is '[…] integrating climatic change and culture change studies' (ibid.: 115) and pleading for historical research to be carried out on communities of the past in order to explore interactions between human lifestyles and climatic change. From a human-ecological perspective, Wisner (2010: 131ff.) views worldwide 'cultural diversity' as being threatened by climate change, for instance when certain population groups are confronted with greater problems than others (ibid.: 134).

As found in the German-speaking sphere, *anthropological approaches* on 'climate and culture' (Crate 2011; Crate and Nuttal 2009) have developed with increased activity since the 2000s. They often connect the term culture (or 'climate culture', Barnes and Dove 2015) with local and/or historical peculiarities of perceiving and dealing with climate change. Again, most of these attempts use qualitative ethnographic research methods to develop theory.[2]

Many English studies at the intersection between sociology, cultural geography, and human ecology have developed a narrower concept of culture, in that they limit culture to specific forms of socially shared knowledge (see Sections 2.1.2 and 2.1.3 and Chapter 5).

Thus, at the turn of the millennium, the first studies on different perceptions of problems and preferences for climate-mitigation measures emerged (Thompson 2003; Pendergraft 1998). An early approach that was applied to climate change was the 'cultural theory of risk' (Adger *et al.* 2013: 113; Douglas and Wildavsky 1982; for more details see Chapter 5) which assumes that communities with collectively shared environmental *beliefs* also produce shared selective risk perceptions. However, cultural space has played a rather subordinate role in these studies, which instead explain 'climate change […] behaviour' (Scannell and Gifford 2013: 66) through various forms of background knowledge.

Bringing to mind once again that Anglo-American research has also undergone those complementary development stages mentioned at the start, from climate-mitigation to *climate-adaptation* research (see for example Biesbroek *et al.* 2010), it is hardly surprising that in this tradition a new wave of studies has been observed, now seeking explanation for climate-adaptation preferences (see also Adger *et al.* 2013: 112).

Karen O'Brien views shared *values* in particular as the basis for different degrees of acceptance of adaptation measures. Theoretical reference points for her are the culture-comparative studies of values by Rokeach, Schwartz, and Inglehart (Adger *et al.* 2013; see also Chapter 5). Together with Neil Adger, Jon Barnett, and other colleagues, she also points to the possible relevance of specific *identity constructions* for local climate adaptation (ibid.: 113ff.). Accordingly, the loss or transformation of places and regions could meet with resistance if the actors concerned exhibit a high level of place identification (ibid.). To date, these authors have principally published theoretical studies in which theories such as these are formulated; in addition, there are studies that check these hypotheses empirically (see for example Amundsen 2015; Scannell and Gifford 2013; Schweizer *et al.* 2013: 57). Summarising the understanding of culture in these studies, it becomes particularly noticeable that 'culture' is being defined as specific knowledge constructions such as values or beliefs that explain local differences in the perceptions or preferred measures.

To conclude, social-science research was thus for a long time characterised by essentialist oriented studies that often took the problem of climate change as their starting point, but without making differing understandings of the problem by social actors a subject of reflection. It also becomes clear that the concept of culture in social-science climate research is often governed by different semantics, embracing the entire lifestyle of local groups on the one hand, through to individual constructions such as values or beliefs on the other. In addition, the relationship between culture and space has only seldom been theorised or methodologically considered. In the following section therefore, the concept of culture will be precisely defined. In connection with this, the aspect of cultural spatiality will also be discussed in more detail.

Observing the state of the research, it furthermore becomes clear that until now there have been relatively few quantitative culture-related studies in the German-speaking sphere aimed at theoretically test explanatory patterns for varying patterns of interpretation and practice concerning climate change. In the German-speaking sphere, studies using qualitative methods have dominated. These are very well suited for the determination of climate-change categories or interpretive patterns, but these procedures reach their limits when describing group characteristics or testing structural hypotheses. It is also clear that in the research field, more theories have so far been generated than systematically verified. Quantitative methods are therefore selected for the present study in order to close a gap in the research.

It is also noticeable that until now the explanatory variables have been kept very general (e.g. 'climate change engagement': Scannell and Gifford 2013: 75). We therefore lack a body of work in which specific, detailed preferences for various climate practices can be examined. In the field of spatial development too there have been hardly any comprehensive studies of this kind. Furthermore, it is apparent that the relationship between climate-mitigation and climate-adaptation has only recently been considered as a complementary one.

For a long time, research was characterised by only one or the other aspect, but as both of these are basic strategies for dealing with climate change, both should therefore be taken into account.

2.1.2 Culture concept I: collectively shared knowledge

The general problem of culture-comparative research also appears in social-science climate research: 'There is hardly another term in the social sciences that has been used so frequently, so inconsistently, and so vaguely as that of culture' (Rippl and Seipel 2008: 13). Already by the middle of the twentieth century, Kroeber and Kluckhohn had identified more than 170 different concepts of culture in social-science literature (Kroeber and Kluckhohn 1952). The appearance of debates about at least one 'cultural turn'[3] in the social sciences indicates that right up to the present semantic diversity has increased rather than diminished. The term can thus justifiably be described as 'altogether one of the most ambiguous and haziest concepts in the social sciences' (Hölscher 2006: 55). Still further delimitation is necessary, therefore, in defining the subject of this study.

It can also be shown, however, that the use of the term is reducible to several central concepts.[4] Thus at the turn of the millennium, Reckwitz compiled a 'typology of the concept of culture' (Reckwitz 2000: 64ff.) offering a classification which consists of four central categories. According to these, one can differentiate between (a) normative, (b) holistic, (c) differentiated, and (d) knowledge-oriented concepts of culture (ibid.). The present study is established on the basis of this typology.

The point of departure for the (a) *normative use of the term* stems from the classical Latin verb *colere* (maintain, cultivate, revere) which was first used in connection with agriculture (*agri cultura*, care of the field, see Goetz 1922). Cicero (106–43 BCE) extended the concept to *cultura animi* (care of the mind), which in the form of education would serve a person's spiritual and moral refinement. It was here that the concept of culture obtained a connotation as something to be valued.

The (b) *holistic use of the term* can be traced back to Johann Gottfried Herder (1744–1803). Herder understood cultures as the historically developed lifestyles of different peoples and emphasises their uniqueness. His understanding of culture had great influence during the nineteenth century on the then developing discipline of anthropology; 'cultural anthropology' undertook the task of systematising the lifestyles of different collective bodies (Reckwitz 2000: 72ff.).[5] Thinking in terms of nations also gained momentum with Herder, so that cultures and nations would even sometimes be equated (Rippl and Seipel 2008: 17f.).

Out of the holistic understanding of culture developed a further usage of the term, manifesting itself in the categorical differentiation between *'nature' and 'culture'*. Human life in its entirety was thus interpreted in contradistinction to the living non-human sphere. In this conception, humans create a so-called cultural sphere as 'second nature' (Gehlen 1993: 87). This binary construction appears in climate discourse, for example when flood and coastal protection efforts are interpreted as a human 'struggle against nature' (see Section 4.3).[6]

In contrast, the (c) *differentiation concept of culture* reduces culture to a limited, functional part of society: 'art and culture'.[7] Meant by this are activities that deal in an intellectual or artistic way with the 'production, sharing, and administration of interpretations of the world' (Reckwitz 2000: 79).

The question arises how suitable these concepts of culture are for actor-oriented, culture-comparative research, and indeed there is already discussion of their weaknesses (Schubert 2012: 136f.; Reckwitz 2000: 71). Schubert claims that the use of the *normative concept* can only be attributed to individual actors, rendering it impossible to comprehensively describe all actors under a single conceptual apparatus (Schubert 2012: loc. cit.).[8] The *differentiation-theoretical concept of culture* appears too narrowly defined, as it reduces culture to nothing more than a certain social area of activity (Schubert 2012: 137; Reckwitz 2000: 82). In contrast, the *holistic concept of culture* takes into consideration a limitless variety of aspects, all of which are simultaneously possible for the description of the groups under consideration. Thus it can hardly meet the quality criterion of precision (Schubert 2012: 137; Schnell *et al.* 2008: 128). Furthermore there is the danger of culture-essentialist bias if, from the outset, cultural affiliation is defined by the observer before any empirical comparison.

Reckwitz therefore finally nominates the (d) *knowledge-oriented concept of culture*, which has prevailed in contemporary social-science theories of culture (Reckwitz 2000: 90). This accordingly 'offers to bring the concept of "culture" in relation to that of "knowledge"' (Reckwitz 2000: 85). Culture is thus understood as a collective phenomenon shared by groups of people (see also Roose 2011: 242). In the words of Alfred Schütz, it is assumed that '*culture can be defined in terms of shared knowledge which is socially derived and socially approved*' (Schütz 1962: 146, emphasis added).

The understanding of culture as shared knowledge (Quinn and Holland 1987: 4) offers the advantage that, first, it allows the determination of all possible forms of knowledge. Second, it makes it possible to investigate with sufficient precision which interactions take place between different forms of socially shared knowledge. Since these appear both as explanatory instances (*explanans*) as well as instances to be explained (*explanandum*), any interactions between different forms of socially shared knowledge can be revealed to be '*orders of knowledge*' (Reckwitz 2000: 90).

In culture-related climate research to date, there have appeared such different forms of knowledge as e.g. problem perception and climate-change practice preferences (e.g. Greschke 2010), images of nature (e.g. Pendergraft 1998), and value orientations (e.g. Adger *et al.* 2013). Reckwitz himself hardly fleshes out the concept of knowledge, instead postulating the 'constitution of the social in sense frameworks' (Reckwitz 2000: 87) as their most important common features (see also Schubert 2012: 137). All these examples make it clear that the concept of knowledge also requires more precise definition.

2.1.3 Culture concept II: the social-constructivist concept of knowledge

As a starting point for developing an understanding of knowledge, the definition chosen in the 1960s by Berger and Luckmann (which has since become a common concept of the sociology of knowledge) serves well. In the first instance, they understand knowledge to mean 'the certainty that phenomena are real and have certain characteristics' (Berger and Luckmann 1966: 13). By using the term 'certainty', they link themselves to the phenomenological tradition after Alfred Schütz, who focused on the subjective consciousness of actors (see also Knoblauch 2010: 357). Their central social-constructivist approach consists of the idea that actors' shared realities and stocks of knowledge emerge through social interaction, or more precisely through the processes of externalisation, internalisation, and objectification.[9] In this way, stocks of shared knowledge are constantly formed and changed by communication processes (Christmann 2010: 34; Berger and Luckmann 1966: 78).

Berger and Luckmann show that shared stocks of knowledge – and thus cultures – are subject to processes of change and can differ (Knoblauch 2010: 20f.; Reckwitz 2000: 152). This understanding of knowledge thus goes far beyond the classical dichotomy, introduced by Plato, of 'knowledge' (*episteme*) vs. 'opinion' (*doxa*), where the concept of knowledge is reduced to positivist factual knowledge that can be tested by sensory experience, in contrast to subjective opinion. The social constructivism of Berger and Luckmann sets aside the positivist concept of knowledge and brings into view, instead, social and actor-specific reality (see also Knoblauch 2010: 158). These can, of course, also contain scientifically formulated knowledge such as socially shared concepts or, more specifically, social constructions related to climate change.[10]

Many representatives of the current generation of sociologists of knowledge in Germany focus above all on aspects of the *communicative construction* of knowledge (Knoblauch 1995, 2010, 2013; Knoblauch and Schnettler 2004), so that there is now talk of a further development towards 'communicative constructivism' (Keller *et al.* 2013). Gabriela Christmann, a disciple of Luckmann, has for example argued that 'the development of culture and identity spaces can be consistently deduced to communicative actions' (Christmann 2010: 32). She describes how actors' individual definitions of space finally emerge as collectively shared, group interpretations, derived from communicative exchange. Christmann once again affirms that communication processes can lead to the generation of commonly shared stocks of knowledge, and thus also to the genesis of cultures (ibid.).

She presented the thesis that the emergence of subjects' shared stocks of knowledge can be dependent on their participation in a given *discourse* (Christmann 2004: 337ff.). Using the city of Dresden as an example, she demonstrates that the topics regularly discussed (*topoi*) in the local media can serve as a reservoir for the development of actors' common knowledge about the city (ibid.; Viehöver 2004: 185). She understands discourse as 'an ensemble of various communicative acts related to, or entangled with, one another thematically' (Christmann 2004: 44). The genesis of socially shared stocks of knowledge can accordingly stand in relation to central topics of discussion and the systems of classification that regularly appear in discourses. Related to climate change, the varying perceptions of opportunities and problems reported in the scientific media and newspapers would be examples (see also Christmann *et al.* 2012 and Chapter 8).

The phrasing 'related to [...] one another thematically' reminds us that different constructions of shared knowledge can form systematic patterns, or 'orders', in their occurrence (see in particular Chapter 5). In the 1970s, Michel Foucault formulated in his 'Archaeology of Knowledge' (Foucault 1972) and 'Orders of Discourse' (Foucault 1971) the idea that the ways in which knowledge is generated and negotiated have changed throughout history, as have the schemes of classification and forms of perception which arise (see also Rosa *et al.* 2007: 282; Christmann 2004: 43). Accordingly, social scientists would be able to identify in particular *period-specific, discursive 'orders of knowledge'*.

A central gain of the authors of communicative constructivism such as Hubert Knoblauch, Reiner Keller, or Gabriela Christmann lies in their having brought Michel Foucault's by and large subjectless, structuralist understanding of discourse together with the additionally subject- or actor-related orientation (Keller 2011: 97ff., 2007: 14ff.; Christmann 2004: 43; Knoblauch 1995). This clear tendency towards the subject or towards actors brings me once again to the idea that *orders of knowledge* can take shape not only *historically* with the passage of time, but also socio-*spatially* at a specific moment in time. In the further course of this study a separate approach will be developed to show how *socio-spatial* (cultural) *orders of knowledge at a certain point in time* can be theoretically and methodically revealed.

The concept of knowledge has experienced still further expansion by many authors of communicative constructivism. In his attempt to summarise the present understanding, the Luckmann disciple Hubert Knoblauch comes to the following conclusion:

> Knowledge is everything that we adopt from other people: For instance, the language that I speak offers us a prefabricated typology of things and processes, just as the material culture that I use (whether it consists of straw huts or high-rise buildings) [...]. If we include embodied knowledge, then we also recognise that emotion itself [...] rests upon knowledge [...]. The handling of objects, materials, and technologies also, of course, belongs to knowledge; their meaning exists indeed in this very handling.
>
> (Knoblauch 2010: 360f.)

This concept of knowledge thus comes to refer to more than cognitive 'certainty' (Berger and Luckmann 1966: 13). For example, shared practices are now also explicitly included, such as specific ways of handling material 'culture objects' (Knoblauch 2013: 36) like houses or dykes. Embodied knowledge, such as specific, habitualised gestures or emotional valuations, also count as knowledge here.[11]

I return to the possible danger that a knowledge-oriented concept of culture might not be able to meet the quality criterion of precision (Schubert 2012: 137; Schnell *et al.* 2008: 128). I shall thus begin here by adopting the principal idea of social and communicative constructivism, i.e. that knowledge is socially constructed. This allows for climate change to be explored not as a given, a priori 'true' phenomenon, but more as the varying perceptions and actions of various actors. In the further course of this study, *knowledge constructions of climate change* will therefore be a topic of discussion.

The broad understanding of knowledge in today's communicative constructivism brings me back to the question of how to approach shared knowledge (or cultures) without reverting to the holistic understanding of culture (see Section 2.1.2). Therefore, a brief look should first be given to general epistemological developments of knowledge-oriented cultural theory in the twentieth century and, second, I would like to consider the general, unavoidable problem of selection in research and recognition procedures.

Among the most important strands of knowledge-oriented cultural theory of the twentieth century up until the present, we encounter at least two main lines of thought, although they should be regarded as two poles which are not necessarily strictly separated from each other (Reckwitz 2000: 542). On the one hand, we meet the *phenomenological tradition*, which seeks to comprehend and explain action primarily through subjective meaning reconstruction and/or interpretation (such as in Berger and Luckmann 1966; Schütz 1962; Geertz 1973, 1974). On the other hand, we encounter the *structuralist tradition* (such as Levi-Strauss, or, for post-structuralism, Bourdieu 1982; Foucault 1971, 1972; see also Moebius 2009: 77ff.; Reckwitz 2000: 173ff.). Here, culture and knowledge emerge as an 'immanently structured, super-subjective symbolic order' (Reckwitz 2000: 180).

Reckwitz describes how the phenomenological and structuralist lines converged during the second half of the twentieth century (Reckwitz 2000: 542ff.) and I would like to add that this convergence can also be traced in the developments from social to communicative constructivism described above. The investigation of culture as 'shared knowledge' would then not necessarily consist only of the comprehension of subjective sense; it might also be necessary to reconstruct even the super-subjective structural orders of knowledge (of which the individual actor needn't be explicitly aware). These considerations propel me to emphasise the high value of integrating *quantitative methodological approaches* into the field of social- and communicative-constructivism-based cultural research – in addition to the already well-established and highly valuable repertoire of the qualitative methodological canon in this field.

Finally, I would like to come to the problem of selection. As already mentioned, culture has here been defined in the social-constructivist tradition as the shared knowledge constructions of actors. What has until now been a shortcoming in the debate about culture as 'shared knowledge' is the fact that the knowledge constructions held by human actors can basically only be shared *aspectually*. Even social-science or everyday observations of collectively shared knowledge only ever consist of fragmented interpretations of social reality. Knowledge ('*Erkenntnis*') is thus always bound to the reduction of complexity. It is therefore necessary at this point to restrict the concept of culture to *aspects of socially shared knowledge*. In doing so, cultures emerge in various forms, such as 'climate cultures', 'food cultures', or 'music cultures' (see also the example in Section 2.3).

To conclude, this study discusses 'climate cultures' as specific forms of aspectually shared knowledge constructions. Its semantic field will be precisely defined in the following Section.

2.1.4 Climate cultures as collectively shared vulnerability and resilience constructions

Culture has here been defined in the social-constructivist tradition as the shared knowledge constructions of actors. Linked to this, the term 'climate cultures' can be more closely specified. Climate cultures should be understood in the present work as aspectual, collectively shared knowledge constructions about climate change. But which aspects should it deal with specifically? In order to specify this concept, it is necessary to first introduce the notions of *vulnerability and resilience constructions.*

On the one hand we might ask what ideas of possible climate-change problems and opportunities are shared by actors. These ideas will here be termed *vulnerability constructions* (see also Christmann *et al.* 2012: 22). On the other hand one might consider what practices for building up resilience are preferred and implemented (ibid.: 21). In this context, the term *resilience constructions* will be used. Figure 2.1 illustrates these concepts graphically.

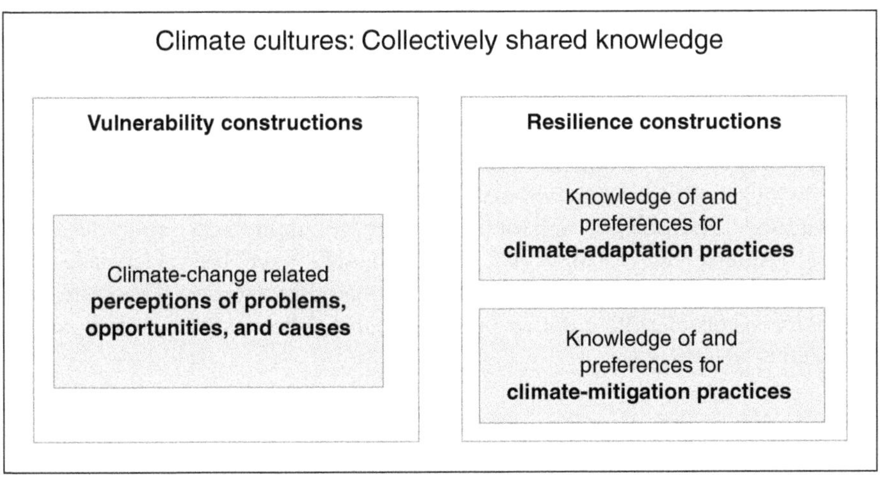

Figure 2.1 Climate cultures as collectively shared vulnerability and resilience constructions.

The concepts of *vulnerability* and *resilience*, now also widespread in the social sciences, were long in use particularly in human-ecological and developing-country research, where they were generally treated as objectively determinable (Christmann *et al.* 2012; Christmann and Ibert 2012: 261). Vulnerability was equated with *factually existing* vulnerability, while resilience was seen as an objectively ascertainable ability to overcome problems (Frommer 2010: 17ff.). From this perspective, various entities such as people, houses, or neighbourhoods could be regarded as objectively vulnerable or resilient, determinable through such variables as age, gender, income, exposure, etc.

There are now many such studies relating to climate change (see for example Adger 2006; Brooks *et al.* 2005; Smit and Wandel 2006; O'Brien *et al.* 2004; Wu *et al.* 2002). My colleagues and I already argued in a 2012 paper that these conceptions can be considered legitimate, previous experience, demonstrating that there are numerous indicators which show with high accuracy that dangers regularly lead to perceptible occurrences of damage (Christmann *et al.* 2012: 21). If one calls to mind that actors behave on the basis of their internalised knowledge and beliefs, however, then this research approach lacks a systematic perspective regarding the question to what extent and why certain knowledge constructions of reality are shared by actors (ibid.). Climate change, vulnerability, and resilience are then no longer only to be comprehended as objectively given; the aim of this research is to systematically explore the constellation of socially shared knowledge constructions pertaining to collective perceptions and dispositions to act (ibid.).

In summary, climate cultures are to be understood as socially shared knowledge constructions about climate change. Vulnerability constructions indicate shared *perceptions* of problems and opportunities presented by climate change. Resilience constructions represent shared preferences for climate-mitigation and adaptation practices, and thus also explicitly include *valuation elements*. In the course of this study the constellation of socially shared vulnerability and resilience constructions will be examined as 'climate cultures'. Subsequently it will be asked which factors contribute to the differentiation of shared knowledge constructions (and climate cultures).

If one asks about the constellation of (climate) cultures, the question inevitably arises as to the spatial references within which these can be investigated. In the previous section we encountered the idea that knowledge constructions are externalised, internalised, and objectified by communication procedures. In recent discourse on globalisation, global changes have been identified that might have led to new spatial formations of shared knowledge. Therefore, it is necessary to consider the question of how possible cultural spatialisations can be explored more closely.

2.2 Space

It may be no coincidence that since the fall of the so-called 'Iron Curtain' in Europe an increased interest in the topic of space has been observable in the social and cultural sciences, soon leading to talk of a 'Spatial Turn' (Döring 2010: 90). At that time, social scientists observing these phenomena were increasingly confronted with various social 'boundary shifts' in Europe. Alongside these politically transformative 'shifts' there also came into view at roughly the same time new processes of globalisation and transnationalisation. At the turn of the new millennium, these finally led to a questioning of 'the idea, deeply anchored in the social sciences, of the isomorphism of territoriality, identity, and culture' (Schroer 2006: 161).

A particular criticism was that the (macro-)sociology of the present could no longer meet the new theoretical and empirical requirements that arose from the shifting social spatialisation phenomena if it only focused on traditional socio-spatial entities, such as the now possibly irrelevant nation state, as a unit of comparison. In this context, demands were accordingly made for new theoretical and methodological approaches in order to investigate those new phenomena (Beck and Grande 2010: 8ff.; Beck 2005: 4; Wimmer and Glick-Schiller 2002).

In the following I shall first look at globalisation discourses that can be viewed as discussions on changed *spatial phenomena* (Section 2.2.1). *Spatial phenomena* relate to perceivable phenomena whose observation is dependent on the knowledge of the researcher, as this is what determines, for example, the categories with regard to units of comparison that are used for the research. I shall therefore take the position that new, experimental paths in sociological theory and empiricism are necessary to better deal with the possible changes discussed, in particular in relation to territorialism and culture.

Concepts of space, on the other hand, are 'idea[s] of space' (Löw 2001: 16), i.e. constructions of the ontology of space. It will be shown that historic concepts of space continue to influence culture-comparative research in the present, and indeed in ways that are often implicitly rather than explicitly dealt with. I therefore argue that a disclosure of spatial concepts can help to approach the new requirements better. This disclosure is given below in Section 2.2.2.

2.2.1 *Spatial phenomena in transition? The deterritorialisation thesis as a challenge for culture-comparative sociology*

Since the end of the twentieth century there have regularly been voices identifying new global structures for physical, social, and cultural spaces.[12] In 1990 Giddens described as a spatial phenomenon the accelerating trend of globalisation taking place before his eyes. By this he meant the 'intensification of worldwide social relations which link distant localities in such a way, that local happenings are shaped by events occurring many miles away' (Giddens 1990: 64). Apart from the question whether social relations, such as trade and transport, were ever not global in nature, he notes a rapid increase in global social connections. Analogous to these observations is the tearing pace of development of communications and transport technologies, the dismantling of barriers in the movement of people and goods, as well as an increase in temporary or permanent migration processes across nations and continents (Massey 1994; Robertson 1992; Harvey 1990). Giddens points to an altered relationship between local and global, in which an incident in Fukushima in Japan, for example, can affect changes in far-away locations, as occurred in Germany, with the proclamation of the so-called '*Energiewende*' (the German national strategy to enhance renewable energy) (see Chapter 6). Robertson names these phenomena the 'compression of the world'[13] (Robertson 1992: 8), as economic, political, technological, social, and cultural conditions now appear to be tightly intermeshed across the globe.[14]

These accounts taken together form the *theory of social deterritorialisation*, according to which society and territory should be seen as ever less congruent (see also Schroer 2006: 161). In connection with the idea of communicative constructivism, the following line of thought has high significance: if communication and migration are increasingly taking place beyond the traditional boundaries of places and regions, then knowledge-related processes, in particular internalisation, externalisation, and objectification, might also be increasingly uncoupled from classical territorial boundaries. Thus the connection of cultural phenomena to historically defined territories would increasingly be called into question. Actors' internalised stocks of knowledge could therefore increasingly be organised according to other criteria, and new forms of cultural spatialisation should therefore also be observable. Authors in the social-science literature continue, however, to reach different conclusions about currently occurring processes and structures (for more detail, see for example Schubert 2012: 173ff.; Leggewie 1999).

Proponents of the *theory of convergence* argue that globalisation and trans nationalisation processes correlate to cultural homogenisation processes. Albrow, for example, sees in globalisation 'all those processes, by which the people of the world are incorporated into a single world society' (Albrow 1990: 9). Robertson assumes an 'intensification of the consciousness of the world as a whole' (Robertson 1992: 8). This would be the case, for instance, if a global appreciation of the problems of climate change emerged and possible challenges were tackled together. The neo-institutionalist John W. Meyer coined the idea of the formation of a 'world culture' (Meyer 2007: 261), in which there would be an increase in so-called cultural universals in the form of globally shared beliefs and values (ibid.).

In contrast, the proponents of *differentiation theory* postulate that traditional cultural boundaries will also continue to exist locally. Huntington, this grouping's most popular and probably also most controversial representative, saw in the 1990s the danger of deepening political faultlines and a 'clash of civilisations' (Huntington 1996). He emphasised in particular the role of historic, religiously characterised knowledge that continued to exist in territorially embedded civilisations (ibid.: 53). Inglehart took up this idea and came to the conclusion that large territorial cultural differences would continue to be observable worldwide and within the EU (Inglehart 2006, 2002).

There is a further position that comes under the term *hybridisation* or *creolisation theory* (Schubert 2012: 195; Leggewie 1999; Pieterse 1994). Creole refers to the language that emerged in the Caribbean from the mixture of the languages of African-origin slaves and the colonial ruling class (Leggewie 1999). Through this intensified exchange, mixtures of this kind and of other elements of knowledge can increasingly arise, and with them 'translocal mélange cultures' (Pieterse 1994: 161).

It is important to state that these various positions were formulated virtually concurrently in the last 25 years. How is it, then, that despite the empirical verification to which almost all the authors mentioned had access, such different conclusions have been reached?

First of all, it is partly because the authors focus on different set pieces of socially shared knowledge (such as values, beliefs, or linguistic structures). In addition to these *aspectual* differences, there is a further issue of significance, namely the *limited reflection on the underlying cultural concept of space*. Thus the two initial positions (homogenisation vs. differentiation) appear contrary to one another because they pursue implicit *container space concepts* of cultures (see Section 2.2.2). Thus the approach from differentiation theory presumes territorially delimitable cultural spheres. By contrast, homogenisation theory views a priori regionally delimitable cultures as coming closer through increasing trans-territorial and communications processes, eventually converging into one world culture.[15] Only in the approach of hybridisation theory is there consideration of theoretical alternatives to the container space, for example in discussion of 'translocal cultures' (Pieterse 1994: 177). However, even here it has been rare for further exploration using the appropriate means and methods of empirical social research to be suggested.

It is therefore hardly surprising that there has recently been an increasing demand for the development of new appropriate approaches in social-science theorisation and empirical analysis. Ulrich Beck, for example, criticises the 'Europe-blindness of sociology' (Beck 2005: 3), turning away from a 'methodological nationalism' (ibid.), and with his choice of phrasing apparently taking issue with many (macro-)sociologists' adherence to the nation state as the prevailing unit of examination (ibid.). Social and cultural changes in intra-, inter-, trans-, or supra-national European spaces have accordingly failed to register on the sociological radar (see also Beck and Grande 2010). Wimmer and Glick-Schiller (2002: 301) also turn away from a too hasty adherence to national and other territorial units of comparison, while at the same time warning of the dangers of a 'methodological fluidism' (ibid.: 326). Possible deterritorialisation processes should therefore certainly be considered, but not at the cost of forgetting to question what significance can still be ascribed to classical territories, including nation states (ibid.).

What is therefore currently lacking is an analytical apparatus able in theory and practice to meet the needs not only of a possible deterritorialisation of culture and its equally likely consequences, but also the continuing existence of territorial boundaries. The question therefore arises how cultural boundaries can be appropriately investigated in the present study.

As already indicated, the central problem appears to be that in the present culture-related literature there has hardly been any thorough consideration of implicit concepts of space. It is this gap that the following will attempt to close.

2.2.2 Conceptualisations of space in the past and present: absolute and relational space – spatial dimensionality – social and physical space

> 'By asking these questions, then, we must raise the whole problem about place – not only as to what it is, but even whether there is such a thing.'
>
> (Aristotle 1984 [original undated]: 356)

Even after more than 2000 years of documented investigation of space, it seems hardly possible to reach a clear ontological definition of it. Anyone concerning themselves with matters of space will regularly seem to reach the limits of their imaginative capabilities and sensory navigability. It therefore seems all the more important to transpose this question from an ontological one to one viewed from the perspective of a sociology of knowledge, where the concern is 'how space has been conceived of until now' (Schroer 2006: 29), and to question what the implications are that then arise for social theory and the empirical analysis of cultural spaces (see also Kajetzke and Schroer 2010: 201).

Concepts or 'idea[s] of space' (Löw 2001: 16) will be systematised below along the three axes of (1) absolute and relational space, (2) spatial dimensionality, and (3) social and physical space.

Absolute and relational space

Looking over the long history of spatial theory it is possible, despite the 'sheer, endless [...] number of accounts of what space is' (Schroer 2006: 30), to recognise the opposition of *absolute and relational concepts* of space as a pattern repeating since antiquity (ibid.).

Absolute conceptions of space have the common feature that they differentiate dualistically between space and material bodies (Kajetzke and Schroer 2010: 193; Schroer 2006: 30; Löw 2001: 17). Einstein would later find fault with what he called *container space concepts* (Einstein 1960: xiii). Space is thus conceived of as an independent, absolute entity, enveloping material and functioning as a container for all objects. The absolute concept of space can be traced back to antiquity and to Aristotle.[16] He justified his reasoning on the possibility of material changes in fixed locations:

> Where water now is, there in turn, when the water has gone out as from a vessel, air is present.
>
> (Aristotle 1984 [original undated]: 355)

In contrast, relational concepts of space monistically assume that space and material bodies are not to be understood as separate from one another. Accordingly, space first of all consists of the mutual *relational* positions of things (i.e. material bodies) (Kajetzke and Schroer 2010: 19; Schmidt 2009: 299; Schroer 2006: 30). An important representative of this concept, Gottfried Wilhelm Leibniz (1646–1716), argued as follows:

> One forms this [relational] concept of space in the following way. One observes that various things exist simultaneously and finds in them a certain order of their coexistence which is, according to their relationship, more or less simple. This is their mutual location or distance from one another.
>
> (Leibniz 1996 [original undated]: 133)[17]

In physics at the turn of the twentieth century, the relational concept of space was taken up and further developed by exponents such as Ernst Mach and Albert Einstein (Löw 2001: 17).[18] At the same time, the idea of the existence of a superordinate space independent of the material world was abandoned in favour of one built upon the relative location of bodies (Blotevogel 2004: 832).

Various authors have been able to show, however, that both concepts of space have had influence until the present in the cultural and social sciences and in the everyday thinking of actors (Schroer 2007, 2006; Günzel 2006; Löw 2001). Martina Löw, for example, comes to the conclusion that virtually all sociological work on space 'with the exception of phenomenological work, is based on the basic premises of one or the other' (Löw 2001: 17). She sees the container space concept as 'to this day the most dominant concept of space in everyday understanding' (Löw 2001: 17). In her own 'spatial sociology' (ibid.: 271), Löw herself speaks for a relational concept of space (ibid.: 271). Markus Schroer also emphasises that in sociological contexts there is much in favour of an integration of the relational concept of space into social theory, as it stresses the active emergence of space instead of assuming an already-existing, absolute space (Schroer 2007: 42f.). At the same time he warns against a new 'absolutisation of the relational concept of space' (ibid.), since in sociological research it is less a question of 'the universal validity of claimed definitions of space than it is of observing the spatial constructions of various actors' (ibid.).[19]

Both conceptions should therefore be considered in social-science analysis, as 'none of these qualities is imperative to space' (Kajetzke and Schroer 2010: 203). It is much more important that empirical investigation 'provides theoretical diversity, and room for thought' (ibid.).

Spatial dimensionality

Alongside the absolute and relative concepts of space, the question of spatial dimensionality has also been investigated in the past (Schmidt 2009: 291). Descartes defined space as occupied by material bodies by its 'length, width, and depth' (Descartes 2006 [1644]: 44). With the development of his two- and three-dimensional Cartesian coordinate system, Descartes combined the Euclidean geometry of antiquity with arithmetic (Löw 2001: 27). In this way he made a multitude of terrestrial spatial phenomena calculable. This coordinate system later met its limits, however, in theoretical and empirical physics. Thus Einstein introduced a further dimension in his relational concept of space, namely that of time (Schmidt 2009: 291). Einstein's combined relational model of space, in which time and matter determine one another, now proves itself more precise in space-technology applications in comparison with classical Euclidean and Cartesian geometry and arithmetic.[20] Thus Schmidt also claims that the question of the determination of space and time in physics is currently 'resolved, in principle' (2009: 292), although dissent continues to prevail with regard to the number of possible spatial dimensions. Thus, for example, current string and M-theory deals with up to 26 spatial dimensions (ibid.).

It is important to emphasise at this point that space does not have to be limited to the three Cartesian dimensions, but can rather be conceived of in many diverse dimensionalities.

Physical space – social space

Space hardly counts among the classic themes of sociology. This may be related to the story of its emergence, since the establishment of the discipline was long characterised by attempts to distinguish itself from other fields such as biology, psychology, and particularly geography, which had proclaimed space as its subject of study (Schroer 2007: 37). Kajetzke and Schroer rightly point to today's much quoted and lamented 'spatial blindness' (Läpple 1991: 163), which served for a long time as a '[self-]prescribed blindness' to disciplinary delimitation (Kajetzke and Schroer 2010: 192). Space can accordingly first become a subject of sociology if it 'isn't dealt with as something naturally given, but socially manufactured' (ibid.). Schroer nonetheless reminds us that this perspective reflects a purely modern 'separation of society and nature' (Schroer 2007: 38).[21] This distinction was very powerful in the twentieth century, however, and led to the differentiation of physical and social space by various authors.

Already by the 1920s, the sociologist Pitirim A. Sorokin (Kajetzke and Schroer 2010: 198f.; Sorokin 1998 [1927]: 3) emphasised: 'Social space is something quite different from geometrical space' (Sorokin 1998 [1927]: 3). He differentiates between distance in the physical/geometric and social senses, for example when he notes that a servant and master might indeed be physically close, but at the same time in social space, with regard to status dispositions, are very far from one another (ibid.). In a similar way, Bourdieu also constructs a multi-dimensional, relational 'space of social positions' (Bourdieu 1982: 212).[22]

Since the 1970s, however, there has been increasing support for the idea that a conceptual separation between 'natural/physical' and 'social' space is illusory. Henri Lefebvre, for instance, was one of the first to break through this binary separation, emphasising the active social construction of *all* socially shared and utilised space (Kajetzke and Schroer 2010: 196; Lefebvre 1991 [1974]: 355).[23] At the turn of the millennium, physical and social space are hardly separated any more in Martina Löw's work:[24] she would much rather explain the creation of spaces 'out of the (active) arrangement of social goods and people' (Löw 2001: 271). Here, then, is a new conceptualisation of the social, in that both people and 'goods' are considered in her relational concept of space.

Finally these approaches – contextualised in climate-change-related research – lead one to the question of the action relevance of physical space itself. In this context, Christmann and Ibert (2012) have argued along the lines of Bruno Latour's actor-network approach that both the human attribution of meaning to physical objects as well as those objects' presence or absence can imply different courses of action *simultaneously* (ibid.: 265ff.). In the analysis of ways of dealing with climate change, it must therefore be assumed from the outset that local and regional physical conditions can also have relevance for actors' shared stocks of knowledge. My supposition here is that natural and social space are not fundamentally separate, and furthermore that 'natural' objects can be ascribed with sociality.[25] This seems of relevance, for instance, when adaptation measures against different natural threats such as flooding or storms should be carried out, while at the same time we have to assume that there is shared knowledge about locally varying landscape conditions (height above sea level, etc.). Where this shows itself to be of relevance, I shall return to it in the course of this study (see Chapters 9 and 10).

2.3 Synthesis: cultural space as relational space

Having discussed the historical concepts of space, the following sections will explore current cultural spatialisations that take into account the possible changes discussed in relation to territoriality and culture. In doing so, relational spatial concepts will be employed. Aspects such as dimensionality and materiality will also be considered. Relational cultural spaces should be both theoretically and empirically justified. Thus I shall now discuss my theoretical (Section 2.3.1) and then empirical foundation (Section 2.3.2), in order to present the further course of action.

2.3.1 *Relational cultural space I: social-theoretical conceptualisation*

Building upon the considerations of the previous sections, a relational concept of cultural space can now be further detailed based on the following idea:

> Relational cultural spaces can be understood as relational configurations of actors that share aspectual knowledge constructions.

At this point one should recall the conceptualisation of culture as 'socially shared knowledge' discussed above. The authors of social and communicative constructivism have described how actors generate certain stocks of common knowledge (see Section 2.1.3). Considering climate change, this implies, for example, shared appreciations of possible problems or similar practices (see Section 2.1.4). These knowledge constructions can be more or less similar. It is therefore important to emphasise that in this conception, *similarities and differences in knowledge constructions* each function as culture defining moments. As carriers of knowledge constructions, actors can be located as more or less distant from one another, consequently forming relational cultural spaces. According to this concept, cultures manifest themselves through actors' shared knowledge constructions and emerge as (relational) differences from the knowledge held by other actors (see also Leggewie 1999).

It can therefore be supposed that actors' knowledge can only be *aspectually* externalised and internalised. Social scientists – and all other cultural 'observers' – can only investigate knowledge aspectually (in particular when social scientists consider meeting the quality criterion of precision, see Section 2.1.3). As different aspects can be focused upon, it follows that *multiple cultural affiliations* are possible for each individual actor. Cultural affiliations are indeed dependent upon which aspect of knowledge is being focused upon. Actors can thus stand culturally near and distant to one another at one and the same time, according to which aspect is being considered by the observer.[26]

If actors' knowledge can be explored aspectually, this has consequences for the *dimensionality of relational cultural spaces*. As there are many conceivable aspects of knowledge, various aspects can *simultaneously* be applicable in the determination of relational cultural spaces, such as when 'food cultures' are defined according to eating-implement preferences, seasonings used, taboos, etc.[27] As a consequence, relational cultural space can occur in multiple dimensions, analogous to the above-mentioned spatial concept of modern physics (see Section 2.2.2).

At the same time, the concept of cultural space as the relational configurations of knowledge shared by actors leads to questions about the relation of *society and matter* (see Section 2.2.2). For the present study, two aspects appear above all others to be of relevance: first, the monistic concept of *'knowledge as space'*, and second the question, also current in climate discourse, about the categorical tension between *knowledge and the material environment*.

Just as Leibniz, Einstein, and Mach adhere to the idea that material and space are not necessarily two separate entities (see Section 2.2.2), so too can the relationship between knowledge and space be *monistically* conceived. As cultures are principally constituted of shared knowledge in the relational concept of cultural space, the constellation of actors' knowledge appears to the observer *'as' space* (and not 'in' space). Accordingly, relational cultural space is above all social space, as the carriers of knowledge are each human, social actors.

The categorical concept of social space brings one back to the now questionable categorical separation, discussed in Section 2.2.2, between *materiality and sociality*. If one accepts Löw's idea, according to which space can be defined as arising 'out of the arrangement of goods and people' (Löw 2001: 271), so it is implied that actors' material environments can be of relevance even to the genesis of cultural spaces. This can be seen particularly in climate cultures, since the material environment can become an object of internalised knowledge. If, for example, shared preferences for climate-adaptation practices (see Section 2.1.4) can be distinguished as climate cultures, then a way must be found not to ignore existing knowledge about the sociality of physical environments, such as the landscape conditions which act as influencing variables (see also Chapters 9 and 10).

To summarise the concept once again: relational cultural spaces appear as relational configurations of actors with shared knowledge constructions. Because cultural space can be considered aspectually, the same actors can be related to one another in many different ways depending on the observed aspects of knowledge. The dimensionality of cultural space is theoretically open, but should be empirically established. Material environments can also be of relevance for the genesis of cultural spaces.

2.3.2 *Relational cultural space II: methodology for investigation*

If the congruence of culture and place/region comes into question in social theory, then ways must also be found to *empirically* investigate new possible spatial constellations of shared knowledge. The theoretical conceptualisations of the previous section should therefore be expanded with considerations about the methodical exploration of relational cultural spaces.

The unreflective implementation of cultural units *ahead of* empirical analysis was criticised (see Section 2.2.1) as a central problem of so-called 'methodological nationalism' (Beck and Grande 2010). This problem of the *cultural a priori* can be resolved by the relational concept of cultural space, as it allows for the application of structure-detecting analytical methods (see for example Backhaus *et al.* 2008). In the course of the analysis itself, it is determined which actors should be grouped together as distinct cultures according to their similarities in knowledge. The relational concept of cultural space thus also enables us to separate territory and culture analytically.[28]

Within the first steps of the empirical analysis, it is thus possible to determine *aspectual climate cultures*, allowing cluster analysis to be used (see for example Bacher *et al.* 2010; Backhaus *et al.* 2008: 389ff.). Cluster analyses summarise the properties of actors as 'geometric points in a multi-dimensional variable space' (Wiedenbeck and Züll 2010: 528) and describe their relational positions in the form of distance ratios. In this way, relational configurations can be detected and distinguished from one another, according to knowledge constructions. As a comprehensive body of literature is available for cluster analyses, the procedure will only be sketched here with a number of fictional examples.

For the investigation, the *number of observed aspects or knowledge constructions* must first be established as variables. These determine the *dimensionality of cultural spaces*. Thus Figure 2.2(a) shows a cultural space in two dimensions, with actors positioned with regard to their preferences for climate-adaptation measures (dimension 1) and climate-mitigation measures (dimension 2). Actors assessed measures on a scale from –5 (total rejection) to +5 (total advocacy). It should be emphasised that further constructions could be introduced, and thus other dimensional arrangements are conceivable.[29]

Although the actors' assessments appear here in a Cartesian coordinate system, they can now be *related relationally* on the basis of the similarity of their preferences for measures. Figure 2.2a shows the similarities between actors in the form of *relational distances*.[30] The actors are thus positioned relative to their proximity to, or distance from, one another. This corresponds to the constellation of the relational concept of space, and can be represented mathematically in the form of a distance matrix (Backhaus *et al.* 2008: 393ff.).

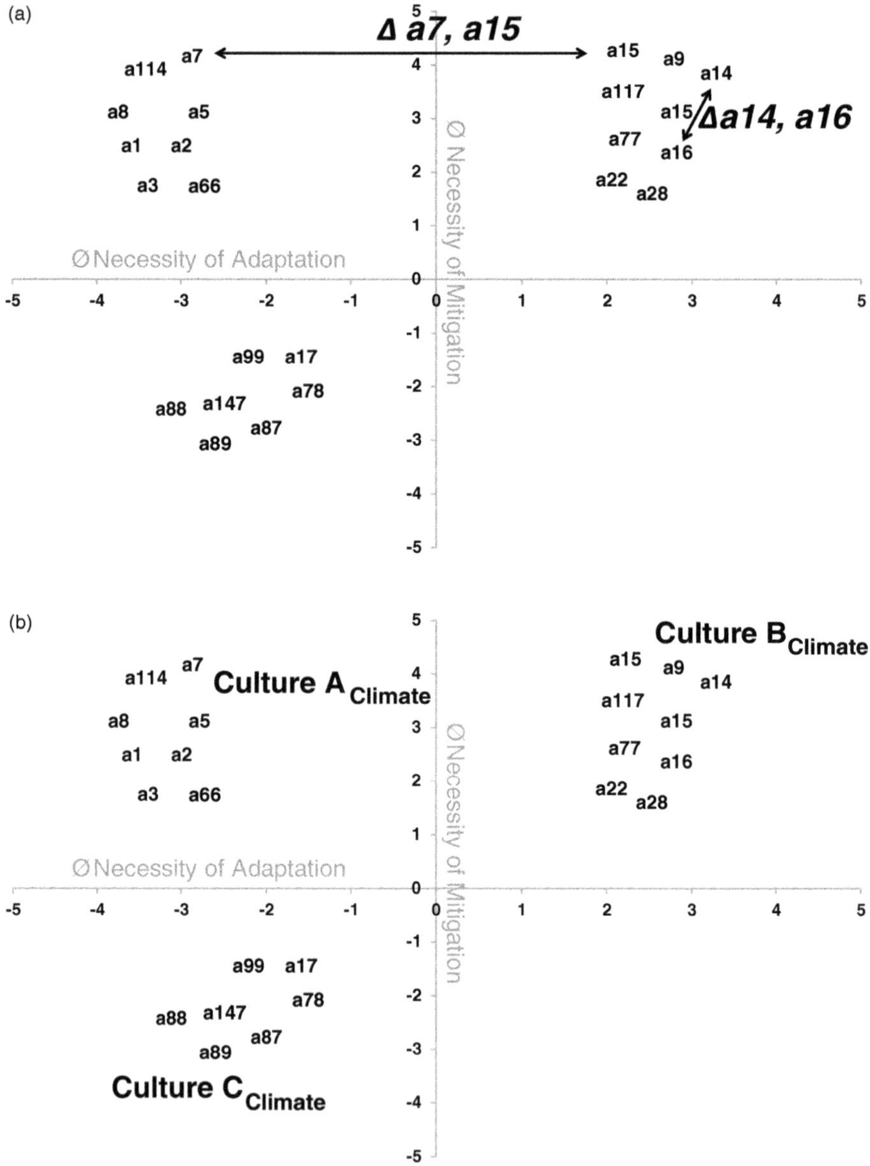

Figure 2.2 Visualisation of relational cultural spaces (fictional examples).

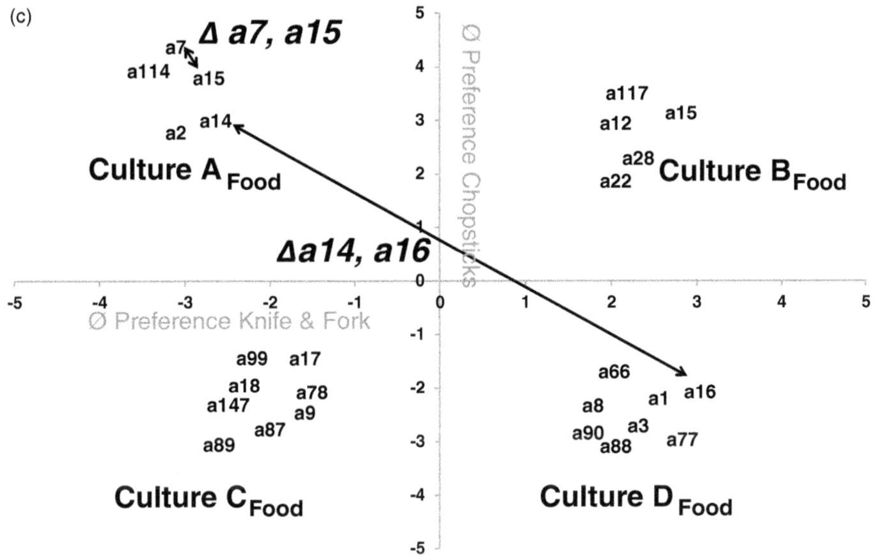

Figure 2.2 Continued.

On the basis of the proximity and distance ratios, determined within the framework of the distance/similarity matrix, actors can in the next step *be grouped into cultures*. The method corresponds, for example, to the grouping often used in hierarchical clustering methods by means of fusion algorithms (Backhaus *et al.* 2008: 411). Actors who are close to one another in their assessments share common patterns of interpretation or knowledge about the aspects considered. They can therefore be interpreted as cultures according to the knowledge-based concept of culture.[31] In the present example, three climate cultures are apparent (see Figure 2.2b): Culture A (a purely climate-mitigation culture), Culture B (a combined climate-mitigation and adaptation culture), and Culture C (a culture of climate-mitigation and adaptation rejection).

These cultures are *aspectually configured* and can be interpreted neither as holistic nor essentialist. This is illustrated in Figure 2.2c. Although the same actors are considered, they now appear, for example, with regard to 'food cultures' in completely different relational configurations. In the example, four food cultures are apparent, whereas only three cultures were found previously.[32] These examples also show that actors can be subject to 'plural affiliations' (Sen 2006: 23) with regard to cultural groupings. Individual actors can thus be regarded, as a rule, as belonging to more than one simple culture.

2.4 Interim conclusion

The aim of this chapter was to contextualise and specify the central concepts 'climate cultures' and 'space'. It has been shown that the concept of culture in social-science climate research has recurred in diverse semantic guises, ranging from Herder's holistic orientation to the knowledge-based conception. For the present position, and following the typology of Reckwitz (2000), further systematic differentiation has been made between the normative, holistic, differentiation, and knowledge-based concepts of culture. The knowledge-based understanding proved to be the most appropriate for the study: cultures can accordingly be considered as actors' socially shared stocks of knowledge. According to the social-constructivist theoretical tradition, knowledge appears to be socially contingent, and is externalised, internalised, and objectified via communication processes.

It has been argued that stocks of knowledge, and therefore cultures, can only ever be considered aspectually. The aim of this study is therefore to provide a systematic description and explanation for the commonalities and differences of socially shared stocks of knowledge with respect to climate change. This common climate-change-related knowledge will be termed climate cultures, whereby distinctions can be made between various vulnerability constructions (i.e. perceptions of climate-related problems and opportunities) and resilience constructions (i.e. knowledge of and preferences for climate-related practices).

This raises the question of how the spatial constellation of climate cultures can be theorised and empirically investigated. According to the theory of deterritorialisation, climate cultures might also emerge in novel constellations. The criticism has been raised that for a long time, social-science analysis has barely reacted in terms of either its theory or method to these new possible conditions, and that there is therefore a lack of applications for the systematic consideration of these potential new constellations. Moreover, it seems problematic that in comparative cultural research, implicitly present concepts of space are seldom reflected upon.

Having thus pursued different historically formed concepts of space, it is apparent that space can be conceived as an absolute (container) and/or a relational structure; that space has, moreover, appeared in the social sciences as social and/or physical; and finally, that space can be constructed in different dimensions.

In order to be able to undertake more precise analysis of climate-culture constellations, a relational concept of cultural space has been proposed, in which cultures are understood as relational configurations of actors with socially shared knowledge constructions. Subsequently, it will be possible to investigate which factors contribute to the separation of the stocks of knowledge and thus provide explanations for climate-culture differences. This approach is especially suitable for cross-sectional analyses at particular points in time.

Finally, I provide a brief overview of the individual analytical steps required in this study for the description and explanation of climate-culture differences.

Step 1: Analysis of discourse (core developments and categories). Before climate cultures can be determined according to the relational concept of space, the question arises about the central knowledge constructions in the research field. Thus the relevant discourses surrounding climate change must first be analysed more closely. Several central developments in particular are to be explored with regard to the topics (*topoi*) which are central to the discourses, as well as the classifications and categories used. This is carried out in Chapters 3 to 6. Building upon this, a survey can be developed to determine what knowledge constructions are internalised or socially shared by the actors, and what positions are represented in the discourse. The empirical method and its operationalisation are described in detail in Chapter 7.

Step 2: Empirical analysis of categories shared by respondents. On the basis of the data obtained, it will then be necessary to ascertain which central categories and classifications are shared by *actors*. To this end, multi-dimensional scaling, factor analysis, and reliability analyses will be utilised.[33] The central categories will then serve to determine the variables for consideration in the cluster analyses, and thus also the dimensionality of the relational cultural spaces.

Step 3: Grouping of respondents into climate cultures. This is the analytical core of determining relational climate cultures: finding groups that are most similar in their perceptions of opportunities and problems, as well as in their preferences for measures. To this end, hierarchical cluster analyses will be carried out, allowing actors to be positioned and grouped according to their proximity and distance.

Step 4: Testing of hypotheses to explain cultural differences. Following the determination of climate cultures according to the relational concept of space, the factors contributing to an explanation of the different group affiliations can be sought. Thus it is only after these have been determined that explanatory factors for differences can be checked. It will also be possible to clarify to what extent classical territorial borders such as nation states or other historical regions serve as (climate-)cultural borders.

Notes

1 The concept of climate will be dealt with in Chapter 3.

2 Crate (2011) offers a comprehensive overview of global, cultural-anthropological research fields surrounding climate and culture.

3 In the late 1980s, Jeffrey Alexander (1988) used the concept of the *cultural turn* to describe earlier socio-theoretical developments. Reckwitz summarises the debates up to the new millennium and describes the *cultural turn* as the 'conceptual shift in the social sciences in the last third of the twentieth century towards perspectives from cultural studies and cultural theory' (2000: 22). Hölscher notes a tenfold increase in usage of the English term 'culture' from the 1960s into the early 2000s in sociological abstracts (Hölscher 2006: 55f.).

4 A concisely edited conceptual history from antiquity to the twentieth century can be found, for instance, in Perpeet (1984). Williams summarises the development of usage in the Anglo-American sphere (Williams 1977). In his typology, Reckwitz (2000: 64ff.) differentiates between four classes of the concept of culture. Rippl and

Seipel describe six different kinds of culture definition (Rippl and Seipel 2008: 15). Moebius's systematic classification enumerates around 13 categories (Moebius 2009, further enumeration of this kind can also be found in Schubert 2012: 136).

5 Thus Tylor defined culture as 'knowledge, belief, art, morals, law, custom, and any other capabilities and habits acquired by man as [a] member of society' (Tylor 1871, quoted in Reckwitz 2000: 74).

6 Recently theorists have again begun to question this conception. Philipp Descola (2013) understands the differentiation between human and non-human culture as specifically western and points to other conceptions, such as Native American totemism, in which animal groups can fall within the frameworks of familial clan kinships. In his actor-network approach, Bruno Latour (1996) conceptualises the relevance of non-human and non-living entities for social action.

7 This concept of culture is closely related to the development of social-science efforts to describe complex, labour-sharing societies. One of the most discussed studies is Talcott Parsons's structural-functionalist conception, in which culture is introduced alongside politics, economics, and social community as one of four functionally necessary subsystems within society as a whole (Reckwitz 2000: 81; Parsons 1961: 38).

8 There is a further problem with the normative concept, which is its unreflective positioning towards the '*Werturteilsproblem*' (question of value judgements in science). In the Weberian understanding, for example, value judgements are only permissible during the selection of research questions and should be avoided in the further course of the research (Weber 1991: 256ff.; Albert 1972). Academics wishing explicitly to follow this postulate, despite much disagreement about value judgements in sociology in the twentieth century, must at least reject a normative conception within the framework of the operationalisation of culture.

9 The concept of *objectification*, linked to Hegel and Marx, indicates the process of reification, in which something becomes an object that is acknowledged as such by members of a common group (Knoblauch 2010: 161; Berger and Luckmann 1966: 106). To reach this point, subjective meaning must first be communicated between actors in the process of *externalisation*. (Of course, language itself already creates a scheme of differentiation that contains such objectifications.) *Internalisation* indicates that this meaning has been appropriated or 'incorporated' by other actors (Berger and Luckmann 1966: 78). Objectification, externalisation, and internalisation thus create the framework for the genesis of socially shared knowledge, and therefore also of cultures.

10 Important lines of development in the genesis of scientific knowledge about climate change will be traced in Section 3.2. Here it will be shown that various technological and social circumstances played an important role in first enabling specific perspectives and problem formulations to form.

At this point it must however be explicitly pointed out that *social constructivism* is not to be equated with *radical constructivism* or even *scepticism*. Under no circumstances does it imply that reality is to be understood as a mere construction of individual consciousness. The central thesis is that reality arises systematically out of agency and consolidates in stable structures (Schnettler 2011: 162). Climate change itself is accordingly neither empirically preconditionless nor arbitrarily constructed. Indeed, systematic observations, problem formulations, and projections about climate change have 'real' systematic consequences for global society.

However, as climate knowledge is to be seen as socially contingent, the systematics of its genesis, as well as of actor-specific differences in knowledge, come into view. The societal consequences of this are also of high relevance when one brings to mind that social conflicts surrounding climate change represent conflicts of knowledge and interpretation. An example of this is the way in which the scientific formulation of problems by central actors is called into question (see also Christmann *et al.* 2012: 26).

11 Reckwitz likewise differentiates between cognitive knowledge, practice, and text as forms of knowledge (Reckwitz 2000: 710).

12 For a more precise definition of the conceptual pair 'physical – social space', see Section 2.2.2.

13 At the same time, in human geography, Harvey adds another dimension and speaks of 'time-space compression' (1990).

14 As early as the 1960s, the media theorist McLuhan formulated his much-cited theory on the emergence of a so-called *global village*, according to which the world would continue to merge as a result of technological and media developments (Kessl and Reutlinger 2010: 146; McLuhan 1994 [1964]: 87).

15 In connection with this, the hegemonic constellation of a presumed 'western' culture is regularly discussed, for example under the slogans of 'Americanisation' or 'McDonaldisation' (Berger 2002; Leggewie 1999).

16 Historic representatives of absolute space include Aristotle, Copernicus, Galileo, and Newton (see also Löw 2001: 17; Newton 1999). Aristotle developed the idea of an ontologically autonomous space as container (Aristotle 1984 [original undated]: 355). In doing so he assumes a fixed, finite space, whose centre is formed by the Earth (see also Löw 2001: 24). His ideas persisted in Europe throughout the Middle Ages (influenced by Christianity) into the Renaissance (ibid.: 26). Isaac Newton (1643–1727) finally dismissed the Aristotelian concept of finitude and postulated an absolute, homogenous, and infinite concept of space.

17 For Leibniz, space itself can have no substantial quality: 'If space [...] were a quality, then it would need to be a quality of a certain substance. But to what substance should empty, finite space [...] belong?' (Leibniz 2006 [1715]: 63).

18 The positivist Mach criticises the notion of superordinate container space as unexperiencable and therefore metaphysical (Löw 2001: 31). Einstein's theory of relativity contradicts the Newtonian concept of homogenous space by showing that Newtonian mechanics only apply in special cases within a much more complex universe (ibid.: 34). He states that space and time phenomena vary depending on an observer's frame of reference, and are therefore not universally or absolutely valid (ibid.; Einstein 2006 [1930]: 99).

19 It should be mentioned that the recent increased importance of the relational understanding of space has not only been limited to sociology. Some developments in geography should especially be noted here, as the discipline has always gained much of its legitimacy and identity through its close relation to spatial issues. Following Werlen (2000: 72) we can state that human geography and sociology converged during the second half of the twentieth century, as seen for example in the development of 'humanistic geography' (e.g. with the integration of Alfred Schütz's phenomenology into geography by Anne Buttimer), 'radical geography' (e.g. David Harvey, Richard Peets using Marx), and 'critical human geography' (e.g. Derek Gregory, Nigel Thrift referring to Anthony Giddens). Coincidently, the 'absolute' understanding of space (e.g. Alfred Hettner, Friedrich Ratzel) lost prominence in geography.

As global changes became more and more apparent by the end of the twentieth century, attention to translocal spatial phenomena increased in importance, and several authors from the field of geography and their ideas (e.g. David Harvey's idea of 'time-space compression' (1990: 260), or Doreen Massey's 'global sense of place' (1994: 146), attracted a great deal of attention in sociology (e.g. Löw 2001). 'Thinking space' (Crang and Thrift 2000) developed as a significant approach due to its greater consideration of different concepts of space, and the explicit use of relational ideas has also recently emerged (see e.g. Bathelt and Glückler, who mention 'knowledge as a relational source'; 2011: 63).

20 For example, satellite Global Positioning Systems (GPS) make adjustments of measurements based upon Einstein's theories (Ley *et al.* 2009: 569; Korb 2008).

21 Early authors such as Simmel and Durkheim often took pains to distinguish themselves from physical-space deterministic thinking (see also Kajetzke and Schroer 2010: 193f.). Simmel states, for example, that 'not the space, but rather the structure and composition of its parts proceeding from the soul have social significance' (Simmel 1992 [original undated]: 688). Durkheim points out that spatial categories such as north or south are inevitably 'of social origin' (Durkheim, 1998 [original undated]: 30).

22 Here, actors can be arranged according to their available stocks of different forms of capital (see Kajetzke and Schroer 2010: 198f.). At the same time, Bourdieu emphasises interactions between positions in social and physical space, such as when places occupied or inhabited by actors in physical space serve as indicators for positions in social space (ibid.: 200).

23 In addition, Lefebvre differentiates between spatial practice (perception and production of places), spatial representation (knowledge and symbols, the instruments of architects and planners), and representational spaces (spaces imagined by actors) (Kajetzke and Schroer 2010: 196; Lefebvre 1991 [1974]).

24 Löw combines Giddens's idea, that spaces arise from the active organisation and positioning of actors (spacing), with Norbert Elias's concept of 'synthesis', which emphasises the active formation and memorisation of spatial perceptions by actors (Kajetzke and Schroer 2010: 201; Löw 2001: 271).

25 The threat from the sea, for example, might be interpreted differently inland than in coastal regions.

26 This can be exemplified by the following aspects: 'Climate cultures: Appreciation of climate-change problems' and 'Food cultures: Preferred eating implements'. Under the first aspect, actors with a high awareness of climate change are closer to one another than to actors with a very low awareness. However, the relative spatial arrangement changes completely if, instead of climate cultures, food cultures are considered. In this case, those who prefer to use chopsticks would be closer to one another than to those who prefer to use a knife and fork. This leads to actors appearing close to one another in one instance, and widely separated in another. This should make it clear that the present understanding of cultural spaces is characterised by the 'recognition of different affiliations' (Sen 2006: 19). Cultural affinities are given not only by the empirical world, but also result from the selective focus of the observer. The observer decides on the aspects of knowledge to be considered, at the same time creating the potential for cultural affiliation.

27 At the same time, Inglehart observed different socially shared political, family, and gender values in comparison with 'cultural spheres' (Inglehart 2006).

28 Only once the relational determination of cultures has been made can it be *tested* whether local or regional boundaries (such as national boundaries) continue to have a discriminatory effect on actors' shared knowledge (or cultures), as well as which other factors are of relevance. This can also be used to check the extent to which container-space oriented cultural concepts are empirically evident.

29 In graphic representation, however, there remains a maximum of three dimensions available to us.

30 In the example it is clear that actors a7 and a15 show much less similarity than actors a14 and a16.

31 In Figure 2.2b a14 and a16, for example, exhibit shared knowledge constructions.

32 Here the same actors have given their preference for dining implements, more specifically for 'knife and fork' vs. 'chopsticks'. Now different actors appear proximal or distant (i.e. similar or dissimilar) in their positions relative to one another, and thus can be grouped as belonging to the same culture (here, for example, a7 and a15). The example indicates four different food cultures, each with collectively shared practices. They differentiate themselves according to the different combinations of commonly preferred practices (see Figure 2.2c).

33 Diverging from the German version of this book, the results of the multi-dimensional scale (MDS) will not be discussed here. All the categories used here have been tested by both MDS and factor analysis (see Heimann 2017). The results can nevertheless be understood by monitoring the factor analysis alone. In addition, not all the results given for the reliability analyses will be discussed. All results can, however, be found in the Appendix of the German version (Heimann 2017). The author will be happy to provide an English version of the German Appendix on request.

References

Adger, W. Neil (2006) 'Vulnerability'. *Global Environmental Change* 16: 268–281.

Adger, W. Neil, Barnett, Jon, Brown, Katrina, Marshall, Nadine, and O'Brien, Karen (2013) 'Cultural Dimensions of Climate Change Impacts and Adaptation'. *Nature Climate Change* 3 (2): 112–117.

Albert, Hans (1972) 'Wertfreiheit (Werturteilsproblem)'. In: Bernsdorf, Wilhelm (ed.): *Wörterbuch der Soziologie*, Volume 3. Frankfurt am Main: Fischer, pp. 914–916.

Albrow, Martin (1990) 'Introduction'. In: Albrow, Martin and King, Elisabeth (ed.): *Globalization, Knowledge and Society. Readings from International Sociology*. London: Sage, pp. 3–13.

Alexander, Jeffrey C. (1988) 'The New Theoretical Movement'. In: Smelser, Neil J. (ed.): *Handbook of Sociology*. Beverly Hills: Sage, pp. 77–101.

Amundsen, Helene (2015) 'Place Attachment as a Driver of Adaptation in Coastal Communities in Northern Norway'. *Local Environment: The International Journal of Justice and Sustainability* 20 (3): 257–276.

Aristotle (1984 [original undated]) 'Physics'. In: Barnes, Jonathan (ed.): *The Complete Works of Aristotle*, Volume 1. Princeton: Bollingen Series LXXI, pp. 315–446.

Bacher, Johann, Pöge, Andreas, and Wenzig, Knut (ed.) (2010) *Clusteranalyse. Anwendungsorientierte Einführung in Klassifikationsverfahren*. München: Oldenbourg Verlag.

Backhaus, Klaus, Erichson, Bernd, Plinke, Wulff, and Weiber, Rolf (2008) *Multivariate Analysemethoden. Eine anwendungsorientierte Einführung*. Berlin/Heidelberg: Springer Verlag.

Barnes, Jessica and Dove, Michael R. (2015) *Climate Cultures. Anthropological Perspectives on Climate Change*. New Haven/London: Yale University Press.

Bathelt, Harald and Glückler, Johannes (2011) *The Relational Economy. Geographies of Knowing and Learning*. Oxford/New York: Oxford University Press.

Beck, Ulrich (2005) 'Europäisierung. Soziologie für das 21. Jahrhundert'. *Aus Politik und Zeitgeschichte* (APuZ) 34–35: 3–11.

Beck, Ulrich and Grande, Edgar (2010) 'Jenseits des methodologischen Nationalismus. Außereuropäische und europäische Variationen der Zweiten Moderne'. *Soziale Welt* 61: 187–216.

Berger, Peter L. (2002) 'Introduction. The Cultural Dynamics of Globalization'. In: Berger, Peter L. and Huntington, Samuel P. (ed.): *Many Globalizations. Cultural Diversity in the Contemporary World*. New York: Oxford University Press.

Berger, Peter L. and Luckmann, Thomas (1966) *The Social Construction of Reality. A Treatise in the Sociology of Knowledge*. Garden City, New York: Doubleday.

Biesbroek, G. Robbert, Rob, J. Swart, Carter, Timothy R., Cowan, Caroline, Heinrichs, Thomas, Mela, Hanna, Morecroft, Michael D., and Rey, Daniela (2010) 'Europe Adapts to Climate Change: Comparing National Adaptation Strategies'. *Global Environmental Change* 20: 440–450.

Blotevogel, Hans-Heinrich (2004) 'Raum'. In: Akademie für Raumforschung und Landesplanung (ed.): *Handwörterbuch der Raumordnung*. Hannover: VSB, pp. 831–841.

Bourdieu, Pierre (1982) *Die feinen Unterschiede. Kritik der gesellschaftlichen Urteilskraft*. Frankfurt am Main: Suhrkamp.

Brooks, N., Adger, W.N., and Kelly, P.M. (2005) 'The Determinants of Vulnerability and Adaptive Capacity at the National Level and the Implications for Adaptation'. *Global Environmental Change* 15 (2): 151–163.

Bryson, Reid A. (1994) 'On Integrating Climatic Change and Culture Change Studies'. *Human Ecology* 22 (1): 115–128.

Christmann, Gabriela B. (2004) *Dresdens Glanz, Stolz der Dresdner. Lokale Kommunikation, Stadtkultur und städtische Identität*. Wiesbaden: Deutscher Universitäts-Verlag.

Christmann, Gabriela B. (2010) 'Kommunikative Raumkonstruktionen als (Proto-) Governance'. In: Kilper, Heiderose (ed.): *Governance und Raum*. Baden-Baden: Nomos, pp. 27–48.

Christmann, Gabriela B. and Ibert, Oliver (2012) 'Vulnerability and Resilience in a Socio-Spatial Perspective'. *Raumforschung und Raumordnung* 70: 259–272.

Christmann, Gabriela B., Heimann, T., Mahlkow, N., and Balgar, K. (2012) 'Klimawandel als soziale Konstruktion?' *Zeitschrift für Zukunftsforschung 1* (urn:nbn:de:0009–32–34126). Available at: www.zeitschrift-zukunftsforschung.de/ausgaben/2012/1/3412/Christmann_et_al (accessed 28 October 2013).

Conrad, Jobst (2010) 'Sozialwissenschaftliche Analyse von Klimaforschung, -diskurs und -politik am Beispiel des IPCC'. In: Voss, Martin (ed.): *Der Klimawandel*. Wiesbaden: VS Verlag für Sozialwissenschaften, pp. 101–115.

Crang, Mike and Thrift, Nigel (2000) *Thinking Space*. Abingdon/New York: Routledge.

Crate, Susan A. (2011) 'Climate and Culture: Anthropology in the Era of Contemporary Climate Change'. *The Annual Review of Anthropology 2011*. 40: 175–194. DOI: 0.1146/annurev.anthro.012809.104925 (accessed 29 October 2013).

Crate, Susan A. and Nuttal, Mark (2009) *Anthropology and Climate Change: From Actions to Transformations*. New York: Routledge.

Descartes, René (2006 [1644]) 'Über die Prinzipien der materiellen Dinge'. In: Dünne, Jörg and Günzel, Stephan (ed.): *Raumtheorie. Grundlagentexte aus Philosophie und Kulturwissenschaften*. Frankfurt am Main: Suhrkamp, pp. 44–57.

Descola, Philipp (2013) *Beyond Nature and Culture*. Chicago: University of Chicago Press.

Döring, Jörg (2010) 'Spatial Turn'. In: Günzel, Stephan (ed.): *Raum. Ein interdisziplinäres Handbuch*. Stuttgart/Weimar: Metzler, pp. 90–99.

Douglas, Mary and Wildavsky, Aaron (1982): *Risk and Culture. An Essay on the Selection of Technological and Environmental Dangers*. Berkeley/Los Angeles/London: University of California Press.

Durkheim, Emile (1998 [original undated]): *Die elementaren Formen des religiösen Lebens*. Frankfurt am Main: Suhrkamp.

Einstein, Albert (1960) 'Vorwort von Albert Einstein'. In: Jammer, Max: *Das Problem des Raumes*. Darmstadt: Wissenschaftliche Buchgesellschaft, pp. xii–xv.

Einstein, Albert (2006 [1930]) 'Raum, Äther und Feld in der Physik'. In: Dünne, Jörg and Günzel, Stephan (ed.): *Raumtheorie. Grundlagentexte aus Philosophie und Kulturwissenschaften*. Frankfurt am Main: Suhrkamp, pp. 94–102.

Foucault, Michel (1971) 'Orders of discourse'. Inaugural lecture at the Collège de France. *Social Science Information* 10 (2): 7–30.

Foucault, Michel (1972) *The Archaeology of Knowledge*. London: Tavistock Publications Limited.

Frommer, Birte (2010) *Regionale Anpassungsstrategien an den Klimawandel. Akteure und Prozess*. Darmstadt: IWAR.

Geertz, Clifford (1973) 'Thick Description: Toward an Interpretative Theory of Culture'. In: Geertz, Clifford: *The Interpretation of Cultures*. New York: Basic Books.

Geertz, Clifford (1974) '"From the Native's Point of View": On the Nature of Anthropological Understanding'. *Bulletin of the American Academy of Arts and Sciences* 28 (1): 26–45.

Gehlen, Arnold (1993) 'Der Mensch. Seine Natur und seine Stellung in der Welt'. In: Gehlen, Arnold: *Gesamtausgabe* Volume 3, Frankfurt am Main: Klostermann.

Giddens, Anthony (1990) *The Consequences of Modernity*. Stanford, CA: Stanford University Press.

Goethe-Institut (ed.) (2011) 'Dem Klima ausgeliefert? – Nico Stehr und Hans von Storch im Interview'. In: *Goethe.de/Klima. Kultur und Klimawandel*. Available at: www.goethe.de/ges/umw/prj/kuk/the/kul/de7184847.htm (accessed 31 August 2015).

Goetz, Georgius (ed.) (1922) M. *Porci Catonis. De Agri Cultura*. Leipzig: B.G. Teubner.

Greschke, Heike (2010) *Jahresbericht 2010 der Nachwuchsforschergruppe 'Klimawelten'*. Bielefeld: Graduate School in History and Sociology.

Grothmann, Torsten, Daschkeit, Achim, Felgentreff, Carsten, Görg, Christoph, Horstmann, Britta, Scholz, Imme, and Tekken, Vera (2011) 'Anpassung an den Klimawandel: Potenziale sozialwissenschaftlicher Forschung in Deutschland'. *Gaia* 20 (2): 84–90.

Grundmann, Reiner and Stehr, Nico (2003) 'Die Natur und die Natur der Gesellschaft'. In: Fischer, Ludwig (ed.): *Projektionsfläche Natur: Zum Zusammenhang von Naturbildern und gesellschaftlichen Verhältnissen*. Hamburg: Hamburg University Press, pp. 261–275.

Günzel, Stephan (2006) 'Physik und Metaphysik des Raums. Einleitung'. In: Dünne, Jörg and Günzel, Stephan (ed.): *Raumtheorie. Grundlagentexte aus Philosophie und Kulturwissenschaften*. Frankfurt am Main: Suhrkamp, pp. 19–43.

Harvey, David (1990) *The Condition of Postmodernity: An Enquiry into the Origins of Cultural Change*. Cambridge/Oxford: Blackwell.

Heimann, Thorsten (2017) *Klimakulturen und Raum. Umgangsweisen mit Klimawandel an europäischen Küsten*. Wiesbaden: Springer VS.

Hölscher, Michael (2006) *Wirtschaftskulturen in der erweiterten EU. Die Einstellungen der Bürgerinnen und Bürger im europäischen Vergleich*. Wiesbaden: VS Verlag für Sozialwissenschaften.

Hulme, Mike (2015) *Climates and Cultures*. Six Volume Set. SAGE Library of the Environment: London.

Huntington, Samuel P. (1996) *The Clash of Civilizations and the Remaking of World Order*. New York: Simon & Schuster.

Inglehart, Ronald (2002) 'Cultural Cleavages in the European Union: Modernisation and Cultural Persistence'. In: Fuchs, Dieter, Rollerl, Edeltraut, and Wessels, Bernhard (ed.): *Bürger und Demokratie in Ost und West*. Wiesbaden: Westdeutscher Verlag, pp. 73–84.

Inglehart, Ronald (2006) 'Mapping Global Values'. *Comparative Sociology* 5 (2/3): 115–136.

Kajetzke, Laura and Schroer, Markus (2010) 'Sozialer Raum: Verräumlichung'. In: Günzel, Stephan (ed.): *Raum. Ein interdisziplinäres Handbuch*. Stuttgart/Weimar: Metzler, 192–203.

Keller, Reiner (2007) *Diskursforschung. Eine Einführung für SozialwissenschaftlerInnen*. Wiesbaden: VS Verlag für Sozialwissenschaften.

Keller, Reiner (2011) *Wissenssoziologische Diskursanalyse. Grundlegung eines Forschungsprogramms*. Wiesbaden: VS Verlag für Sozialwissenschaften.

Keller, Reiner, Knoblauch, Hubert, and Reichertz, Jo (ed.) (2013) *Kommunikativer Konstruktivismus. Theoretische und empirische Arbeiten zu einem neuen wissenssoziologischen Ansatz*. Wiesbaden: Springer VS.

Kessl, Fabian and Reutlinger, Christian (2010) 'Ökonomischer Raum. Megacities und Globalisierung'. In: Günzel, Stephan (ed.): *Raum. Ein interdisziplinäres Handbuch*. Stuttgart/Weimar: J.B. Metzler, pp. 145–161.

Knoblauch, Hubert (1995) *Kommunikationskultur. Die kommunikative Konstruktion kultureller Kontexte*. Berlin/New York: De Gruyter.

Knoblauch, Hubert (2010) *Wissenssoziologie*. Konstanz: UVK Verlagsgesellschaft.

Knoblauch, Hubert (2013) 'Grundbegriffe und Aufgaben des kommunikativen Konstruktivismus'. In: Keller, Reiner, Knoblauch, Hubert, and Reichertz, Jo (ed.): *Kommunikativer Konstruktivismus. Theoretische und empirische Arbeiten zu einem neuen wissenssoziologischen Ansatz*. Wiesbaden: Springer VS.

Knoblauch, Hubert and Schnettler, Bernt (2004) 'Vom sinnhaften Aufbau zur kommunikativen Konstruktion'. In: Gabriel, Manfred (ed.): *Paradigmen der akteurszentrierten Soziologie*. Wiesbaden: VS Verlag für Sozialwissenschaften.

Korb, Kathrin (2008) *Satellitennavigation in der Schule. Integration moderner Technik in den Physikunterricht*. Mainz: no publisher's imprint. Available at: www.physik.uni-mainz.de/lehramt/lehramt/Vortraege/Anleitung/Kathrin_Korb_StEx.pdf (accessed 21 June 2015).

Krauss, Werner (2007) *Küstenforschung. Ethnologie einer Wissenslandschaft*. Projektbericht GKSS. Geestacht: Küstenforschungszentrum.

Krauss, Werner, Schäfer, Mike S., and von Storch, Hans (2012) 'Introduction: Post-Normal Science'. *Nature and Culture* 7 (2, Summer): 121–132. Available at: http://sciencepolicy.colorado.edu/students/envs-geog_5100/krauss_2012.pdf (accessed 28 August 2015).

Kroeber, Alfred L. and Kluckhohn, Clyde (1952) *Culture. A Critical Review of Concepts and Definitions*. Cambridge, MA: Peabody Museum. Available at: https://archive.org/stream/papersofpeabodymvol.47no1peab#page/n11/mode/2up (accessed 23 January 2016).

Läpple, Dieter (1991) 'Essay über den Raum. Für ein gesellschaftswissenschaftliches Raumkonzept'. In: Häußermann, Hartmut and Ipsen, Detlev (ed.): *Stadt und Raum. Soziologische Analysen*. Pfaffenweiler: Centaurus, pp. 157–207. Available at: www.rainer-rilling.de/gs-villa07-Dateien/laepple.pdf (accessed 25 June 2015).

Latour, Bruno (1996) 'On Actor-network Theory. A Few Clarifications'. *Soziale Welt* 47 (4): 396–381.

Lefebvre, Henri (1991 [1974]) *The Production of Space*. Oxford/ Malden: Blackwell.

Leggewie, Claus (1999) 'Kulturen der Welt – Weltkultur. Zur Einführung'. *Transit – Europäische Revue* Volume 17. Available at: www.iwm.at/transit/transit-17/ (accessed 7 August 2013).

Leibniz, Gottfried Wilhelm (2006 [1715]) 'Briefwechsel mit Samuel Clarke'. In: Dünne, Jörg and Günzel, Stephan (ed.): *Raumtheorie. Grundlagentexte aus Philosophie und Naturwissenschaften*. Frankfurt am Main: Suhrkamp, pp. 58–73.

Leibniz, Gottfried Wilhelm (1996 [original undated]) 'Streitschriften zwischen Leibniz und Clarke'. In: Cassirer, Ernst (ed.): *Philosophische Werke in vier Bänden. Volume 1: Hauptschriften zur Grundlegung der Philosophie*. Hamburg: Felix Meiner Verlag, pp. 81–182.

Ley, Wilfried, Wittmann, Klaus, and Hallmann, Willi (2009) *Handbook of Space Technology*. Chichester: John Wiley & Sons.

Löw, Martina (2001) *Raumsoziologie*. Frankfurt am Main: Suhrkamp.

Massey, Doreen (1994) 'A Global Sense of Place'. In: Massey, Doreen: *Space, Place, and Gender*. Cambridge: Polity Press.

McLuhan, Marshall (1994 [1964]) *Understanding Media. The Extensions of Man*. Cambridge/Boston: MIT Press.

Meyer, John W. (2007) 'Globalization. Theory and Trends'. *International Journal of Comparative Sociology* 48 (4): 261–273.

Moebius, Stephan (2009) *Kultur*. Bielefeld: Transcript Verlag.

Newton, Issac (1999) [Translated by Cohen, I. Bernard and Whitman, Anne]: *The Principia: Mathematical Principles of Natural Philosophy*. Berkeley, CA: University of California Press.

O'Brien, Karen, Leichenko, Robin, Kelkar, Ulka, Venema, Henry, Aandahl, Guro, Tompkins, Heather, Javed, Akram, Bhadwal, Suruchi, Barg, Stephan, Nygaard, Lynn, and West, Jennifer (2004) 'Mapping Vulnerability to Multiple Stressors: Climate Change and Globalization in India'. *Global Environmental Change* 14 (4): 303–313.

Parsons, Talcott (1961) 'An Outline of the Social System'. In: Parsons, Talcott, Shils, Edward, Naegele, Kaspar D., and Pitts, Jesse R. (ed.): *Theories of Society. Foundations of Modern Sociological Theory*. New York: The Free Press, pp. 30–79.

Pendergraft, Curtis A. (1998) 'Human Dimensions of Climate Change: Cultural Theory and Collective Action'. *Climatic Change* 39: 643–666.

Perpeet, Wilhelm (1984) 'Zur Wortbedeutung von "Kultur"'. In: Brackert, Helmut and Wefelmeyer, Fritz (ed.): *Naturplan und Verfallskritik. Zu Begriff und Geschichte der Kultur*. Frankfurt am Main: Suhrkamp, pp. 21–28.

Pieterse, Jan Nederveen (1994) 'Globalisation as Hybridisation'. *International Sociology* 9 (2, June): 161–184.

Quinn, Naomi and Holland, Dorothy (1987) 'Culture and Cognition'. In: Quinn, Naomi and Holland, Dorothy (ed.): *Cultural Models in Language and Thought*. Cambridge: Cambridge University Press, pp. 3–40.

Reckwitz, Andreas (2000) *Die Transformation der Kulturtheorien*. Weilerswist: Velbrück.

Renn, Ortwin, Arnold, Annika, Schetula, Viola, and Schweizer, Pia-Johanna (2011) 'Das Ringen der Sozialwissenschaften um ihre Rolle in der Klimawandeldebatte'. *Soziologische Revue* 34: 463–472.

Rippl, Susanne and Seipel, Christian (2008) *Methoden kulturvergleichender Forschung*. Wiesbaden: VS Verlag für Sozialwissenschaften.

Robertson, Roland (1992) *Social Theory and Global Culture*. London/Thousand Oaks/New Delhi: Sage.

Roose, Jochen (2011) 'Jürgen Gerhards: Quantifizierende Kultursoziologie'. In: Moebius, Stephan and Quadflieg, Dirk (ed.): *Kultur. Theorien der Gegenwart*. Wiesbaden: VS Verlag für Sozialwissenschaften, pp. 241–250.

Rosa, Hartmut, Strecker, David, and Kottmann, Andrea (2007) *Soziologische Theorien*. Konstanz: UVK.

Scannell, Leila and Gifford, Robert (2013) 'Personally Relevant Climate Change: The Role of Place Attachment and Local Versus Global Message Framing in Engagement'. *Environment and Behavior* 45 (1): 60–85.

Schmidt, Jan C. (2009) 'Physik'. In: Günzel, Stephan (ed.): *Raumwissenschaften*. Frankfurt am Main: Suhrkamp, pp. 291–307.

Schmitt, Lea (2015) *Klima, Raum und Zeit im Wandel. Eine ethnographische Untersuchung von Adaptationen und Konflikten auf der westfriesischen Insel Ameland*. München: Oekom.

Schnell, Rainer, Hill, Paul B., and Esser, Elke (2008) *Methoden der empirischen Sozialforschung*. München: Oldenbourg Verlag.

Schnettler, Bernt (2011) 'Thomas Luckmann: Kultur zwischen Konstitution, Konstruktion und Kommunikation'. In: Moebius, Stephan and Quadflieg, Dirk (ed.): *Kultur. Theorien der Gegenwart*, pp. 212–226.

Schroer, Markus (2006) *Räume, Orte, Grenzen. Auf dem Weg zu einer Soziologie des Raums*. Frankfurt am Main: Suhrkamp.

Schroer, Markus (2007) 'Raum als soziologischer Begriff. Programmatische Überlegungen'. In: Wehrheim, Jan (ed.): *Shopping Malls. Interdisziplinäre Betrachtungen eines neuen Raumtyps*. Wiesbaden: VS Verlag für Sozialwissenschaften, pp. 35–53.

Schubert, Sophia (2012) *Die globale Konfiguration politischer Kulturen. Eine theoretische und empirische Analyse*. Wiesbaden: Springer VS.

Schütz, Alfred (1962) 'Husserl's Importance for the Social Sciences' In: *Collected Papers I. Problems of Social Reality*. Den Haag: Martinus Nijhoff, pp. 162–173.

Schweizer, Sarah, Davis, Shawn, and Thompson, Jessica Leigh (2013) 'Changing the Conversation about Climate Change: A Theoretical Framework for Place-Based Climate Change Engagement'. *Environmental Communication* 7 (1): 42–62.

Sen, Amartya (2006) *Identity and Violence. The Illusion of Destiny*, 1st edn. New York: Allen Lane.

Simmel, Georg (1992 [original undated]) 'Der Raum und die räumlichen Ordnungen der Gesellschaft'. In: Rammstedt, Otthein (ed.): *Georg Simmel. Gesamtausgabe. Soziologie. Untersuchungen über die Formen der Vergesellschaftung*. Frankfurt am Main: Suhrkamp, pp. 687–790.

Smit, Barry and Wandel, Johanna (2006) 'Adaptation, Adaptive Capacity and Vulnerability'. *Global Environmental Change* 16: 282–292.

Sorokin, Pitirim Aleksandrovich (1998 [1927]) *Social Mobility*. London/New York: Routledge [Harper & Row].

Stehr, Nico (1996) 'The Ubiquity of Nature: Climate and Culture'. *Journal for the History of the Behavioural Sciences* 32: 151–159.

Stehr, Nico and von Storch, Hans (1999) *Klima, Wetter, Mensch*. München: C.H. Beck.

Thompson, Michael (2003) 'Cultural Theory, Climate Change and Clumsiness'. *Economic and Political Weekly* 38 (48).

Tylor, Edward B. (1871) *Primitive Culture. Researches into the Development of Mythology, Philosophy, Religion, Language, Art and Custom*. London: John Murray.

Viehöver, Willy (2004) 'Workshop Diskurs–Wissen–Kultur'. *Historical Social Research* 29 (4): 178–189.

Von Storch, Hans and Krauss, Werner (2005) 'Culture Contributes to Perceptions of Climate Change'. *Nieman Reports* 59 (4): 30–35. Available at: www.nieman.harvard.edu/reportsitem.aspx?id=100600 (accessed 10 November 2010).

Voss, Martin (2010) 'Einleitung: Perspektiven sozialwissenschaftlicher Klimaforschung'. In: Voss, Martin (ed.): *Der Klimawandel*. Wiesbaden: VS Verlag für Sozialwissenschaften, pp. 9–40.

Weber, Max (1991) 'Der Sinn der "Wertfreiheit" der soziologischen und ökonomischen Wissenschaften'. In: *Schriften zur Wissenschaftslehre*. Stuttgart: Reclam, pp. 176–236.

Welzer, Harald, Soeffner, Hans-Georg, and Giesecke, Dana (2010) 'KlimaKulturen'. In: Welzer, Harald, Soeffner, Hans-Georg and Giesecke, Dana (ed.): *Klimakulturen. Soziale Wirklichkeiten im Klimawandel*. Frankfurt/New York: Campus Verlag, pp. 7–19.].

Werlen, Benno (2000) *Sozialgeographie. Eine Einführung*. Bern/Stuttgart/Wien: UTB.

Wiedenbeck, Michael and Züll, Cornelia (2010) 'Clusteranalyse'. In: Wolf, Christof and Best, Henning (ed.): *Handbuch der sozialwissenschaftlichen Datenanalyse*. Wiesbaden: VS Verlag für Sozialwissenschaften.

Williams, Raymond (1977) 'Culture'. In: Williams, Raymond: *Keywords. A Vocabulary of Culture and Society*. Glasgow: William Collins, Sons & Co. Ltd., pp. 76–82.

Wimmer, Andreas and Glick-Schiller, Nina (2002) 'Methodological Nationalism and beyond: Nation-State Building, Migration and the Social Sciences'. *Global Networks* 2 (4): 301–334. Available at: www.princeton.edu/~awimmer/B52.pdf (accessed 5 August 2015).

Wisner, Ben. (2010) 'Climate Change and Cultural Diversity'. *International Social Science Journal* 61: 131–140.

Wu, Shuang-Ye, Yarnal, Brent, and Fisher, Ann (2002) 'Vulnerability of Coastal Communities to Sea-level Rise: A Case Study of Cape May County, New Jersey, USA'. *Climate Research* 22: 255–270.

3 Climate-change constructions in science

Basic concepts and their historical development

The previous chapter considered the theoretical basis for the analysis of climate cultures. In the present study I ask which forms of climate-related knowledge are shared by actors and what factors can explain differences. To answer this, a social-constructivist research perspective will be adopted using phenomenological consideration of actor-specific and socially shared realities. This also applies to scientifically formulated knowledge. With reference to Central Europe, Weingart *et al.* (2002) have described through discourse analysis how climate change knowledge has increasingly been diffused from the natural sciences into other societal fields during the latter half of the twentieth century.

It can therefore be concluded that specific forms of socially shared climate change knowledge can be observed in European coastal areas today. But what specific forms can be shared socially? And how has this knowledge itself developed historically? These questions will be discussed further in this chapter. The focus will be on vulnerability constructions of climate change such as the basis terminology concerning climate change as well as its causes and impacts. Discourse on practices to create resilience will be the subject of the next chapter (Chapter 4).

The basic terminology used today will be considered first (Section 3.1), in order to better understand what is meant by climate, climate change, and related terms from the natural sciences. Historical developments in scientific climate knowledge will subsequently be examined (Section 3.2). It will become clear that scientific knowledge about climate change is also subject to social conditionality, being dependent on technological, economic, and other social conditions. One should therefore speak of 'climate-change constructions in science'.[1] This chapter will also give an introduction to the current global institutional context in which scientific knowledge about climate change is produced.

Following this, I shall then be able to discuss the climate change impacts and geographical features in current scientific discourse, in particular with a focus on the research area in Europe (Section 3.3).

3.1 Current concepts of climate

In order to discuss globally encountered scientific concepts of climate change, the central terminology used in the field must first be introduced.

Climate and climate elements. Schönwiese provides a comprehensive summary of scientific climate definitions used since the nineteenth century (Schönwiese 2008: 55ff.). Building upon this, his own definition is as follows:

> The terrestrial climate is the local, regional, or, as the case may be, global **statistical description of relevant climate elements** [...] that give a sufficiently detailed description of [...] the conditions and variations of the Earth's atmosphere [...] over a not too brief period of time.
>
> (Schönwiese 2008: 56, emphasis added)

Schönwiese thus defines climate as an analytical construct of climate science based on regular observation of the Earth's atmosphere. Climatology divides atmospheric events into various *climate elements*, such as air temperature, precipitation, humidity, air pressure, wind, or cloud cover. It attempts to describe and explain their constellations at various spatial and temporal scalings. It is thus the application of statistical methods that makes climate distinguishable.

Territorial scale of climate. In relation to *territorial scale*, the concept of climate is applied to areas of very different sizes. The local climate of an individual weather station describes the smallest possible unit size, while at the opposite extreme the largest, macro-scalar unit provides the worldwide climate in the form of data collected from stations across the globe. All possible spatial scalings in between are conceivable. The resolution of climate models can currently be scaled to no less than 100 km, and it should be noted that, due to this, local judgements made by actors about future climate developments are regularly associated with high levels of uncertainty (compare Kappas 2009: 5).

Temporal scale and climate change. The term climate has relatively clear temporal limitations in its use. Atmospheric phenomena are considered as *weather* when they last for between one hour and approximately 30 days (e.g. low-pressure areas and storms). Atmospheric events are only described as climate phenomena when observed over relatively long observation periods (determined variously, depending on the issue in question). In order to be able to identify and compare climate uniformly, the World Meteorological Organisation (WMO) has defined so-called standard reference periods, each of which covers a period of at least 30 years (Kappas 2009: 3; Schönwiese 2008: 51). At this point it should be made clear that, according to this definition, a change in climate conditions can be described statistically only after 30 years have elapsed, although phenomenon-related events connected to such a change can already be observed during shorter time periods. Thus Rahmstorf and Schellnhuber emphasise that changes in climate can also occur very abruptly (Rahmstorf and Schellnhuber 2007: 28). The Intergovernmental Panel on Climate Change (IPCC) defines climate change as follows:

> Climate change in IPCC usage refers to a change in the state of the climate that can be identified (e.g. using statistical tests) by changes in the mean

> and/or the variability of its properties and that persists for an extended period, typically decades or longer. It refers to any change in climate over time, whether due to natural variability or as a result of human activity.
> (IPCC 2007d: 30)

Climate factors and climate systems. In current investigations of climate elements, many of the various possible influencing factors that are taken into account are not solely attributable to the atmosphere itself (Kappas 2009: 86, Jacobeit 2007). Climate is regarded as 'a result of the physico-chemical processes [...] and interactions within the climate system, as well as external influences on this system' (Schönwiese 2008: 56). The term *climate factor* refers to processes and states that have a decisive influence on the constellations of climate elements (e.g. geographical position).

These ideas result in the current concept of the *climate system*, whose diverse interactions and feedback effects must therefore be made observable and describable (Kappas 2009: 86). The climate system comprises the different geosystems of the atmosphere (air masses), the geosphere (land masses), the hydrosphere (oceans), the biosphere (living creatures), and the cryosphere (ice masses). External influences such as the activity of the sun or variations in orbit around it are also absorbed as *factors* within the climate models (see also Section 3.2).

The idea of a global climate system and related notions of the variability of its properties result principally from the development of a globally organised climatology. The scientific genesis of knowledge about climate change will be detailed below. It will be shown that core ideas are historically contingent. Scientific climate knowledge seems to depend constantly on technological and social circumstances.

3.2 Historical constructions of scientific climate knowledge and their global institutions

The term 'climate' derives etymologically from the ancient Greek '*κλινω*' (klino: 'I incline') (Schönwiese 2008: 11). In around 500 BCE, Parmenides of Elea introduced an early systematic description of climate zones. He defined these as areas with similar temperature conditions then observable within the known ancient world (warm climates, temperate climates, cold climates). He explained different conditions between regions on the basis of the locally varying average angle of inclination of the sun at its zenith and associated variations in local irradiation intensities. His conclusions were possible only on the basis of the temporal and spatial limitations of his surveillance, which may also have contributed to Parmenides's static understanding of climate. Aside from daily and annual variations, he regarded climate as a largely unchanging phenomenon, a view that would dominate the sciences for centuries (ibid.).

Various technical and social developments were of relevance in the emergence of scientific concepts about changing climate conditions. Important preconditions were the attainment of appropriate spatial and temporal ranges for the observation horizons. One can distinguish between three basic types of climate information source: (1) *historical sources*, (2) *directly measured observations* (neoclimatology), and (3) *indirectly inferred information* (paleoclimatology) (Schönwiese 2008: 283).

1 The earliest written *historical sources* date back to circa 3000 BCE. Information about precipitation and river levels were recorded for agricultural planning by the civilisations based along the Nile and Ganges. Many further sources were created around the world in the course of the millennia, which later made it possible to interpret historical weather and climate conditions and contributed to ideas about how they changed. The reception of ancient Greek and Roman literature thus led to the first notions of regional climate change among European authors of the seventeenth and eighteenth centuries such as Abbé Dubos and David Hume (see in more detail Fleming 1998: 11ff.).
2 *Directly measured observations* (neoclimatological data) first became available from the 1650s onwards. This can be seen as the beginning of modern meteorology, where climatology forms a sub-discipline. An important prerequisite for this was the invention of the technology that could take physical measurements of atmospheric conditions, such as the thermoscope for temperature (Galileo in 1611) or the barometer for air pressure (Torricelli in 1643) (Schönwiese 2008: 9f., 283ff.; Behringer 2007: 25).

Scientific understanding of *global* climatic conditions is based upon the ability to compare global data in the form of synchronous time series measurements in different locations. Thus of equal significance for present-day science is the emergence of possible institutionalised coordination for the comparison of data, although it took several more centuries before these became globally established. The development of worldwide scientific networks for systematic data collection led in the late nineteenth century to the founding of the International Meteorological Organisation (IMO, today WMO) (WMO 2009: 3; Fleming 1998: 33ff.).

During the nineteenth century, the foundation of public and private scientific associations dedicated to the gathering of climate and weather data began to increase in the western hemisphere.[2] The invention of the telegraph in 1833 allowed data from different locations to be centrally and speedily collated for the first time, enabling the calculation of short-term weather forecasts. Permanent national institutions for meteorology and climatology arose (Schönwiese 2008: 13). But since atmospheric conditions are hardly subject to national boundaries, several conferences were held from the 1850s onwards to institutionalise a permanent, internationally standardised system of observation and coordination. In 1873, as already noted, this resulted in the establishment of the International Meteorological Organisation (IMO), which became the World Meteorological Organisation (WMO) – a part of the United Nations – in 1951 (WMO

2009: 3) and today sets uniform standards for the international coordination of meteorological and climatological information systems. *Globally* unified coordination possibilities for the direct recording of weather and climate data have thus existed for almost one and a half centuries. In addition, there are also long, almost unbroken series of observations from a large number of locations, many of which reach back between 200 and 250 years (Schönwiese 2008: 12f.).[3]

3 Palaeoclimatology seeks to gather *indirectly inferred data* (proxy data) in order to access periods even further back in time. Its development is closely linked to the idea of long-term changes in climate conditions. In the 1830s, Karl Friedrich Schimper formulated the idea of an earlier 'ice age' on the basis of observations of the landscape (Schimper 1837; see also Kühlmann 2011: 363f.). However, major leaps in innovation did not take place until the mid-twentieth century. An essential milestone was the discovery of the oxygen-isotope method in 1951, after which geological materials could be utilised 'as a thermometer' (Urey *et al.* 1951: 399). At present, up to 800,000 years of past data have been deduced (Behringer 2007: 22).

On the basis of these data, it is now assumed that periodic changes in the global average temperature have repeatedly occurred during the Earth's history (Behringer 2007: 44). According to this, climate changes have occurred over periods of a few centuries or millennia. Long-term climate change has thus been an established feature of scientific discourse since the nineteenth century at the latest, and the issue of its possible causes has since gained momentum and led to numerous theories. The development of scientific discourse on the causes of climate change should therefore be examined more closely.

By the 1820s, Joseph Fourier had theoretically dealt with the issue of precisely how terrestrial temperatures occur (Behringer 2007: 243). What is today known as the *greenhouse effect* is related to the conclusion that the Earth's temperature would be significantly lower without the presence of certain atmospheric gases such as water vapour, carbon dioxide, or methane (Kappas 2009: 82ff.; Schönwiese 2008: 116ff.; Rahmstorf and Schellnhuber 2007: 30ff.). John Tyndall provided empirical foundations in 1861 by determining the light-absorption properties of these gases. He concluded that different atmospheric proportions of water vapour and carbon dioxide lead to different temperature outcomes (Tyndall 1861: 28), but had to admit that he could not determine the effect size of different gas components on temperature (Tyndall 1861: 29).

In 1896, the Swedish scientist Svante Arrhenius took up these considerations and sought to determine the extent of varying carbon dioxide (CO_2) and water vapour levels on ground temperature. Using an early climate model, he came to the conclusion that a fall in the atmospheric proportion of CO_2 of slightly less than half could lead to global temperature differences of between 4 and 5 degrees Celsius, initiating an ice age (Arrhenius 1896: 268). Without the aid of digital technology, these climate-sensitivity calculations would have taken several months, hence placing technological and economic limits upon the analyses. Arrhenius also discussed the human influence on the climate through the then rapidly increasing industrial emissions, and considered warming to be a possibility. Indeed, he saw this leading to improved living conditions in cold regions such as Scandinavia, as well as opportunities for agriculture through higher harvest yields (Fleming 1998: 74).

By the mid-twentieth century, many other theories that are still discussed today had been developed to explain climate change. According to Fleming, three important theoretical traditions can be distinguished: the above-mentioned *molecular* theories, *astronomical* theories, and *terrestrial* theories (Fleming 1998: 109). Astronomical theories establish climate change based on extraterrestrial variables. The mathematician Milankovitch thus explained regular cold periods via the long-term cycles of the Earth's orbit and its position in relation to the sun (Rahmstorf and Schellnhuber 2007: 20ff.; Milankovitch 1930). Meteorite impacts, alterations in sun activity, and rotations of the Milky Way were also taken into consideration (Schönwiese 2008: 328). Among the terrestrial theories are Alfred Wegener's considerations of continental drift, which have also been deemed as climate relevant, or the notion that 'albedo' (the reflection of sunlight from the Earth) influences climate.

Initially these various methods were often considered separately and they partly competed against one other. Increasing attempts have been made, however, to integrate them into a common model and, additionally, to consider interactions between atmospheric, terrestrial, and astronomical elements. Nonetheless, for a long time, reliable data and techniques for testing these hypotheses have often been lacking (Fleming 1998: 83ff.).

The question whether industrial CO_2 has a large effect on global temperatures once more gained in significance from the end of the 1930s. Callendar observed an increase in the average temperatures of the North Atlantic regions (Callendar 1938: 232ff.). For him, anthropogenic climate change had already begun.

In 1955 Hans Suess observed that the origin of carbon isotopes had changed in North American trees and was increasingly fossil in nature. 'Artificial coal and oil combustion' were the only sources that came into question for him (Suess 1955: 415): this change would have to have been anthropogenic. Shortly afterwards came efforts to directly measure CO_2 time-series data, and in 1957 Charles Keeling succeeded in laying the foundations for a measurement series on Mauna Loa, Hawaii. He observed an increasing annual trend in atmospheric CO_2 levels. As of 2016, the figures have continued to increase from year to year (IPCC 2013b: 10). The Kyoto Protocol efforts do not appear to have made any difference.

A connection between global warming and increased levels of fossil greenhouse gases in the atmosphere remained highly controversial, however. Observations suggested otherwise, as the measured average global temperatures fell between the 1940s and 1970s, despite an observed increase in CO_2. In fact, a theory of 'global cooling' (Fleming 1998: 129) and of an imminent ice age appeared just as plausible.

High uncertainty persisted about the strength of the effects of different conceivable variables on global temperature. At the same time, important new techniques of observation and analysis were developed. From the mid-1960s the first observation satellites were used. The rapid development of the computer following the Second World War enabled climate-model calculations of increasing complexity to be carried out in an ever decreasing period of time (Fleming 1998). The aim was not only to make statements about the strength of the effects and interactions between the observed variables, but also to enable projections of different potential futures. Using their models, Manabe and Wetherald calculated in 1967 that a doubling of CO_2 in the atmosphere would lead to a global increase in temperature of around 2.3 degrees Celsius (Manabe and Wetherald 1967: 254).

At this time, particularly in western societies, momentum was gathering in the public discourse about possible environmental problems caused by industrialisation processes. Climate discussion increasingly took place beyond the scientific sphere, in politics and the media (Rahmstorf and Schellnhuber 2007: 30; Behringer 2007: 246; Weingart *et al.* 2002). To address possible environmental challenges and uncertainties, the United Nations Environment Programme (UNEP) was adopted at the United Nations Conference on the Human Environment in Stockholm in 1972. Among other things, a global monitoring system was introduced to provide new insights into atmospheric changes and their interdependencies (GEMS, Global Environmental Monitoring System) (ibid.). Based on these and other new observation and modelling methods, the 1970s saw growing consensus within scientific discussion that future anthropogenic global warming could become a major problem (Behringer 2007: 252). The first World Climate Conference met in 1979 for further consultation (WMO 2013), and as a result the World Climate Programme (WCP) and further global research programmes were created (ibid.). Some countries subsequently saw the creation of the first national research programmes on global warming (Weingart *et al.* 2002: 30).

From the mid-1970s, average global temperatures again demonstrated an upward trend. Droughts also increased, such as in the Sahel zone. The possibility of anthropogenic climate change thus gained further attention beyond science (Weingart *et al.* 2002: 253). In the 1980s, the focus was directed towards a further problem when chlorofluorocarbon emissions (CFCs) were held responsible after a rapid, possibly life-threatening decline in the stratospheric ozone layer was observed. Political representatives worldwide were very quick to respond. In the 1987 Montreal Protocol, UN member states committed to national limits. The share of CFCs in the atmosphere has since been reduced (Schönwiese 2008: 376f.). A year later, in order to better coordinate both the observation of atmospheric events and possible policy measures, the WMO and UNEP founded the Intergovernmental Panel on Climate Change (IPCC).

The IPCC can be classified as a 'hybrid or border organisation, lying between science and politics' (Conrad 2010: 102). Its tasks consist of summarising the state of knowledge on climate change and strategies for coping with it, and separately to derive policy-relevant conclusions (ibid.). The three main Working Groups (WG) of the IPCC consist of scientists from different disciplines whose reports are peer reviewed and published every five years (Behringer 2007: 254). Working Group I (Physical Science Basis) summarises the current state of scientific research. Working Group II (Climate Change Impacts, Adaptation, and Vulnerability) addresses the possible impacts of climate change. Working Group III (Mitigation of Climate Change) processes the research on possible measures to reduce climate change (Conrad 2010: 103; Schönwiese 2008: 379; IPCC 2007a).

The founding of the IPCC may well have been the most important step in the genesis of discourse on the subject of anthropogenic climate change as a global challenge, as the need to deal with it on a worldwide scale has now been institutionalised. Already in 1990, in its first climate report, the IPCC describes observable changes in the climate system (IPCC 1990: 195ff.) and emphasises the influence of humans on the world's climate:

> Emissions resulting from human activities are substantially increasing the atmospheric concentrations of the greenhouse gases carbon dioxide, methane, chlorofluorocarbons (CFC's) and nitrous oxide. These increases will enhance the greenhouse effect, resulting on average in an additional warming of the Earth's surface. The main greenhouse gas, water vapour, will increase in response to global warming and further enhance it.
>
> (IPCC 1990: xii)

In 2007 the authors added: 'Evidence of a human influence on the recent evolution of the climate has accumulated steadily' (IPCC 2007c: 669). Six years later they then came to the conclusion '[...] science now shows with 95 percent certainty that human activity is the dominant cause of observed warming since the mid-20th century' (IPCC 2013c: v). The IPCC thus regularly communicated the consensus, strong within global climate research, that global climate changes are taking place and that these are anthropogenic in origin.

3.3 Vulnerability constructions in recent discourse: changes in climate and the debated impacts for the coastal regions of Europe

Scientific observers of climate change conclude that many more phenomena other than temperature change will occur. As described, the present understanding of climate systems is characterised by a multitude of possible interactions between sub-systems (see Section 3.1). The IPCC lists so-called climate impacts accompanying global warming (see for example, IPCC 2013c), such as changes in the atmosphere, sea level, cryosphere, or geochemical circulation (IPCC 2013b: 2ff.). The authors give descriptions and explanations of developments in the past, and make projections of future outcomes on the basis of climate models.

The first question that arises is which scale to take into consideration. Projections of climate research at a global level are frequently attributed higher levels of validity than those on a regional or local level (Rahmstorf and Schellnhuber 2007: 54). Changes in climate, such as a global sea-level rise, can also be ascribed different consequences depending on varying local geophysical conditions. Nevertheless, it must be noted that atmospheric observations are initially made locally, and that climate projections are also discussed at the regional and local level (Section 3.1). The following sections will therefore discuss the observed and projected impacts of climate change relevant to the areas of this study at the global and European scale. Regional particularities discussed at the scale of the countries under consideration will be examined subsequently.

3.3.1 *Climate-change observations and projections at the global and European scale*

Observations of recent climate developments and future projections at the global and European scale can be summarised as follows.

Atmospheric changes

- **Temperature:** The IPCC describes a linearly rising trend in average global temperatures over land and ocean surfaces during the period since the mid-nineteenth century (IPCC 2013b: 3). The three decades since 1980 are considered *likely* to have been the warmest for 1,400 years in the northern hemisphere (IPCC 2013b: 3, 2013a). In Europe, in the 10 years from 2002 to 2011, the average temperature was 1.3 degrees Celsius higher than in the period before industrialisation (EEA 2012: 59). Should levels of greenhouse gases and aerosols in the atmosphere remain the same as in 2000, the IPCC projects further warming by at least 0.1 degree Celsius per decade until and beyond the century's end (IPCC 2013c: 125). Since greenhouse gas levels have so far continued to rise (IPCC 2013b: 10), higher increases should be expected.
- **Number of warm and cold days:** Days are considered warm or cold if they exceed or fall below a locally determined average threshold. It is *very likely* that, since 1950, the annual number of warm and cold days have increased and decreased, respectively, worldwide (IPCC 2013b: 5). In Europe, the number of warm days has apparently tripled since 1880, and since the 1960s the number of cold days has declined by between 5 and 11 days each decade (EEA 2012: 62). These trends are expected to continue in the twenty-first century (IPCC 2013b: 5).
- **Heat waves:** Heat waves are periods of at least three days during which the temperature exceeds an extreme value averaged for the respective station (Zebisch *et al.* 2005: 36). Since 1950 both the frequency and duration of heat waves have increased with *medium confidence* in most land areas (IPCC 2013b: 5). In western Europe the average length of summer heat waves has

doubled since 1880 (EEA 2012: 62). It is *very likely* throughout Europe and worldwide that these figures will continue to increase in degree and duration until the end of the century (ibid.; IPCC 2013b: 5).

- **Rainfall:** Since 1950 both the frequency and intensity of rainfall have *likely* increased globally over most land areas (IPCC 2013b: 5). Stronger regional fluctuations can be observed in Europe. While precipitation in northern Europe has increased by more than 14 mm per decade, it has decreased in southern Europe. On average, the changes in Europe balance out, so that no significant trends are identifiable at this scale. Heavy rainfall has increased in western Europe (EEA 2012: 19). Future projections indicate that these developments will continue and intensify in most European regions (EEA 2012: 69). This would have consequences for local and river-level dependent flooding (see below), although the data are ascribed high uncertainty: since warmer air can absorb more water, it seems plausible that global warming would lead in total to larger quantities being transported in the water cycle (see also Section 4.3.1). However, this is not yet clearly observable (Schönwiese 2008: 324).
- **Droughts:** The intensity and duration of droughts has *likely* increased in many regions since 1970 (IPCC 2013b: 5). In Europe this has been observed most prominently in the central and southern regions (EEA 2012: 17). Projections for Europe suggest that these developments will become more acute in the future, especially in the central, eastern, and southern regions (EEA 2012: 25).
- **Storms:** Tropical storms have *likely* increased since 1970 in some regions. On the other hand, there is *low confidence* relating to the temporal and spatial observational horizon in the tropics (IPCC 2013b: 5). The last two hundred years have seen no significant trends for storms in Europe, despite some period-related variance (EEA 2012: 71; Matulla *et al.* 2007). In Europe, apart from in Mediterranean regions, a future increase in wind speeds is projected (EEA 2012: 28). However, as there are no clear trends at present, there is high regional uncertainty.

Changes in the cryosphere (ice and snow cover)

- **Mass loss:** In the last two decades the ice masses at the poles and worldwide on glaciers have with *high confidence* reduced in volume (IPCC 2013b: 7). This is also true for almost all European glacial regions. Since 1850 the Alps have lost two thirds of their ice mass (EEA 2012: 80). As an increase in average temperatures is projected even at constant greenhouse gas levels, it can be assumed that these developments will continue for most glaciers in Europe as well as ice-covered areas throughout the world (IPCC 2013c: 345f.; EEA 2012: 86ff.).
- **Snow cover:** Decreasing snow cover in spring has been observed with *high confidence* in the northern hemisphere. The phenological transition of the seasons is thus taking place earlier. Winter appears to be shortening (IPCC 2013b: 7). There are major regional differences in Europe, however (EEA

2012: 75f.). The annual duration of snow cover will *likely* decrease further. At the same time, an increase in snow levels during cold winter months is projected, especially for northern Europe (EEA 2012: 76f.), meaning that precipitation intensities will increase, while at the same time the duration of snow periods will decrease.

Oceanic and sea-level changes

- **Sea-level rise:** Sea level has with *high confidence* increased since the mid-nineteenth century, and more than in the preceding 2000 years. Between 1901 and 2010 this rose by a global average of 19 cm (IPCC 2013b: 9). The relative sea-level rise is not the same everywhere: in Europe, for example, land along the southern North Sea and Baltic Sea coasts has sunk since the last Ice Age, adding to the process (Woth and von Storch 2008). The global rise in recent years is explained mainly by thermal expansion, that is, the change in physical expansion due to temperature changes. The melting of glaciers and polar ice also contributes to the increase (IPCC 2013b: 9). Nevertheless, the exact processes are not fully understood, and projections are also subject to a high level of uncertainty (Vermeer and Rahmstorf 2009: 1). The 2007 IPCC report (AR4) projected an increase of between 18 cm and 59 cm to the year 2100, depending on emission levels (IPCC 2007c: 13). The most recent report increases these figures to between 26 cm and 98 cm (IPCC 2013c: 25).

Changes in geochemical circulation

- **Proportion of atmospheric greenhouse gases:** The proportions of carbon dioxide, methane, and nitric oxide are with *high confidence* at the highest concentrations seen in the last 800,000 years. Since the onset of industrialisation the amount of atmospheric carbon dioxide has increased by 40% (IPCC 2013b: 9). Future concentrations will depend largely on economic and technological developments. The IPCC therefore bases projections on various scenarios, defined in the latest IPCC report as four 'Representative Concentration Pathways (RCPs)', corresponding to different atmospheric proportions (IPCC 2013c: 29). The associated radiation balances can then be integrated into complex models in order to calculate the respective possible consequences.
- **Proportion of oceanic carbon dioxide:** With *high confidence*, oceanic carbon dioxide content has seen a large increase, leading to changes in pH value. Since carbon dioxide in the water reacts with carbonic acid, acidification processes have begun (IPCC 2013b: 10). This process could result in changes to flora and fauna, and pose a particular threat to limestone-forming organisms such as coral. Increasing carbon dioxide emissions are expected to lead to prolonged acidification processes (IPCC 2013c: 27).

In addition to these geophysical changes, the IPCC also discusses climate-change impacts for coasts arising from the complex interplay of various factors.

It distinguishes between the effects on 'natural and human systems' (IPCC 2014: 4) and thus assumes a worldview of separation between humans and nature (see also Chapter 5). Working Group II of the IPCC is dedicated to both of these aspects, summarising the global state of climate-impact research. Here, only a selection of impacts will be discussed, namely those that are important for the spatial development of coastal regions.

- **Flooding:** Flooding is normally perceived as a disaster when areas of human land use and/or built infrastructure are affected. At the global scale on average there is little confidence regarding an observed trend of increasing frequency and extent of flooding events. For Europe, on the other hand, it is reported with *high confidence* that floods have occurred with higher average water levels in the second half of the twentieth century than in the first (IPCC 2013c: 112). This could also be attributed to altered rainfall conditions. Nevertheless, it has been noted that flooding may also be increasing due to changes in patterns of human settlement (IPCC 2013c: 214), for example through processes of urbanisation (IPCC 2014: 18). A distinction should be made in regard to coastal regions between *sea flooding* and *inland flooding* (in which rivers and lakes overflow after heavy rainfall or when snow melts inland). From this it should be clear that flooding in coastal regions can also be caused by the interchange between inland and sea water (see also Heimann and Christmann 2013: 39). More serious problems can occur when inland and sea water meet in larger volumes (see also IPCC 2007b: 14).
- **Storm surges:** Storm surges arise from the interplay of wind and water causing land surfaces to be flooded by seawater. Due to changing wind conditions, but also sea-level rise or changes in the current, storm surges can occur with new characteristics. The IPCC reports altered patterns of surface waves in the Atlantic and Pacific Oceans (IPCC 2013c: 277). No changes have been observed so far in Europe, however. According to some scenarios, storm surges might become significantly higher in the future, although other projections do not suggest much change (von Storch and Woth 2011: 12f.).
- **Coastal erosion/Land loss:** As a consequence of rising sea levels and increased storms, the risk to European coastal regions of additional coastal erosion is also being discussed (EEA 2006: 43). 'The changing faces of Europe's coastal areas' (EEA 2006: 1) could also be characterised in the longer term by even greater land loss than has previously been seen (EEA 2006: 12).
- **Salination of drinking water/Changes in freshwater resources:** In connection with sea-level rise, the extent of sea-water penetration of existing groundwater catchments near the coast has also been considered (for more detail see Heimann and Christmann 2013: 33; EEA 2006: 44). Freshwater areas can be affected by salination and changes in precipitation patterns (EEA 2006: 49f.).

3.3.2 Peculiarities of the coastlines of Germany, the Netherlands, Denmark, and Poland

Beyond the above-mentioned global and European scales, observations and projections of climatic changes are also made for national and local scales. As a rule, the topics are linked to those at the larger scales. In order to avoid redundancy, only the most important features regarding the European coasts are addressed here, in relation to specific nation states.

German coastal regions

The German coastal zones can be divided into those on the North Sea and on the Baltic Sea (Jensen and Schwarzer 2013: 108ff.). Physical height variations have been described for both areas in the geomorphological literature. According to this, 70% of coastal zone areas on the North Sea lie no more than five metres above sea level, compared to 12% on the Baltic Sea coast. The land masses on the German Baltic Sea are thus to a large extent comparatively higher than on the North Sea (PRC 2009a: 1ff.).

A summary of recent studies on the 'Vulnerability of Germany's Coastal Zones to Climate Change' (ibid.) commissioned by the European Environment Agency (EEA) concluded in 2009 that both the North Sea and Baltic Sea regions are sensitive to sea-level rise. The probability of extreme water levels is considered higher for the North Sea regions than for those on the Baltic Sea. Although storms occur on both coasts with approximately the same frequency, water levels on the North Sea can be substantially higher due to stronger tides and lower land masses (PRC 2009a: 2).

Intensified coastal erosion is also being discussed as a serious problem for the Baltic Sea shores (ibid.; Jensen and Schwarzer 2013: 113ff.). Land loss could increase in many areas by as much as 40cm annually (PRC 2009a: 2). Threats to the Wadden Sea have also been described: those areas that at present are only partially flooded in the rhythm of the tides could become permanently flooded. The Wadden Sea would thus lose its fundamental characteristics (PRC 2009a: 3; Bundesregierung 2008: 25f.).

The vulnerability assessment commissioned in 2005 by the German Bundesumweltamt (Federal Environment Agency) also concludes that days reaching temperatures above 30 degrees Celsius and longer heat waves increased significantly in Germany during the twentieth century and especially since the 1980s (Zebisch *et al.* 2005: 36f.). Several models project a further increase (ibid.: 38; Bundesregierung 2008: 48). Further, the study describes an increase in heavy rainfall (Zebisch *et al.* 2005: 39). Projections indicate that further changes in rainfall could occur until 2100.

There has also been discussion of the possibility of an increase in dry periods in eastern Germany (PRC 2009a: 3). The 2005 Vulnerability Assessment concluded that in Germany only about a quarter of the available water resources are being utilised. The authors thus view there to be sufficient coping potential in the case of drought (Zebisch *et al.* 2005: 48).

Dutch coastal regions

The Netherlands is located on the North Sea. Given its physical characteristics, the European Environment Agency concluded in 2006 that the Netherlands and Belgium were the European Union's 'most vulnerable countries' (EEA 2006: 42) in regard to sea-level rise. Only 15% of coastal areas lie higher than 5 m above sea level (ibid.). The Netherlands is additionally characterised by the fact that about one-third of its total surface area lies below sea level (Van Koningsveld *et al.* 2008: 367). Many of these so-called 'polder areas' have been wrested from the sea since the Middle Ages, giving rise to a common proverb: 'God created the world, but the Dutch created the Netherlands' (Kunze 2009: 9).

Enhanced technological measures against flooding have been implemented since the thirteenth century, among them drainage canals and windmills, which were used to drive pumps regulating the water content of the polder areas. Windmills and canals are closely linked today to collective knowledge about the Netherlands. Place names such as Amster*dam* and Rotter*dam* also indicate the importance of flood barriers in the foundation of these settlements (Van Koningsveld *et al.* 2008: 371). A large section of the population lives in these polder areas and in recent years approximately 70% of the country's GDP was produced there (NEAA 2009: 11; see also Van der Meulen *et al.* 2013: 136ff.).

Regarding future sea-level rise, the Dutch Delta Commission calculates even more drastic projections than the IPCC. It sees a potential local sea-level rise of between 0.65 m and 1.35 m by the end of the century. By the year 2200 it even expects an increase of 2.0–4.0 m compared to current levels (Deltacommissie 2008: 10).

The Netherlands National Environment Agency emphasises that a large part of the Netherlands forms a river delta, with four major rivers meeting there: the Rhine, the Meuse, the Scheldt, and the Ems (NEAA 2009: 11). The Dutch climate is characterised by intensive periods of rain in winter and longer dry periods in summer. Projections by the Royal Meteorological Institute of the Netherlands (KNMI) suggest an intensification of these conditions. Not only is sea-level rise identified as a major problem, so too are the additional water volumes of the rivers.

An intensification of hot spells is also projected. Referring to heat and drought, a serious scarcity of freshwater resources during the summer has been observed. By contrast, the KNMI provides no evidence of existing or future changes in regard to extreme storm weather (PRC 2009b: 2f.; Gupta *et al.* 2006: 18).

Danish coastal regions

As in Germany, the Danish coastal regions can be divided roughly into those on the North Sea and those on the Baltic Sea, whereby the proportion on the Baltic Sea predominates. The country also consists of numerous islands. It has been calculated that 22% of all coastal areas lie less than 5 m above sea level (Sørensen 2013: 96ff.; EEA 2006: 42).

The European Environment Agency's meta-analysis of studies on the impacts of climate change for European coastal regions concludes that 'Denmark's vulnerability to coastal flooding and erosion is rather limited' (PRC 2009c: 1). Although many areas are threatened by flooding, it is regularly emphasised that many of these low-lying regions are barely inhabited. An exception, however, are some areas on the North Sea, and municipalities such as Lolland and Greve, as well as Copenhagen on the Baltic Sea (ibid.).

The study above all names increasing rainfall and storm weather as potential threats. Thus several coastal regions saw repeated evacuations in the first decade of the new millennium due to flooding after heavy rainfall (PRC 2009c: 2). An increase in precipitation in winter, and a decrease in summer, could increase the water supply accordingly for some areas, especially in West Jutland, while diminishing it in others (ibid.: 3).

In relation to heat, several meta-studies conclude that average temperatures have increased in Denmark and will continue to do so (KEMIN 2009: 242; The Danish Government 2008: 15). In the 2008 Danish Adaptation Strategy heat is identified as a possible problem for the big cities (The Danish Government 2008: 25), but overall the authors see few significant problems beyond that.

Polish coastal regions

The Polish coastal regions are located exclusively on the Baltic Sea, with 30% of coastal areas lying less than 5m above sea level (Furmanczyk 2013: 81ff.; EEA 2006: 42). A rough physical distinction is made between three regions: first, the estuary region of the Odra River in the West Pomeranian Voivodship, characterised mainly by the interplay between the Odra and the sea; second, a middle segment on the border between West Pomerania and Pomerania, with a long coastline running from the south to the east and including numerous lakes near the coast; and third, the area south of the Hel Peninsula to the east of the Pomeranian Voivodship, and in northern Warmian-Masurian with the Gdańsk Bay (ibid.).

A risk of increased coastal erosion has been identified for all regions in connection with sea-level rise (Kundzewicz and Matczak 2012: 302; PRC 2009d: 2f.). The entire Polish coast is characterised by soft soil (Labuz 2013: 34f.). In recent years, annual land losses of about 50–90 cm have been observed along large stretches of the coast (Pruszak and Ostrowski 2005: 24). The risk of flooding is estimated to be particularly high in the area of the Odra estuary. The problem of storm surges has also been discussed.

The threat of pollution of the coastal lakes has been identified as a problem in the middle segment between West Pomerania and Pomerania (PRC 2009d: 3). Because there are numerous industrial facilities less than one metre above sea level in Gdańsk Bay, the authors of the European Commission's 2009 analysis see an increased threat potential resulting from sea-level rise (ibid.). The possible loss of the Hel Peninsula has even been discussed. This would lead to waves arriving at Gdańsk Bay in altered form, and the entire transportation of sediment undergoing major change (ibid.).

The danger of water scarcity has been highlighted (ibid.; Kundzewicz and Matczak 2012: 302), and larger problems caused by reduced rainfall or heat waves have been noted. Attention has also been drawn to the presence of fresh water in the coastal regions in areas which are now under threat of infiltration by salt and brackish water (PRC 2009d: 3).

3.4 Interim conclusion

This chapter has, first of all, introduced the basic terminology of climate-change-related science. In the study of historical scientific developments, some aspects of the 'cultural situatedness and historical contingency of our climate knowledge' (Fleming 1998: 9) have been highlighted. This showed that the idea of changing climate only arose once actors obtained the technology that enabled observations of longer periods and larger spatial areas and regions.

Ideas about a change in climate conditions first emerged from the study of historical sources and only later from directly measured data. Today climate change is indicated by statistical reference variables, such as shifting averages and variations. The phenomenon seems not to be directly perceptible, but it has also been noted that, alongside statistical description, climatic changes can also be linked by actors to concrete events.

During the last few centuries there has been an increasing spatial expansion and institutionalisation of scientific observation, along with the social treatment of weather and climate phenomena. Since the first joint attempts at observation during the economic prosperity of the Italian Renaissance, efforts have become increasingly globalised. With the founding of the WMO in the late nineteenth century there emerged the first globally active, transnational organisation. The idea of anthropogenic climate change created further diversity in the global institutions, leading to the formation of the IPCC and the World Climate Conferences. Ways of dealing with climate change therefore present exceptional research potential for exploring the global constellations of (climate)-cultural spaces further. According to the latest social-science theory of deterritorialisation, it appears that national spaces (often reduced to container spaces) are nowadays of little significance for the separation of climate science *per se*, or for the resulting forms of climate change knowledge. This theory will be examined in more detail later.

Climate change is also associated in both the natural and social sciences with various problems and opportunities for the world and for specific places and regions. On the basis of historical observations, changes taking place in the last centuries and decades have been described. Additionally, climate projections have been calculated for possible future changes, including sea-level rise, increased precipitation and an increase in flooding, increasing heat waves, and changes in storm intensity.

The regions under investigation might be affected to varying degrees owing to differences in their physical landscapes. Actor-specific knowledge of vulnerability and resilience could therefore also differ according to local peculiarities. I will outline below in more detail how this fundamental problem in describing climate cultures can be reduced.

Notes

1 The choice of the concept of constructions must not, however, lead one to consider all knowledge as arbitrarily constructed. Luckmann himself warned against confusing social constructivism with scepticism or radical constructivism (Luckmann 1999: 18). The aim of this study is to systematically take account of actor-specific knowledge and its genesis, not to question all possible phenomena.

2 Such as the Meteorological Society of London (1823), the Annuaire météorologique de la France (1849), the Army Medical Department in the USA (1819), and the Anstalten für Witterungskunde in Saxe-Weimar-Eisenach (Fleming 1998: 43).

3 For some English weather stations, directly recorded data are available from 1659 onwards.

References

Arrhenius, Svante (1896) 'On the Influence of Carbonic Acid in the Air upon the Temperature of the Ground'. *Philosophical Magazine and Journal of Science*, Series 5, Volume 41 (April): 237–276. Available at: www.rsc.org/images/Arrhenius1896_tcm18-173546.pdf (accessed 29 November 2013).

Behringer, Wolfgang (2007) *Kulturgeschichte des Klimas. Von der Eiszeit bis zur Globalen Erwärmung*. München: C.H. Beck.

Bundesregierung (ed.) (2008) *Deutsche Anpassungsstrategie an den Klimawandel*. Berlin: no publisher's imprint.

Callendar, G.S. (1938) 'The Artificial Production of Carbon Dioxide and its Influence on Temperature'. *Quarterly Journal of the Royal Meteorological Society* 64 (275): 223–240. Available at: http://onlinelibrary.wiley.com/doi/10.1002/qj.49706427503/pdf (accessed 6 December 2013).

Conrad, Jobst (2010) 'Sozialwissenschaftliche Analyse von Klimaforschung, -diskurs und -politik am Beispiel des IPCC'. In: Voss, Martin (ed.): *Der Klimawandel*. Wiesbaden: VS Verlag für Sozialwissenschaften, pp. 101–115.

Deltacommissie (2008) *Working Together with Water. A Living Land Builds for its Future. Findings of the Deltacommissie 2008* [*Samen werken met water. Een land dat leeft, bouwt aan zijn toekomst. Bevindingen van de Deltacommissie 2008*]. Alkmaar: Hollandia Printing. Available at: www.deltacommissie.com/doc/deltareport_full.pdf (accessed 15 May 2012).

EEA European Environment Agency (ed.) (2006) *The Changing Faces of Europe's Coastal Areas*. Copenhagen/Luxembourg: Office for Official Publications of the European Communities. Available at: www.eea.europa.eu/publications/eea_report_2006_6 (accessed 1 May 2014).

EEA European Environment Agency (ed.) (2012) *Climate Change, Impacts and Vulnerability in Europe. An Indicator Based Report*. Copenhagen/Luxembourg: Office for Official Publications of the European Union. Available at: www.eea.europa.eu/publications/climate-impacts-and-vulnerability-2012 (accessed 25 April 2014).

Fleming, James Rodger (1998) *Historical Perspectives on Climate Change*. New York: Oxford University Press.

Furmanczyk, Kazimierz (2013) 'Poland'. In: Pranzini, Enzo and Williams, Allan (ed.): *Coastal Erosion and Protection in Europe*. Abingdon/New York: Routledge, pp. 81–95.

Gupta, Joyeeta, van Asselt, Harro, Amelung, Bas, Bessembinder, Janette, Bogaardt, Marc-Jeroen, Bolwidt, Leonie, Buiteveld, Hendrik, Huynen, Maud, Leemans, Rik, Martens, Pim, Verhagen, Jan, and van der Wegen, Mick (2006) *Assessing Dangerous Climate Impacts for the Netherlands. Climate Change Scientific Assessment and Policy Analysis*. Bilthoven: Netherlands Environmental Assessment Agency (MNP).

Heimann, Thorsten and Christmann, Gabriela B. (2013) *Klimawandel in den deutschen Küstenstädten und –gemeinden. Befunde und Handlungsempfehlungen für Praktiker*. Available at: www.irs-net.de/publikationen/sonderpublikationen/index.php (accessed 11 April 2013).

IPCC Intergovernmental Panel on Climate Change (1990) *Climate Change. The IPCC Scientific Assessment*. Cambridge: Cambridge University Press. Available at: www.ipcc.ch/publications_and_data/publications_ipcc_first_assessment_1990_wg1.shtml (accessed 13 December 2013).

IPCC Intergovernmental Panel on Climate Change (ed.) (2007a) 'Organization. Structure'. IPCC website. Available at: www.ipcc.ch/organization/organization_structure.shtml (accessed 7 December 2013).

IPCC Intergovernmental Panel on Climate Change (ed.) (2007b) *Climate Change 2007: Impacts, Adaptation and Vulnerability*. Contribution of Working Group II to the Fourth Assessment Report of the Intergovernmental Panel on Climate Change. Cambridge/New York: Cambridge University Press.

IPCC Intergovernmental Panel on Climate Change (2007c) *Climate Change 2007: The Physical Science Basis*. Contribution of Working Group I to the Fourth Assessment Report of the Intergovernmental Panel on Climate Change. Cambridge/New York: Cambridge University Press.

IPCC Intergovernmental Panel on Climate Change (ed.) (2007d) *Climate Change 2007: Synthesis Report*. Contribution of Working Groups I, II and III to the Fourth Assessment Report of the Intergovernmental Panel on Climate Change. Geneva: IPCC.

IPCC Intergovernmental Panel on Climate Change (ed.) (2013a) *Approved Summary for Policymakers*. Twelfth Session of Working Group I. IPCC WGI AR5. Available at: www.climatechange2013.org/images/uploads/WGIAR5-SPM_Approved27Sep2013.pdf (accessed 30 October 2013).

IPCC Intergovernmental Panel on Climate Change (ed.) (2013b) *Climate Change 2013. The Physical Science Basis*. WGI I AR5. Summary for Policy Makers. Available at: www.climatechange2013.org/images/uploads/WGI_AR5_SPM_brochure.pdf (accessed 6 December 2013).

IPCC Intergovernmental Panel on Climate Change (ed.) (2013c) *Climate Change 2013: The Physical Science Basis*. Contribution of Working Group I to the Fifth Assessment Report of the Intergovernmental Panel on Climate Change, Cambridge/New York: Cambridge University Press. Available at: www.climatechange2013.org/images/report/WG1AR5_ALL_FINAL.pdf (accessed 27 April 2014).

IPCC Intergovernmental Panel on Climate Change (ed.) (2014) *Climate Change 2014: Impacts, Adaptation, and Vulnerability. Part A: Global and Sectoral Aspects*. Contribution of Working Group II to the Fifth Assessment Report of the Intergovernmental Panel on Climate Change. Cambridge/New York: Cambridge University Press.

Jacobeit, Jucundus (2007) 'Zusammenhänge und Wechselwirkungen im Klimasystem'. In: Endlicher, Wilfried and Gerstengarbe, Friedrich-Wilhelm (ed.): *Der Klimawandel. Einblicke, Rückblicke und Ausblicke*. Potsdam: G&S, pp. 1–16.

Jensen, Jürgen and Schwarzer, Klaus (2013) 'Germany'. In: Pranzini, Enzo and Williams, Alan (ed.): *Coastal Erosion and Protection in Europe*. Oxford/New York: Routledge, pp. 108–135.

Kappas, Martin (2009) *Klimatologie. Klimaforschung im 21. Jahrhundert. Herausforderung für Natur- und Sozialwissenschaften*. Heidelberg: Springer/Spektrum Akademischer Verlag.

KEMIN The Danish Ministry of Climate and Energy (ed.) (2009) *Denmark's Fifth National Communication on Climate Change. Fifth Report on the National Implementation of the United Nations Framework Convention on Climate Change and the Kyoto Protocol*. Available at: http://unfccc.int/resource/docs/natc/dnk_nc5rev.pdf (accessed 10 March 2015).

Kühlmann, Wilhelm (2011) 'Schindler, Karl Friedrich'. In: *Killy Literaturlexikon*. Berlin/Boston: De Gruyter, pp. 363–364.

Kundzewicz, Zbigniew W. and Matczak, Piotr (2012) 'Climate Change Regional Review: Poland'. *WIREs Climate Change* 3: 297–311. DOI: 10.1002/wcc.175.

Kunze, Rolf-Ulrich (2009) 'Autobahn-Landschaftsbild: Die Niederlande als geschütztes Landschaftsartefakt'. *Journal of New Frontiers in Spatial Concepts* 1: 8–10. Available at: http://ejournal.uvka.de/spatialconcepts/archives/495 (accessed 22 February 2015).

Labuz, Tomasz A. (2013) 'Polish Coastal Dunes: Affecting Factors and Morphology'. *Landform Analysis* 22: 33–59. Available at: http://dx.doi.org/10.12657/landfana.022.004 (accessed 4 May 2014).

Luckmann, Thomas (1999) 'Wirklichkeiten: individuelle Konstitution und gesellschaftliche Konstruktion'. In: Hitzler, Ronald, Reichert, Jo and Schröer, Norbert (ed.): *Hermeneutische Wissenssoziologie. Standpunkte zur Theorie der Interpretation*. Konstanz: UVK, pp. 17–28.

Manabe, Syokuro and Wetherald, Richard T. (1967) 'Thermal Equilibrium of the Atmosphere with a Given Distribution of Relative Humidity'. *Journal of the Atmospheric Sciences* 24 (3): 241–259.

Matulla, C., Schöner, W., Alexanderson, H., von Storch, H., and Wang, X.L. (2007) 'European Storminess. Late Nineteenth Century to Present'. In: *Climate Dynamics* 31 (2–3): 125–130. Available at: www.image.ucar.edu/idag/Papers/storch.matulla.estorms%5B1%5D.pdf (accessed 25 April 2014).

Milankovitch, Milutin (1930) 'Mathematische Klimalehre und astronomische Theorie der Klimaschwankungen'. In: Köppen, Wladimir Peter and Geiger, Rudolf (ed.): *Handbuch der Klimatologie* 1. Berlin: Gebr. Bornträger, pp. 1–176.

NEAA Netherlands Environmental Assessment Agency (PBL) (2009) *Roadmap to Climate Proof Netherlands*. Bilthoven: PBL.

PRC Policy Research Corporation (2009a) *The Economics of Climate Change Adaptation in EU Coastal Areas. Country Report: Germany*. Brussels: European Commission. Directorate-General for Maritime Affairs and Fisheries. Available at: http://ec.europa.eu/maritimeaffairs/documentation/studies/documents/germany_climate_change_en.pdf (accessed 12 December 2013).

PRC Policy Research Corporation (2009b) *The Economics of Climate Change Adaptation in EU Coastal Areas. Country Report: The Netherlands*. Brussels: European Commission. Directorate-General for Maritime Affairs and Fisheries. Available at: http://ec.europa.eu/maritimeaffairs/documentation/studies/documents/netherlands_climate_change_en.pdf (accessed 12 December 2013).

PRC Policy Research Corporation (2009c) *The Economics of Climate Change Adaptation in EU Coastal Areas. Country Report: Denmark*. Brussels: European Commission.

Directorate-General for Maritime Affairs and Fisheries. Available at: http://ec.europa.eu/maritimeaffairs/documentation/studies/documents/denmark_climate_change_en.pdf (accessed 12 December 2013).

PRC Policy Research Corporation (2009d) *The Economics of Climate Change Adaptation in EU Coastal Areas. Country Report: Poland.* Brussels: European Commission. Directorate-General for Maritime Affairs and Fisheries. Available at: http://ec.europa.eu/maritimeaffairs/documentation/studies/documents/poland_climate_change_en.pdf (accessed 12 December 2013).

Pruszak, Zbigniew and Ostrowski, Rafal (2005) 'Shrinking Shores. Our Beaches'. *Academia. The Magazine of the Polish Academy of Sciences* 5: 24–25.

Rahmstorf, Stefan and Schellnhuber, Hans Joachim (2007) *Der Klimawandel. Diagnose, Prognose, Therapie.* München: C.H. Beck.

Schimper, Karl Friedrich (1837) *Die Eiszeit.* Available at: http://gedichte.xbib.de/Schimper_gedicht_Die+Eiszeit.htm (accessed 26 November 2013).

Schönwiese, Christian Dietrich (2008) *Klimatologie.* Stuttgart: Eugen Ulmer KG.

Sørensen, Per (2013) 'Denmark'. In: Pranzini, Enzo and Williams, Allan (ed.): *Coastal Erosion and Protection in Europe.* Abingdon/New York: Routledge, pp. 96–107.

Suess, Hans (1955) 'Radiocarbon Concentration in Modern Wood'. *Science* 122 (3166): 415–417. Available at: www.sciencemag.org/content/122/3166/415.2.full.pdf (accessed 6 December 2013).

The Danish Government (ed.) (2008) *Danish Strategy for Adaptation to a Changing Climate.* Copenhagen: Schultz.

Tyndall, John (1861) 'The Bakerian Lecture: On the Absorption and Radiation of Heat by Gases and Vapours, and on the Physical Connexion of Radiation, Absorption, and Conduction'. *Philosophical Transactions of the Royal Society of London* 151: 1–36.

Urey, H.C., Lowenstam, H.A., Epstein, S., and McKinney, C.R. (1951) 'Measurement of Paleotemperatures and Temperatures of the Upper Cretaceous of England, Denmark and the Southeastern United States'. *Bulletin of the Geological Society of America* 62: 399–416.

Van der Meulen, Frank, Van der Valk, Bert, and Arens, Bos (2013) 'The Netherlands'. In: Pranzini, Enzo and Williams, Allan (ed.): *Coastal Erosion and Protection in Europe.* Oxford/New York: Routledge, pp. 137–157.

Van Koningsveld, D.M., Mulder, J.P.M., Stive, M.J.F., Vandervalk, L., and Vanderweck, A.W. (2008) 'Living with Sea-level Rise and Climate Change: A Case Study of the Netherlands'. *Journal of Coastal Research* 24 (2): 367–379.

Vermeer, Martin and Rahmstorf, Stefan (2009) 'Global Sea Level Linked to Global Temperature'. In: *Proceedings of the National Academy of Sciences of the United States of America* (PNAS), 7 December 2009.

Von Storch, Hans and Woth, Katja (2011) 'Sturmfluten'. *Historische Sozialkunde: Geschichte – Fachdidaktik – Politische Bildung* 41 (3). Available at: www.hvonstorch.de/klima/pdf/hammerl.woth.pdf (accessed 16 May 2014).

Weingart, Peter, Engels, Anita, and Pansegrau, Petra (2002) *Von der Hypothese zur Katastrophe. Der anthropogene Klimawandel im Diskurs zwischen Wissenschaft, Politik und Massenmedien.* Opladen: Leske & Budrich.

WMO (2009) *The World Meteorological Organization at a Glance. Working Together for Monitoring, Understanding and Predicting Weather, Climate and Water.* Geneva: WMO. Available at: www.wmo.int/pages/about/documents/WMO990.pdf (accessed 26 November 2013).

WMO (2013) *World Climate Conferences*. Available at: www.wmo.int/pages/themes/climate/international_wcc.php (accessed 13 December 2013).

Woth, K. And von Storch, H. (2008) 'Klima im Wandel. Mögliche Zukünfte des Norddeutschen Küstenklimas'. In: *Dithmarschen. Landeskunde – Kultur – Natur* 1: 20–31.

Zebisch, Marc, Grothmann, Torsten, Schröter, Dagmar, Hasse, Clemens, Fritsch, Uta, and Cramer, Wolfgang (2005) *Klimawandel in Deutschland. Vulnerabilität und Anpassungsstrategien klimasensitiver Systeme*. Dessau: Umweltbundesamt. Available at: www.umweltbundesamt.de/publikationen/klimawandel-in-deutschland-vulnerabilitaet (accessed 1 May 2012).

4 Resilience constructions

Introduction to climate-mitigation and adaptation strategies

The previous chapter presented the scientific discussion of climate change and projections for further changes in the future. Such changes could have numerous implications for different areas of human life. Various forms of climate-related action are thus also conceivable. A distinction can be drawn between climate-change adaptation and climate-change mitigation. Climate-mitigation measures are designed to prevent or reduce the effects of climate change (IPCC 2014a). Climate adaptation measures serve to help adjust to changing climate conditions (IPCC 2014b).

Climate-mitigation and adaptation measures can find application in virtually all areas of society (Leggewie 2010: 40f.; Institut der deutschen Wirtschaft 2009). This study will focus on the area of spatial development. This has the advantage that both climate-mitigation and adaptation measures are available to the actors working in this area (BMVBS 2010). The concept of spatial development will therefore be examined more closely (Section 4.1), before the discourses surrounding various possible measures are introduced (Sections 4.2 (mitigation) and 4.3 (adaptation)). As many of these discourses have their origin outside the field of spatial development, it will occasionally be necessary to present debate from beyond spatial development.

4.1 On the concept of spatial development

There is no uniform understanding of the term 'spatial development'. Nevertheless, common semantic references can be identified. The European Union has sought to develop a uniform glossary on spatial development, and this study will therefore orient itself towards this terminology. Here, the concept of spatial *development* must be differentiated from the concept of spatial *planning* (COMMIN 2014):

> Spatial planning and spatial development are related terms that describe the conscious shaping of people on the development of an area (a 'space').

These terms therefore refer to human action[1] aimed at the design of space. By 'space', the authors understand the 'natural, built, and social environment of an area' (ibid.). This reveals that spatial planning and development seek to influence not only the physical, material, and ecological design of an area, but also social formations.

The terms can be differentiated roughly according to their focus on specific actors and institutions. *Spatial planning* traditionally refers above all to arrangements by state-sovereign institutions. The more recent concept of *spatial development*, by contrast, refers to the more general development processes of areas, places, or regions (COMMIN 2014). It thus includes spatial planning of the state, but also refers back to the stronger involvement of various other social actors in the development of space (see also Section 2.2).

Since the 1980s, a 'shifting pattern in styles of governing' (Jann 2009: 3) has been discussed, with much attention given to western democracies. In contrast to the classical, hierarchical mechanisms of state control, new forms of communication and coordination between the state, the economy, and civil society have become increasingly important, such as in the form of new political networks, participatory negotiation models, or local partnerships between state and non-state actors (ibid.; Benz *et al.* 2007: 13).

As also in other policy areas, the 'conscious influence of people on the development of an area' (COMMIN 2014) has become ever less defined by a top-down hierarchy of legitimated political and administrative representatives. Instead, diverse cooperative and participatory forms have gained in significance (Mayntz 2008: 45). In this context, Nuissl has spoken of a paradigm shift 'towards a discursive planning culture' (Nuissl 2007: 44), placing a stronger focus on the participation of various social actors in the development of space (ibid.). More recently, the concept of *spatial development* has increasingly been used to emphasise the 'constitutive, dynamic character of the development of a space' (Birkmann *et al.* 2013: 17).

Spatial development can be understood as an umbrella term referring to the conscious human influence on the design of areas that regularly takes place in the interplay between various political, administrative, economic, and civil society actors. The concrete climate-related measures for spatial development that are already being discussed and/or implemented will now be considered in more detail.

4.2 Climate mitigation

The scientific development of the concept that human action can lead to large-scale changes in the atmosphere was followed by numerous ideas for the shaping of their mitigation. Working Group III of the IPCC deals with ways in which anthropogenic climatic change can be mitigated. Climate mitigation is thus defined as follows: 'Mitigation is a human intervention to reduce the sources or enhance the sinks of greenhouse gases' (IPCC 2014a: 4). The central focus is therefore on measures to influence the proportion of atmospheric greenhouse gases. The example of carbon dioxide (CO_2) shows a regular fluctuation over one year in the globally measured annual atmospheric content (IPCC 2013b: 10,

Keeling curve). In the annual balance, the values currently show an asymmetry. The annual *import* of CO_2 into the atmosphere, such as through incineration or animal respiration, does not correspond to the *export*, such as by plant photosynthesis or chemical interactions with land masses and oceans. Consequently, two basic climate mitigation strategies are conceivable:

- The **reduction of existing greenhouse gas emissions** (such as by lowering energy consumption or switching to non-fossil fuels).
- The **conservation or increase of greenhouse gas sinks** (such as forests, fenlands, and carbon capture and storage (CCS) technology) to absorb greenhouse gases from the air (IPCC 2014a: 4f.; Birkmann *et al.* 2013: 14).

To date, numerous ideas have been taken up and developed within various societal fields of application, among them spatial development (see for example IPCC 2013b: 18ff.; Fleischauer *et al.*, 2013: 90ff.; Birkmann *et al.*, 2010: 7ff.; Stock 2005: 500ff.). In the next section, the reduction strategies available in this field will be considered.

4.2.1 Reduction of greenhouse gas emissions: sufficiency, efficiency, and consistency strategies

The IPCC estimates that worldwide emissions of greenhouse gases (GHG) currently result from energy and heat generation (25%), agriculture and forestry (24%), industrial production (21%), transport (14%), and building use (7%) (IPCC 2013a: 8). The burning of fossil fuels is considered to be the main source of emissions, especially for carbon dioxide, which accounts for almost 80% of total emissions (ibid.: 6f.). The International Energy Agency states that in 2012 about 82% of global energy needs was generated from burning oil, coal, and natural gas (IEA 2014: 6) In the European Union, the fossil component was comparatively lower according to Eurostat, accounting for 75% of primary energy consumption (Eurostat 2014).

The reduction of GHG emissions can be assessed for both the primary energy-source type ('input') and for energy utilisation ('output'). Three 'substrategies' emerge from this technical-procedural perspective (Reusswig 2010: 90): (1) *sufficiency* and (2) *efficiency strategies* aim at a reduction in the energy input; (3) *consistency strategies* focus on the migration between different types of energy generation (see also Fleischauer *et al.* 2013: 95; Huber 2011: 58ff.; Reusswig 2010: 90, 161ff.; Wilson and Piper 2010: 22).

The *sufficiency discourse* is probably the oldest ecological discourse of the present day, being traceable back to the very beginning of industrialisation in Europe in the eighteenth and nineteenth centuries. At that time critics of civilisation, such as Jean-Jacques Rousseau and Thomas Robert Malthus, gained increasing attention for their ideas (Huber 2011: 135). In the mid-eighteenth century Rousseau rejected the European civilisation of his time and, in exaltation of the supposedly more authentic 'noble savage' (Berger 2004: 357), sought a path 'back to nature' (ibid.). In 1826 Malthus foresaw in the ceaseless population growth a real danger for further economic development, anticipating a collapse of civilisation and widespread famine should these developments continue along their existing paths. One solution he proposed was moral abstinence (Malthus 1798).

A century and a half later, his ideas gained new impetus when a research team gathered together by the 'Club of Rome' and headed by Donella and Dennis Meadows modelled the global 'limits of growth' (Meadows *et al.* 1972). In the 1970s, ecological and economic boundaries could now be identified using system-dynamic computer models. In the context of the globally perceived oil and growth crises, broad support for sufficiency strategies was also found among the burgeoning environmental movements. These ideas later found their way into the climate-change discourse (Huber 2011: 161ff.).

The key question for sufficiency discourse is: 'How much is enough?' (ibid.). To be sure, nowadays comparatively few radical suggestions come to the fore, such as considering the phasing out of industry or the adoption of social prosperity and growth models as a model by which to solve ecological problems. Some ideas, however, have survived into modern climate discourse (ibid.). As a climate-mitigation strategy, Reusswig defines sufficiency as the 'reduction of the absolute level of energy demand by changes in mentality and behaviour' (Reusswig 2010: 90). Sufficiency aims for lower energy use overall, either through changes in use or reductions in consumption.

In the field of spatial development, ideas following the logic of sufficiency are also still represented. These can be found, for example, in relation to *traffic* and *settlement structures* (Fleischauer *et al.* 2013: 95; IPCC 2013a: 22ff.). Motorised private transport (MPT) has been the preferred means of transport in central Europe for years (BMVBS/BBSR 2009: 32). The IPCC also takes into consideration new spatial planning structures which allow for a return to non-motorised transport, such as going on foot or by bicycle (IPCC 2013a: 23). Spatial development can therefore target the *reduction of individual motorised traffic*. Possible contributions to this are zones with traffic calming, the restriction of other spatial and functional uses in preference for the development of cycle routes, and the promotion of collective transport concepts (public transport and car sharing) (BMVBS/BBSR 2009: 34). It is also possible to *limit land use*, such as reducing plot sizes or the space between building spaces, in order to implement *redensification* (following the guiding principle: 'internal development before external development') (ibid.). For the reconstruction of *settlement structures*, a return to *mixed-use development* has been proposed, such as the kind of development common before industrialisation, with housing and work spaces combined (BMVBS/BBSR 2009: 33).

Ecological discourse, and later also climate discourse, have not been restricted to sufficiency strategies. While in the years following the appearance of the 'limits of growth' large differences between ecologically and economically oriented actions existed, this perspective began to disappear by the early 1980s. Under the motto of 'ecological modernisation' (Jänicke 2000, 1984), approaches were developed which sought to use the logics of economic and industrial-technological rationalisation and innovation for ecological ends. Hence today, representatives of ecological-modernisation theories such as Joseph Huber, Martin Jänicke, Arthur Mol, and Gert Spaargaren reject the idea that technological modernisation processes must come to an end in order to solve ecological problems (Huber 2011: 143; Jänicke 2000: 1ff.; Mol and Spaargaren 2000: 18ff.).

An early approach created the idea of *eco-efficiency* (Huber 2011: 166), which attempts to achieve the 'saving of energy per production or consumption unit through technological and/or behavioural changes' (Reusswig 2010: 90). The same outputs are to be obtained while using less input in order to reduce emissions (Huber 2011: 166). *Energy efficiency strategies* continue to be discussed as measures for climate mitigation (see for example IPCC 2013a: 23ff.). This is also true with regard to spatial development. Thus the IPCC points out that the *energy consumption of buildings can be optimised* (IPCC 2013a: 24). New buildings can be equipped with improved thermal insulation, and existing buildings should be retrofitted so that less energy needs to be consumed to achieve similar room temperatures (BMVBS/BBSR 2009: 24).

Efficiency and sufficiency approaches have, however, both been criticised for the fact that these rarely offer long-term solutions to resource and emission problems, since only consumption levels are reduced (Huber 2011: 171). They extend the time frame until ecological limits are met – e.g. tipping points, such as the complete melting of the Greenland ice sheet or the exhaustion of fossil mineral deposits – but the problematic developmental trajectory would not itself be eliminated (ibid.). In addition, increases in efficiency have regularly seen rebound effects, leading not to a reduction in the consumption of resources and/or emissions, but to the use of the additional efficiency to increase performance. This eventually means that even more resources are consumed than before, missing the original target of lowering emissions (see also Heimann 2013: 64).

Ecological modernisation discourse ultimately resulted in the strategy of ecological and metabolic *consistency* (Huber 2011: 172). Consistency strategies no longer aim only at an improvement in the efficiency of fossil and other energy sources, but at a complete structural change in favour of new technologies, products, and practices designed in such a way that the relevant materials cycles can be anchored within the ecological environment sustainably and largely without causing any form of disturbance (ibid.: 172).

Connected to this is the idea that energy use is not necessarily linked to the burning of fossil fuels. From the consistency strategic perspective, a *migration to renewable energy* would enable the complete decoupling of social and economic developments from fossil resources, hailing a 'post-carbon society' (Reusswig 2010: 90). Former patterns of use and development ('outputs') could continue to be pursued *consistently* with the same or even higher energy input, but free of any globally relevant greenhouse gas emissions. Potential renewable or regenerative energy sources are being discussed that are 'inexhaustible within human temporal horizons' (Quaschning 2007: 34). Among these are sources yielding energy primarily from the sun (e.g. photovoltaics and solar thermal energy), gravitational or planetary energy sources (e.g. hydropower and wind power), bioenergy from renewable resources (e.g. wood and biodiesel), and geothermal energy (Fleischauer *et al.* 2013: 98; Destouni and Frank 2010; Quaschning 2007: 79ff.).

The migration from fossil to renewable energies can also be associated with physical-spatial changes, since all forms of energy generation have different requirements of these spaces (Fleischauer *et al.* 2013: 102). Wind turbine rotors, for example, have been and are being widely installed on- and offshore. Photovoltaic plants can be installed either in private or public buildings or in exposed, open spaces. Hydroelectric power plants require transformations of marine or inland expanses of water. Space is also required for the production of biomass (ibid.). Fleischauer *et al.* (2013: 118) argue that changing patterns of demand with regard to land use could lead to new competing spatial requirements, acceptance problems, and conflicts. It follows from this that the migration to regenerative energies can also be explicitly proposed or prevented by spatial development actors, who must make and/or approve decisions regarding the *protection of the land used by renewable energies* (ibid.; BMVBS 2011: 23ff.). In addition, renewable energy generation is associated with the decentralisation of the production sites (Donnerbauer 2012; BMVBS 2011: 12ff.) An increasing number of providers can offer power stations relatively smaller packages (in the form of kilowatt hours), such as when solar panels are installed on the roofs of smaller households. This means that *energy generation by smaller public and private households and companies* is gaining in importance. The actors involved must thus clarify whether and to what extent local space-related patterns of use need to be formalised (ibid.).

4.2.2 Absorption of greenhouse gases from the atmosphere: ecological and technological sinks

Sinks are reservoirs on the Earth's surface which are able temporarily or permanently to absorb carbon dioxide. The concept of natural sinks is closely linked to the understanding of global carbon cycles, in which regular carbon flows take place between the geosystems – the atmosphere, land, and water masses – and the biosphere (see for more detail Schönwiese 2008: 346ff.). The German Advisory Council on Global Change (Wissenschaftlicher Beirat der Bundesregierung Globale Umweltveränderungen, henceforth WBGU) works on the assumption that approximately two to five gigatonnes of carbon dioxide are extracted from the atmosphere every year by means of so-called natural sinks (WBGU 2003: 53). On the Earth's surface, both the soil and the vegetation act as natural carbon sinks. The biosphere has an important role, since in the world's rural ecosystems alone there exists about three times as much carbon as is present in the form of atmospheric carbon dioxide (Fleischauer *et al.* 2013: 108). Due to their current surface area, forests have the most important part to play, particularly those in the tropics and boreal zones. It is also known that fenland can absorb the largest portion of carbon per unit area, as organic material is very highly compressed in these areas (ibid.).

The WBGU argues that owing to the varying spatial distribution of natural sinks, global differences in the annual carbon balance can be observed. Accordingly, about half of CO_2 emissions generated by the United States are absorbed across that country, while only about one-third of European emissions are re-assimilated in Europe (WBGU 2003: 54). Land use thus has an important influence on carbon sinking. Changes in local and global land use, for example through forestry and conversion to urban or agricultural areas, can contribute to a reduction in the sinking capacity, and also to increased emissions. According to the IPCC, almost a quarter of global greenhouse gas emissions are attributable to agricultural and forestry activities, in particular clearing, fertilisation, and animal husbandry (IPCC 2013a: 25). But the carbon storage capacity of ecosystems itself depends on human patterns of use. Thus more intensive forest management, for example, is associated with a decline in the carbon storage capacity of these areas, while unused forests, including wilderness reserves such as national parks, are attributed far greater storage capacities (Fleischauer *et al.* 2013: 109; SRU 2008: 155). The targeted drainage of fenlands can similarly lead to the degradation of organic substances and release the carbon stored therein (ibid.).

Out of these considerations arises the idea that spatial development actors can actively influence the global carbon balance through the planning and implementation of specific land use. The IPCC thus emphasises the *expansion of carbon sinks* as an important task in climate mitigation (IPCC 2013a: 4).

In addition to the targeted promotion of natural sinks, possibilities for the *technological storage of* CO_2 (carbon capture and storage technology, henceforth CCS) have also been proposed (IPCC 2005; Fleischauer *et al.* 2013: 110). In this process, CO_2 is separated and stored in artificial, mostly underground storage sites. Although diverse applications have been theorised for this technology, the combustion of fossil fuels, particularly in coal-fired power plants, has been the main focus of discussion (ibid.). CCS would serve as a 'bridge technology' (EC 2013: 16), as long as the 'dependence on fossil fuels continues to exist' (ibid.: 3). The technology would be marketable if high-cost obligations for meeting global emissions targets (such as emissions trading) were implemented, and if renewable energies carried much higher kilowatt-hour prices than those for fossil fuels.

The main task of spatial development actors would in future be to achieve consensus on the use of CCS practices, as well as on the regulation of transport routes and storage sites (Fleischauer *et al.* 2013: 110). The use of CCS technology is currently largely in the testing phase, however. After the publication of an IPCC special report on CCS technology (IPCC 2005) several pilot projects were launched worldwide by research institutes and energy companies. In 2009 the European Union also adopted a directive to create a legal framework for CCS deployment in Europe (EC 2009). However, their further political and practical implementation has been very slow in most of the 27 participating countries, not least due to their low acceptance among local populations (Hirschhausen *et al.* 2012: 3; European Commission 2011).

From developments it can be concluded for now that the future commercial introduction of CCS technology in Europe remains uncertain and is likely to be associated with major conflicts between various stakeholder groups.

4.3 Climate adaptation

Alongside the systematic reappraisal and projection of the observed and potential impacts of climate change, Working Group II of the IPCC deals with adaptation measures for responding to these changes. The authors define climate adaptation as follows:

> The process of adjustment to actual or expected climate and its effects. In human systems, adaptation seeks to moderate or avoid harm or exploit beneficial opportunities. In some natural systems, human intervention may facilitate adjustment to expected climate and its effects.
>
> (IPCC 2014b: 5)

The Working Group emphasises the relevance of climate and weather phenomena to human activity. The use of the word 'adjustment' indicates that climate changes may necessitate modifications of or breaks from previous routines. Adaptation measures are meant to either moderate or prevent harm or target potential opportunities. The dual reference to the possibility of interventions in 'human' and 'natural systems' hints at the variety of possible adaptation measures. Climate adaptation deals with both the currently observed and anticipated future phenomena interpreted as having an impact on climate change. Measures can be reactive (where pre-existing damage is perceived) or proactive (anticipatory) (Eichhorst and Madry 2013; Frommer 2010: 17).

Chapter 3 already considered a variety of currently discussed climate-change impacts. The term 'adjustment' implies a reaction to this specific knowledge. The current study examines the spatial development of European coastal regions. Therefore, so as not to expand the scope of this study too far, it will be limited to the crucial problem constructions within the discourse, namely water-related challenges (such as sea and inland flooding), heat, and storms.

4.3.1 Measures against sea and inland flooding: protection, accommodation, retreat

When considering the challenges to coastal regions in Chapter 3, it was revealed that several causes of flooding appear in the discourse. On the one hand, new formations of sea-water masses are regarded as posing a challenge to coastal regions, while the authors also identify new problems caused by changing inland water conditions (Knieling *et al.* 2009: 15ff.; IPCC 2007: 14). Since potential challenges have regularly been met in the past with varying courses of action, these possible measures will each be considered separately. In doing so, a distinction needs to be made between adaptation measures which target sea flooding and inland flooding, respectively.

Sea flooding

The authors of the IPCC's first report in 1990 distinguished between three possible basic strategies for spatial development in the context of increasing sea levels, namely protection, accommodation (integration), and retreat (IPCC 2014b: 387, 2007: 342, 2001: 367, 1996: 312ff., 1990: xliv, 147ff.; Tol *et al.* 2008: 435).

- **Protection:** The aim of this strategy is the complete preservation of land areas by the sea (ibid.).
- **Accommodation:** Land areas near the sea continue to be used by people, even if they are subject to altered conditions such as regular or total flooding by sea water. To this end, alterations are made to infrastructure or to land use patterns (ibid.).
- **Retreat:** No efforts are made to maintain the land. In extreme cases, areas are abandoned and resettlement is undertaken (ibid.).

In order to be able to completely preserve landscapes within the framework of the *protection* strategy, various defensive measures (IPCC 1990: 148) are identified in relation to the sea. The IPCC distinguishes between 'hard' and 'soft' structures (ibid.). 'Hard structures' are technologically constructed infrastructures, such as the new installation or extension of dykes, dams, or flood walls, and the construction of flood gates or wave breakers (ibid.). 'Soft structures' do not require the installation of technological infrastructure, and include beach nourishment or the cultivation of vegetation to prevent erosion (ibid.: 149). In recent years, alongside the option of land conservation ('hold[ing] the line', IPCC 2007: 342), even land reclamation ('advanc[ing] the line', ibid.) has been considered as a protection strategy, such as through the technological closure of estuaries (ibid.).

Considering the descriptions of protection strategies, terms which carry military connotations occur frequently (e.g. 'defence', 'hold the line', 'advance the line', ibid.). A strict categorical distinction is made between land used by humans and the sea that threatens it. Spatial development here primarily entails 'defence' and 'demarcation' against the sea. In the context of *accommodation strategies*, however, the idea of struggle shifts to the background. The sea is no longer viewed as a supposed opponent against which the land must be completely protected. Rather, the new intention is 'to live with the water' (Deltacommissie 2008). Land used by people near the sea should still be developed, but with modifications being made to temporarily or permanently integrate the increasing water volumes. Houses can be erected on piles or equipped with flexible protective devices, for example (IPCC 1996: 313, 1990: 147). Technological floating devices for buildings have also been seriously discussed and implemented (so-called 'floating houses', Albers *et al.* 2013: 17; Knieling *et al.* 2009: 24). In addition, land modification in agriculture and forestry is being considered, such as the cultivation and farming of more salt-resistant plants or animals in areas near the sea, or their conversion into fish farms in the event of permanent flooding (IPCC 1990: xliv, 147f.).

Accommodation strategies are based on the possibility of combining human land use with the expansion of sea water. The *retreat strategy*, on the other hand, aims to abandon land used by humans to the sea. The retreat can be actively planned, for example, by deconstructing dykes and/or deliberately moving the dyke further inland ('retreat[ing] the line', IPCC 2007: 342). In extreme cases, existing settlements might also be abandoned (IPCC 1990: 147f.). Renaturation measures, leaving larger areas exposed to the effects of water, flora, and fauna, are also conceivable in this context ('wetland restoration', IPCC 2007: 342).

Inland flooding

Inland flooding can pose a threat to coastal regions, and in addition, higher or increasingly fluctuating water levels can be expected for inland bodies of water such as rivers, lakes, and streams (see Chapter 3; IPCC 2007: 550). As new relationships between areas used by humans and changing water levels might occur here, similar adaptation strategies are conceivable to those discussed above for sea flooding, i.e. *protection*, *accommodation*, *or retreat*.

There are some peculiarities relating to inland flooding strategies, however, which are likewise evident in the causes attributable to inland flooding. First we must recall the concept of the water cycle (or hydrological cycle) (Kappas 2009: 187ff.; Schönwiese 2008: 152ff.). Namely, (rain) water masses which are not stored and do not evaporate inevitably make their way to the sea (ibid.: 2011: 19).

Should unusually large volumes of water meet human settlements and farmland for a certain period, this is usually experienced as flooding. Staying with the idea of the water cycle, we can see, on the one hand, that inland flooding may lead to problems along coastal regions directly at the site of the precipitation and, on the other hand, that the starting point of the flooding may be distant inland rainfall. To prevent problematic situations developing, different starting points for flooding events should be considered as positions of leverage to prevent floods at various stages in the water cycle.

The first point to consider relates to human *settlements in the floodplains* of rivers. For Europe, which is relatively densely populated, the IPCC does not consider anthropogenic climate change a principal cause of the increase in flood disasters in recent decades, so much as the use of areas that were already regularly flooded in the past (IPCC 2014c: 1280). Thus a comparatively recent anthropocentric view has emerged in the IPCC discourse on inland flooding, which identifies human activity as the main cause of flood disasters (Garrelts *et al.* 2008). A paradigm shift has also been discussed in connection with inland flooding, in which similar narratives appear as have already been described in the context of sea flooding. The 'battle against water' (ibid.: 72f.) should thus be undertaken less by means of technological infrastructures 'as was customary in central Europe over the centuries' (ibid.), being replaced by the principle of 'giving rivers more space' (ibid.). This approach currently dominates, at least in the German-language engineering science literature, and the strategy used until the 1970s of promoting human land use in the vicinity of water and protecting it with measures such as walls and dams is now regarded as highly problematic (see also Krohn 2013: 606; Patt and Gonsowski 2011: 159; Garrelts *et al.* 2008: 73).[2]

Viewing human activity near the water as the actual cause of flood disasters need not necessarily lead to a preference for local strategies of retreat. As water travels long distances from the highlands to estuaries, measures can be implemented at any site beginning from the sources of precipitation inland, along rivers, and up to the estuaries. A different starting point therefore relates to *measures for increasing water reserves*. As already discussed, the size of the water volumes that ultimately cause streams, rivers, and lakes to overflow is not only associated with the amount of precipitation, but also with evaporation and retention factors (Patt and Gonsowski 2011: 19). The creation of retention facilities should lead to a slowing down of the water and thus reduce the maximum level of watercourses (Jüpner 2013: VII). *Ground surface properties* could be the first aspect to consider, since different soils have varying water-retention capacities. Surface sealing through the construction of buildings, roads, squares, and so on, can reduce water-retention capacities, causing water to drain away faster (MKULNV 2011: 19). Measures to expose these surfaces, so-called *unsealing*, should help to once again improve water-retention potential (ibid.: 27). Where unsealing is not possible, for example where buildings already exist or are being constructed, the use of specialised surface designs can also help to slow rain runoff, for example the *installation of roof and facade greening* (ibid.).

Greater retention capabilities can also be created through spatial planners identifying areas to be left unbuilt, creating *retention areas* which can be flooded in case of need (UBA 2008: 2).

Technological flood protection can also focus on retention options through the installation of associated infrastructure. *Water retention basins* can thus be built, as well as *reservoirs* in mountain regions, in which water can be diverted or collected when it reaches higher volumes (see in more detail Patt and Gonsowski 2011: 170ff.).

In the context of technological strategies, classic *object or area protection measures* such as dykes, dams, and fixed or mobile flood-protection walls and barriers can also be considered (see for example Brombach *et al.* 2013: 333). These are each aimed at protecting physical objects and important areas against water masses (ibid.; UBA 2008: 3).

4.3.2 Measures against heat: building optimisation, restoration, surface design

The IPCC has identified spatial development measures which can be used to mitigate potential problems arising in the context of hot spells (see for example IPCC 2014b: 18, 2007: 70). These have been discussed at the European and national level within politics and administration (see for example MKULNV 2011; EPA 2005). An important role is played by the observation that construction activities can have an effect on microclimatic conditions (IPCC 2007: 375). The IPCC reports that temperatures within settlements are higher than outside them (IPCC 2007: 372). Depending on site conditions, the level of effects which construction work can have might range from small differences that are barely measurable up to the emergence of clearly demarcable 'urban heat islands' (Schönwiese 2008: 336).

In accordance with the IPCC, a distinction can be drawn in regard to measures against heat between (1) *building design* (IPCC 2007: 375) and (2) *settlement form* (IPCC 2007: 381), although transitions between these categories are fluid, since (3) *surface design* is also of significance to the whole.

Buildings can be protected against heat development through *energy-efficient design*, to minimise heating and radiation effects internally and externally (IPCC 2007: 375). Measures can be implemented either during the design and construction of new buildings or retroactively through the use of energy-relevant materials. Steel and glass, for example, have a higher heat capacity than wood (MKULNV 2011: 39). These materials heat up faster during the day and return the energy to the ambient air at night. The installation of suitable materials should aid thermal insulation and overall energy optimisation (ibid.).

Surface design aims to influence the local radiation effects both of incoming solar energy, as well as of the buildings themselves (IPCC 2007: 372, 381). As described in Chapter 3, the albedo effect summarises how surface properties have a direct effect on the reflection of incoming solar energy and, therefore, on temperature conditions on site. Since this is relevant for surface colour design, the installation of lighter surfaces has been discussed (MKULNV 2011: 39; EPA 2005: 8ff.). *Roof and facade greening through plants* should also prevent local heat development (MKULNV 2011: 35ff.; IPCC 2007: 70), contributing to temperature regulation through evaporation processes and the provision of shade (ibid.).

The above-mentioned surface design measures relate to previously constructed and thus sealed-up areas. Since, on the other hand, non-sealed surfaces often reflect less energy, the *removal of sealed surfaces* has also been discussed as an adaptation measure against heat (ibid.: 35). The re-exposure of sealed surfaces should counteract local heating effects and restore the properties of the original soil.

Finally, if these measure against heat are implemented in large-scale formats, such as in urban development and binding site plans, then they can be considered measures for the design of *settlement forms* (IPCC 2007: 381). In urban areas, spaces characterised by vegetation, such as forests, green areas, and parks, are considered together with water-covered areas as cold- and fresh-air producing areas, due to their metabolic and surface properties (MKULNV 2011: 37; BMVBS/BBSR 2009: 20). These can therefore contribute to moderate temperatures in the form of *fresh- and cold-air corridors* (MKULNV 2011: 40). The relationship between sealed and non-sealed areas within an area is also influential to temperatures within settlements. *Reductions in building density* are therefore also being considered (MKULNV 2011: 37).

4.3.3 Measures against storms: technological reinforcement of buildings and shelters

Existing buildings and infrastructure in the coastal regions of Europe might sometimes last several centuries. The decisions made by spatial development actors can therefore be of very long-term significance. The IPCC highlights this situation with regard to the development of measures against storms (IPCC 2007: 384). As already discussed, the intensity and frequency of storms has not yet seen any significant change along European coastlines, but some projections suggest future increases (see Section 3.3). This fact has already led to the consideration of suitable measures by spatial development actors in Europe.

Buildings and other infrastructure can be damaged or destroyed by high wind loads during severe storms. To address this situation, technical minimum standards for building design have been defined in the EU for specific regions, each on the basis of *wind load assumptions* (see European Committee for Standardization 2005; on the situation in the individual countries, see Scia Group 2014). The tightening of design standards in general has been considered to counter possible long-term changes (IPCC 2007: 384). In 2008 the German government stated that *changes to the wind load regulations*, combined with higher loadings for buildings, could contribute to their survival in the face of long-term storm events (Bundesregierung 2008: 19f.). The *technical adaptation of buildings* has therefore also been considered for both existing building stock and new buildings, with comprehensive refurbishment and modernisation work in the former case, or corresponding construction practices in the latter, which are thus able to contribute to their higher resistance (ibid.: 20).

Beyond the European coastal regions travelling in geographical latitude, in areas that have been subjected to high wind loads, the *setting up of shelters* is already a common measure for maintaining the safety of the population during high winds, for example cyclone shelters in Bangladesh (Mallick 2013: 20) or hurricane wind shelters in the USA (Yazdani *et al.* 2006). In the meantime, shelter rooms (UBA 2012: 2) have also been considered for European coastal regions. These could be designed so as to provide protection not only from storms but also from flooding (ibid.). It remains to be seen how these discourses will evolve.

4.4 Interim conclusion

This chapter has examined the globally discussed climate-change mitigation and adaptation measures available in the field of spatial development.

First, two climate-mitigation strategies were described, namely GHG reduction and sinks. Mitigation measures for the reduction of greenhouse gases can be sub-divided into sufficiency, efficiency, and consistency strategies. In addition, a distinction can be made between ecological and technological sinks. Furthermore, in this study, adaptive practices for four different problem constructions are considered: sea flooding, inland flooding, heat, and storms. Within the flooding-related measures one can distinguish between protection, retreat, and accommodation (integration) strategies. Against heat, buildings can be optimised and the surface design of an area can be altered. Alterations to settlements are also possible, in particular through the renovation of buildings. Against storms, buildings can be reinforced, and the creation of shelters has been discussed in this context.

The concept of spatial development has also been introduced, which in this work refers to a conscious influence on the design of areas in the interplay between different actors from administration and politics, the economy, and civil society. The concept of space relates here to both physical and social spaces.

A central challenge within spatial development in contemporary Europe thus already appears. The interplay between actors in spatial development is likely to be influenced by their perspectives and preferences; in other words, by the actors' knowledge. The previous chapter introduced the most important topics (*topoi*) of climate-system and climate-impact research and explored discourses on possible challenges. This chapter has subsequently related the most important strategies under discussion. Differing views on climate change and variations in practice preferences can lead to conflicts and complicate the implementation of measures. Until now, it has been largely unclear which groups of spatial development actors can be distinguished, sharing common problem constructions and/or common practice preferences. These 'zones of cultural friction' (Crate 2011: 188), which are referred to as 'climate cultures' in this study, will be investigated later. Before this, there is the question of how differences between actors can be explained. This will be discussed in the next chapter.

Notes

1 Since this definition emphasises the 'conscious transactions of people', action can be understood as meaningful in the Weberian sense. The actors associate the designing of space with concrete, subjective meaning (Weber 2010: 3).

2 A brief historical introduction to the development of flood strategies in central Europe is also provided by Jüpner (Jüpner 2013: 11ff.).

References

Albers, Meike, Hasibovic, Sanin, and Deppisch, Sonja (2013) *Klimawandel und räumliche Planung. Rahmenbedingungen, Herausforderungen und Anpassungsstrategien in Stadtregionen im Ostseeraum*. Working Paper No. 13. Hamburg: HafenCity Universität.

Benz, Arthur, Lütz, Susanne, Schimank, Uwe, and Simonis, Georg (2007) *Handbuch Governance. Theoretische Grundlagen und empirische Anwendungsfelder*. Wiesbaden: VS Verlag für Sozialwissenschaften.

Berger, Johannes (2004: '"Über den Ursprung der Ungleichheit unter den Menschen". Zur Vergangenheit und Gegenwart einer soziologischen Schlüsselfrage'. *Zeitschrift für Soziologie* 33 (5): 354–374.

Birkmann, Jörn, Böhm, Hans Reiner, Büscher, Dirk, Fleischauer, Mark, Frommer, Birte, Janssen, Gerold, Overbeck, Gerhard, Schanze, Jochen, Schlipf, Sonja, Stock, Manfred, and Vollmer, Maike (2010) *Planungs- und Steuerungsinstrumente zum Umgang mit dem Klimawandel.* Discussion Paper 8, July 2010. Berlin: Berlin Brandenburgische Akademie der Wissenschaften.

Birkmann, Jörn, Böhm, Hans Reiner, Buchholz, Frank, Büscher, Dirk, Daschkeit, Achim, Ebert, Sebastian, Fleischhauer, Mark, Frommer, Birte, Köhler, Stefan, Kufeld, Walter, Lenz, Susanne, Overbeck, Gerhard, Schanze, Jochen, Schlipf, Sonja, Sommerfeldt, Petra, Stock, Manfred, Vollmer, Maike, and Walkenhorst, Oliver (2013) *Glossar Klimawandel und Raumentwicklung.* ARL E-Paper No. 10. Hannover. Available at: http://nbn-resolving.de/urn:nbn:de:0156-73571 (accessed 15 February 2013).

BMVBS (ed.) (2010) *Klimagerechte Stadtentwicklung – Planungspraxis.* Latest version of the BBSR Online Publication No. 25/2009. BMVBS-Online-Publikation November 2010. Availableat: www.bbsr.bund.de/cln_016/nn_629248/BBSR/DE/Veroeffentlichungen/BBSROnline/2009/ON252009.html (accessed 12 November 2010).

BMVBS (ed.) (2011) 'Klimawandelgerechte Stadtentwicklung. Ursachen und Folgen des Klimawandels durch urbane Konzepte begegnen'. *Forschungen* No. 149. Berlin: BBR.

BMVBS/BBSR (ed.) (2009) *Ursachen und Folgen des Klimawandels durch urbane Konzepte begegnen. Skizzierung einer klimawandelgerechten Stadtentwicklung.* Berlin: Bundesinstitut für Bau-, Stadt- und Raumforschung (BBSR). Available at: www.bbsr.bund.de/BBSR/DE/Veroeffentlichungen/BBSROnline/2009/DL_ON222009.pdf?__blob=publicationFile&v=2 (accessed 30 October 2014).

Brombach, Hansjörg, Jüpner, Robert Müller, Uwe, Patt, Heinz, Richwien, Werner, and Vogt, Reinhard (2013) 'Hochwasserschutzmaßnahmen'. In: Patt, Heinz and Jüpner, Robert (ed.): *Hochwasser-Handbuch. Auswirkungen und Schutz.* Berlin/Heidelberg: Springer, pp. 313–381.

Bundesregierung (ed.) (2008) *Deutsche Anpassungsstrategie an den Klimawandel.* Berlin: no publisher's imprint.

COMMIN (Promoting Spatial Development by Creating COMmon MINdscapes) (2014) *Raumordnung, Raumentwicklung und Raumplanung. Spatial Planning and Spatial Development.* Available at: http://commin.org/en/bsr-glossaries/national-glossaries/germany/raumordnung-raumentwicklung-und-raumplanung.html (accessed 20 May 2014).

Crate, Susan A. (2011): 'Climate and Culture: Anthropology in the Era of Contemporary Climate Change'. *The Annual Review of Anthropology 2011.* 40: 175–194. DOI: 0.1146/annurev.anthro.012809.104925 (accessed 29 October 2013).

Deltacommissie (2008): *Working Together with Water. A Living Land Builds for its Future. Findings of the Deltacommissie 2008 [Samen werken met water. Een land dat leeft, bouwt aan zijn toekomst. Bevindingen van de Deltacommissie 2008].* Alkmaar: Hollandia Printing. Available at: www.deltacommissie.com/doc/deltareport_full.pdf (accessed 15 May 2012).

Destouni, Georgia and Frank, Harry (2010) 'Renewable Energy'. *AMBIO. A Journal of Human Environment* 39: 18–21.

Donnerbauer, Robert (2012) 'Strukturwandel im Strom- und Wärmemarkt'. *VDI Nachrichten* 10/2012. Available at: www.vdi-nachrichten.com/Technik-Wirtschaft/Strukturwandel-im-Strom-Waermemarkt (accessed 20 December 2014).

Eichhorst, Urda and Madry, Thomas (2013) *Die Kehrseite der Medaille. Anpassung an den Klimawandel.* Dossier. Bundeszentrale für Politische Bildung Online. Available at: www.bpb.de/gesellschaft/umwelt/klimawandel/38484/anpassung-an-den-klimawandel (accessed 14 May 2015).

European Commission (ed.) (2011) *Special Eurobarometer 364. Public Awareness and Acceptance of CO_2 Capture and Storage*. Brussels: European Commission/TNS.

EC European Commission (ed.) (2009) 'Directive 2009/31/EC of the European Parliament and of the Council of 23 April 2009 on the Geological Storage of Carbon Dioxide and Amending Council Directive 85/337/EEC, European Parliament and Council Directives 2000/60/EC, 2001/80/EC, 2004/35/EC, 2006/12/EC, 2008/1/EC and Regulation (EC) No 1013/2006'. In: *Official Journal of the European Union*. Available at: http://eur-lex.europa.eu/legal-content/EN/TXT/?uri=CELEX:32009L0031 (accessed 8 January 2015).

EC European Commission (ed.) (2013) 'Communication from the Commission to the European Parliament, the Council, the European Economic and Social Committee and the Committee of the Regions on the Future of Carbon Capture and Storage in Europe'. COM/2013/0180 final. Available at: http://eur-lex.europa.eu/legal-content/EN/TXT/?uri=COM:2013:0180:FIN (accessed 8 January 2015).

European Committee for Standardization (ed.) (2005) EN 1991-1-4 Eurocode 1. 'Actions on Structures – Part 1–4: General Actions – Wind Actions, Authority: The European Union per Regulation 305/2011, Directive 98/34/EC, Directive 2004/18/EC'. Available at: https://archive.org/details/en.1991.1.4.2005 (accessed 30 October 2014).

Eurostat (2014) *Energy Balances*. Available at: http://ec.europa.eu/eurostat/web/energy/data/energy-balances (20.11.2014).

EPA United States Environmental Protection Agency (2005) *Reducing Urban Heat Islands. Compendium of Strategies: Cool Pavements*. Washington DC: EPA. Available at: www.epa.gov/heatisland/resources/pdf/CoolPavesCompendium.pdf (accessed10 July 2014).

Fleischauer, Mark, Overbeck, Gerhard, Janssen, Gerold, and Kufeld, Walter (2013) 'Raumplanung und Klimaschutz – ein Überblick'. In: Birkmann, Jörn, Vollmer, Maike, and Schanze, Jochen (ed.): *Raumentwicklung im Klimawandel. Herausforderungen für die räumliche Planung*. Hannover: ARL, pp. 90–119.

Frommer, Birte (2010) *Regionale Anpassungsstrategien an den Klimawandel – Akteure und Prozess*. Darmstadt: IWAR.

Garrelts, Heiko, Lange, Hellmut, and Flitner, Michael (2008) 'Anpassung an den Klimawandel: Siedlungsplanung in Flussgebieten. Wandel und Herausforderungen im Politikfeld Hochwasserschutz'. *Raumplanung*, No. 2 (April): 72–76.

Heimann, Thorsten (2013) 'Konsens oder Dissens. Herausforderungen des Klimawandels für die deutschen Küstenstädte und –gemeinden aus Expertensicht'. In: Heimann, Thorsten and Christmann, Gabriela B.: *Klimawandel in den deutschen Küstenstädten und –gemeinden. Befunde und Handlungsempfehlungen für Praktiker*, pp. 22–72.

Hirschhausen, Christian von, Herold, Johannes, Oei, Pao-Yu, and Haftendorn, Clemens (2012) 'CCTS-Technologie ein Fehlschlag: Umdenken in der Energiewende notwendig'. *DIW Wochenbericht* June 2012, pp. 3–9.

Huber, Joseph (2011) *Allgemeine Umweltsoziologie*. Wiesbaden: Springer VS.

IEA International Energy Agency (ed.) (2014) *Key World Energy Statistics 2014*. Paris: OECD/IEA.

Institut der deutschen Wirtschaft (2009) *Klimaschutz und Klimaanpassung. Ergebnisse des Klimazwei-Förderschwerpunktes*. Köln: Institut der deutschen Wirtschaft.

IPCC Intergovernmental Panel on Climate Change (1990) *Climate Change. The IPCC Response Strategies*. Geneva: WMO/UNEP. Available at: www.ipcc.ch/publications_and_data/publications_ipcc_first_assessment_1990_wg3.shtml (accessed 28 May 2014).

IPCC Intergovernmental Panel on Climate Change (1996) *Climate Change 1995. Impacts, Adaptations and Mitigation of Climate Change: Scientific-Technical Analyses*. Cambridge/New York/Melbourne: Cambridge University Press.

IPCC Intergovernmental Panel on Climate Change (2001) *Climate Change 2001: Impacts, Adaptation, and Vulnerability*. Cambridge: Cambridge University Press. Available at: www.grida.no/publications/other/ipcc_tar/ (accessed 29 May 2014).

IPCC Intergovernmental Panel on Climate Change (2005) *IPCC Special Report on Carbon Dioxide Capture and Storage*. New York: Cambridge University Press.

IPCC Intergovernmental Panel on Climate Change (ed.) (2007) *Climate Change 2007: Impacts, Adaptation and Vulnerability*. Contribution of Working Group II to the Fourth Assessment Report of the Intergovernmental Panel on Climate Change. Cambridge/New York: Cambridge University Press.

IPCC Intergovernmental Panel on Climate Change (ed.) (2013a) *Climate Change 2013. The Physical Science Basis*. WGI I AR5. Summary for Policy Makers. Available at: www.climatechange2013.org/images/uploads/WGI_AR5_SPM_brochure.pdf (accessed 6 December 2013).

IPCC Intergovernmental Panel on Climate Change (ed.) (2013b) *Climate Change 2013: The Physical Science Basis*. Contribution of Working Group I to the Fifth Assessment Report of the Intergovernmental Panel on Climate Change, Cambridge/New York: Cambridge University Press. Available at: www.climatechange2013.org/images/report/WG1AR5_ALL_FINAL.pdf (accessed 27 April 2014).

IPCC Intergovernmental Panel on Climate Change (ed.) (2014a) *Climate Change 2014: Mitigation of Climate Change*. Contribution of Working Group III to the Fifth Assessment Report of the Intergovernmental Panel on Climate Change. Cambridge/New York: Cambridge University Press.

IPCC Intergovernmental Panel on Climate Change (ed.) (2014b) *Climate Change 2014: Impacts, Adaptation, and Vulnerability. Part A: Global and Sectoral Aspects*. Contribution of Working Group II to the Fifth Assessment Report of the Intergovernmental Panel on Climate Change. Cambridge/New York: Cambridge University Press.

IPCC Intergovernmental Panel on Climate Change (ed.) (2014c) *Climate Change 2014: Impacts, Adaptation, and Vulnerability. Part B: Regional Aspects*. Contribution of Working Group II to the Fifth Assessment Report of the Intergovernmental Panel on Climate Change. Cambridge/New York: Cambridge University Press.

Jänicke, Martin (1984) 'Umweltpolitische Prävention als ökologische Modernisierung und Strukturpolitik'. In: Internationales Institut für Umwelt und Gesellschaft. Wissenschaftszentrum Berlin (WZB) (ed.): IIUG Discussion Papers No. 84–1, Berlin: WZB.

Jänicke, Martin (2000) 'Ökologische Modernisierung als Innovation und Diffusion in Politik und Technik: Möglichkeiten und Grenzen eines Konzepts'. In: Forschungsstelle für Umweltpolitik (ed.): *FFU-Report*, January 2000, pp. 1–18.

Jann, Werner (2009) 'Regieren als Governance Problem: Bedeutung und Möglichkeiten institutioneller Steuerung'. In: Jann, Werner and König, Klaus (ed.): *Regieren zu Beginn des 21. Jahrhunderts*. Tübingen: Mohr Siebeck Velag, pp. 1–28.

Jüpner, Robert (2013) 'Hochwasserschutzstrategien'. In: Patt, Heinz and Jüpner, Robert (ed.): *Hochwasser-Handbuch. Auswirkungen und Schutz*. Berlin/Heidelberg: Springer, pp. 11–15.

Kappas, Martin (2009) *Klimatologie. Klimaforschung im 21. Jahrhundert. Herausforderung für Natur- und Sozialwissenschaften*. Heidelberg: Springer/Spektrum Akademischer Verlag.

Knieling, Jörg, Schaerffer, Mareike, and Tressl, Stephan (2009) *Klimawandel und Raumplanung. Flächen- und Risikomanagement überschwemmungsgefährdeter Gebiete am Beispiel der Hamburger Elbinseln*. Coastline Reports 14, Hamburg: HafenCity Universität.

Krohn, Wolfgang (2013) 'Versicherung von Hochwasserschäden'. In: Patt, Heinz and Jüpner, Robert (ed.): *Hochwasser-Handbuch. Auswirkungen und Schutz*. Berlin/Heidelberg: Springer, pp. 553–607.

Leggewie, Claus (2010): 'Futur Zwei. Klimawandel als Gesellschaftswandel'. *Aus Politik und Zeitgeschichte* (ApuZ) 32–33, 40–46.

Mallick, Bishawjit (2013) *Der gesellschaftliche Umgang mit zunehmender Verwundbarkeit. Eine Analyse der sozialen Bedingungen für vulnerabilitätsorientierte räumliche Planung in den Küstenzonen von Bangladesch.* Karlsruhe: KIT.

Malthus, Thomas Robert (1798) *An Essay on the Principle of Population, as it Affects the Future Improvement of Society.* London: John Johnson.

Mayntz, Renate (2008) 'Von der Steuerungstheorie zu Global Governance'. In: Schuppert, Gunnar Folke and Zürn, Michael: *Governance in einer sich wandelnden Welt: PVS – Politische Vierteljahresschrift*, Special Issue, 41/2008. Wiesbaden: VS Verlag für Sozialwissenschaften, pp. 44–60.

Meadows, Donella H., Meadows, Dennis L., Randers, Jorgen and Behrens, William W. (1972) *The Limits to growth A Report for the Club of Rome's Projection on the Predicament of Mankind.* New York: Universe Books.

MKULNV Ministerium für Klimaschutz, Umwelt, Landwirtschaft, Natur- und Verbraucherschutz des Landes Nordrhein-Westfalen (ed.) (2011) *Handbuch Stadtklima. Maßnahmen und Handlungskonzepte für Städte und Ballungsräume zur Anpassung an den Klimawandel.* Düsseldorf: MKULNV.

Mol, Arthur P.J. and Spaargaren, Gert (2000): 'Ecological Modernization Theory in Debate: A Review'. *Environmental Politics* 9 (1): 17–49.

Nuissl, Henning (2007) 'Entwicklung der raumbezogenen Planung in Deutschland'. In: Weiland, Ulrike and Wohlleber-Feller: *Einführung in die Raum- und Umweltplanung.* Paderborn: Verlag Ferdinand Schöningh, pp. 39–47.

Patt, Heinz and Gonsowski, Peter (2011) *Wasserbau. Grundlagen, Gestaltung von wasserbaulichen Bauwerken und Anlagen.* Heidelberg/Dordrecht/London/New York: Springer.

Quaschning, Volker (2007) *Regenerative Energiesysteme. Technologie – Berechnung – Simulation.* München: Carl Hanser Verlag.

Reusswig, Fritz (2010) 'Klimawandel und Gesellschaft. Vom Katastrophen- zum Gestaltungsdiskurs im Horizont der postkarbonen Gesellschaft'. In: Voss, Martin (ed.): *Der Klimawandel. Sozialwissenschaftliche Perspektiven.* Wiesbaden: VS Verlag für Sozialwissenschaften.

Schönwiese, Christian Dietrich (2008) *Klimatologie.* Stuttgart: Eugen Ulmer KG.

Scia Group (ed.) (2014) *Eurocodes. Legal Situation per Country.* Available at: www.eurocodes-online.com/index.php?page=legal-situation-per-country (accessed 30 October 2014).

SRU Sachverständigenrat für Umweltfragen (2008) *Umweltgutachten 2008 – Umweltschutz im Zeichen des Klimawandels.* Berlin: Erich Schmidt Verlag.

Stock, Manfred (2005) 'Klima'. In: ARL – Akademie für Raumforschung und Landesplanung (ed.): *Handwörterbuch der Raumordnung.* Hannover: ARL, pp. 495–505.

Tol, Richard S.J., Klein, Richard J.T., and Nicholls, Robert J. (2008) 'Towards Successful Adaptation to Sea-Level Rise along Europe's Coasts'. *Journal of Coastal Research* 24 (2, March): 432–442.

UBA Umweltbundesamt (ed.) (2008) *Hochwasserschutz. Klimawandel & Klimafolgen in Deutschland. Themenblatt: Anpassung an Klimaänderung in Deutschland.* Dessau: Umweltbundesamt. Available at: www.dfld.de/Presse/PMitt/2008/081215a4.pdf (accessed 25 June 2014).

UBA Umweltbundesamt (ed.) (2012) *Themenblatt: Anpassung an Klimaänderung in Deutschland. Küstenschutz und Raumplanung*, 12/2008. Dessau: Umweltbundesamt. Available at: www.umweltbundesamt.de/publikationen/anpassung-an-den-klimawandel-kuestenschutz (accessed 30 October 2014).

WBGU Wissenschaftlicher Beirat der Bundesregierung Globale Umweltveränderungen (2003) *Über Kyoto hinaus denken – Klimaschutzstrategien für das 21. Jahrhundert*. Special Report. Berlin: WBGU.

Weber, Max (2010) *Wirtschaft und Gesellschaft*. Frankfurt am Main: Verlag Zweitausendeins.

Wilson, Elizabeth and Piper, Jake (2010) *Spatial Planning and Climate Change*. New York: Routledge.

Yazdani, Nur, Townsend, Tanya, and Kilcollins, Danny (2006) 'Construction Aspects of In-Home Hurricane Wind Shelter Rooms'. *Journal of Coastal Research* 22 (4, July): 862–871.

5 Explanations for climate-culture differences

Derivation of the Values–Beliefs–Identities model

In the previous chapters I have taken a close look at climate-related vulnerability and resilience constructions. Actors might internalise these varying climate-change-related knowledge constructions systematically. This leads us to the question of how these differences could be explained. In other words: what are the factors that contribute to the existence of different climate cultures? This question is the subject of the following section.

5.1 Overview of the theoretical framework for explaining climate-culture differences

Figure 5.1 provides an overview of the theoretical framework for explaining climate-culture differences. As discussed in Chapter 2, my interest here is in the constellation of actors' knowledge constructions. As a first step, therefore, we need to look at the possible *orders of knowledge*. This refers to regular patterns of socially shared climate knowledge and their systematic relation to other forms of internalised knowledge. In addition, the extent to which *further social factors* might explain differences in actors' knowledge will be investigated, in particular field affiliation, national affiliation, and socio-demographic factors.

The concept of knowledge was previously discussed in Chapter 2 in connection with the Foucauldian theoretical consideration of historically varying orders of knowledge. By focusing on historical developments, Foucault highlighted in particular the emergence of orders of knowledge over time. In Chapter 2 I discussed how, with additional reference to the sociology of knowledge and social-constructivist theories, it is also possible to carry out cross-sectional analysis for the investigation of how knowledge is ordered at a specific point in time (see Chapter 2).

Using the concept of *knowledge*, Andreas Reckwitz's 'culture-theoretical model of explanation for action' (Reckwitz 2000: 129) can be introduced here. The essence of this theory contends that there are specific 'effects of knowledge' (ibid.: 149). According to Reckwitz 'the cognitive organisation of actors' reality becomes comprehensible if one assumes that their knowledge possesses an order' (ibid.: 148). From that point of view, socially shared climate knowledge can thus also be embedded in orders of knowledge, e.g. by depending on other

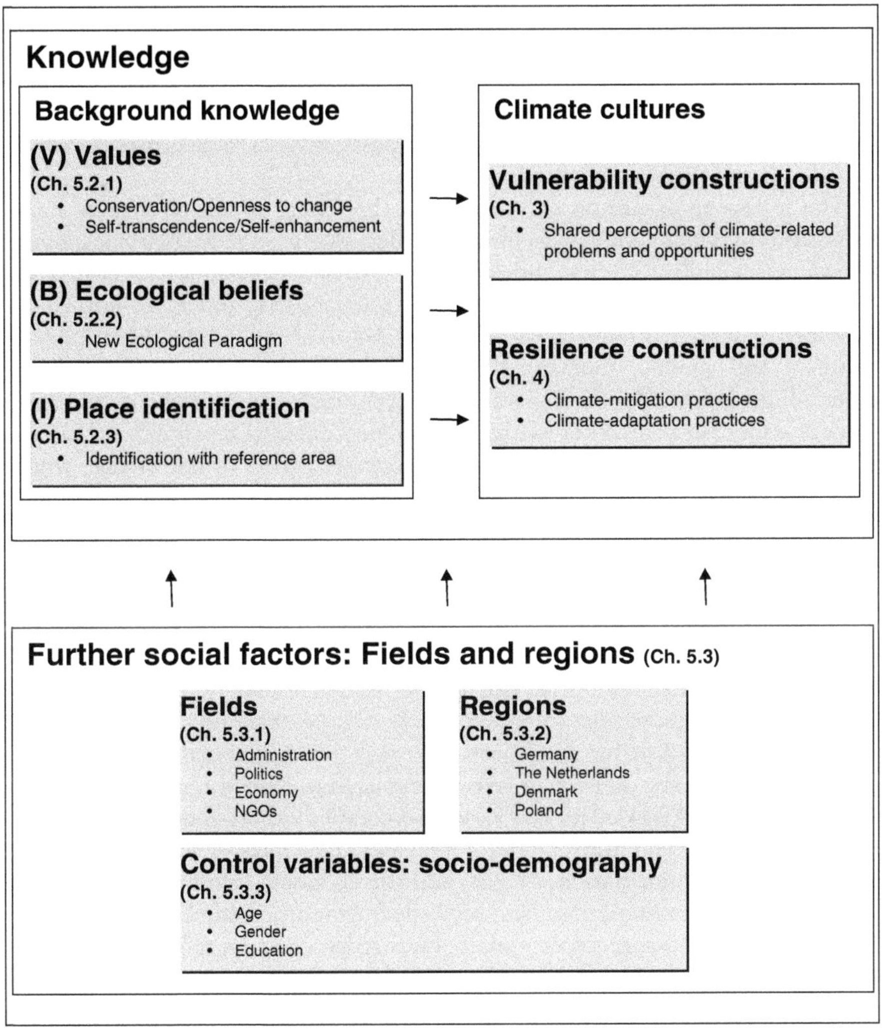

Figure 5.1 Theoretical framework for the explanation of climate-culture differences.

general forms of shared knowledge which in turn can be structured by other social factors – for example, affiliation with fields such as administration, the economy, and civil society, or actors' local and regional affiliations. In the following it will be assumed that the social scientist is able to theoretically comprehend and empirically investigate these orders.

Reckwitz himself understands orders of knowledge as the 'cognitive-symbolic organisation of reality through meaningful schematisation' (ibid.: 161). From any 'schema used by an actor, there can be derived a virtually endless number of compatible beliefs and actions' (ibid.: 166). According to this understanding, individual and collectively shared forms of knowledge regularly form logical relationships to other forms of knowledge. For this reason, formerly internalised knowledge leaves only limited room for future interpretation and action: 'The cognitive-symbolic structures allow certain forms of relation, while excluding others as "inconceivable"' (Reckwitz 2000: 130). This therefore means that in what follows it is necessary to formulate the possible observable orders of climate knowledge and then test them empirically.

The following discussion will show that the extent to which specific internalisations of climate knowledge might be 'dependent on particular background knowledge' (ibid.: 162) has already been widely discussed in social-science climate research. Social-science discourse suggests that orders of knowledge exist between shared climate knowledge in the narrower sense (here: climate cultures) and general, thematically related forms of knowledge (here: background knowledge). Concrete possible relationships between background knowledge and climate-related knowledge will be the subject of Section 5.2. This will lead to a derivation of hypotheses about concrete orders of knowledge, whose empirical investigation will follow in Chapters 9 and 10.

Of further interest to this study is the relevance of *additional social factors* explaining climate-culture differences (Section 5.3). As previously described, climate-culture differences will here be explored according to a relational understanding of cultural space (see Chapter 2). This will not, however, reveal the contribution of social factors to these differences. The extent to which actors' local and regional affiliation (in the form of national affiliation) explain differences in climate knowledge should therefore also be examined (Section 5.3.2). On the other hand, it has long been observed that functionally differentiated societies based on the division of labour tend to form differentiated (sub-)group-specific orders of knowledge, and thus also create cultural boundaries that cut across local, regional, and national boundaries. Examples of such field-specific affiliations are administration, NGOs, and the economy (see Section 5.3.1). In addition, the literature discusses socio-demographic factors as explanatory factors for differences in actors' climate knowledge (Section 5.3.3). The following provides a detailed look at the theoretical framework.

5.2 Background knowledge

The idea that existing background knowledge can provoke different conceptions of and ways of handling the environment has been intensively discussed among various research disciplines since, at the latest, the 1970s. These debates will be looked at in more depth in the course of this section. In the literature, individual and collectively shared values, beliefs, and identities are particularly prevalent as different forms of background knowledge.

5.2.1 *Values*

The authors of the IPCC have, since the 1990s, regularly presumed values to be a relevant variable in the selection of climate-change measures. To date, however, they have not delved very much into the issue of the relevance of concrete constructions, or even named any comprehensive social-science studies (IPCC 2014: 5, 2007: 342, 1996: 292). This does not mean, though, that there is no such work that can be drawn on. The following will therefore first provide a more precise definition of the concept of 'values'. Next, relevant central discourses on the role of values in forming constructions of climate-related problems and measures will be introduced. Connected to this, current gaps in the research can be shown and hypotheses for the present study can be derived.

5.2.1.1 *On the concept of 'values'*

Research into values has a long tradition and has been 'in and out of fashion within sociology' several times (Hitlin and Piliavin 2004: 359). Many studies use the definition formulated in the 1950s by Clyde Kluckhohn. According to this: 'A *value is a conception [...] of the desirable* which influences the selection from available modes, means and ends of action' (Kluckhohn 1967 [1951]: 395, emphasis added).

The term needs to be differentiated from similar social-science concepts, in particular from beliefs, attitudes, and norms (see also Hitlin and Piliavin 2004: 345f.). *Beliefs* are concrete individual or collective concepts of reality, such as that changes in climate are occurring and that these changes have a specific cause (Dietz *et al.* 2005: 346). Values, by contrast, and as the term already implies, include judgements about the content of beliefs. *Attitudes* also include judgements, but relate to concrete objects, while values imply general ideals that are superordinate to individual attitudes and therefore usually appear more consistent over time (Hitlin and Piliavin 2004: 361; Van Deth and Scarbrough 1995: 32). *Norms* are also 'concepts of the desirable', but are linked to social sanctions in the case of non-compliance (Schäfers 2006: 33).

5.2.1.2 *The state of research: values as explanations for climate-related action*

Meanwhile, a certain number of studies have been made concerned with the influence of values on (1) climate-change-related knowledge of opportunities and problems, and (2) climate-mitigation and (3) climate-adaptation preferences. Here the theories and constructions discussed share many similarities, and generally refer back to only a few research traditions. Two such traditions stand out in particular: first of all, environmental values research, associated mainly with the development of environmental sociology that has developed since the 1970s and, in this case, has been dealing with interactions between values and environmental action (including environmental consciousness and practices).[1] Second, a risk-oriented research tradition can be identified, largely founded on the anthropology and political-science influenced studies of Mary Douglas and Aaron Wildavsky. In the early 1980s, within the framework of their 'cultural theory of risk', Douglas and Wildavsky formulated the constructivist-oriented theory that social groups with shared values and beliefs often developed similar, selective concepts of ecological problems and ways of dealing with them (Thompson 2003; Douglas and Wildavsky 1982).

The two theoretical traditions taken together provide the idea of the influence of values on socially shared perceptions and actions, meaning that this basic concept can be made use of in the present study. There are, however, differences in the value constructions used. Each of those that are relevant to climate will be discussed in the following. The influence of values on (1) problem constructions, and (2) climate-mitigation and (3) climate-adaptation preferences will be differentiated.

ON THE INFLUENCE OF VALUES ON VULNERABILITY CONSTRUCTIONS

In relation to perceptions of risk, Douglas and Wildavsky summarise their central idea thus: 'Common values lead to common fears' (Douglas and Wildavsky 1982: 8). According to this idea, internalised values lead to the systematic prioritisation of different bodies of information and knowledge, so that actors with similar values ultimately share common ideas about (environmental) risk.

Since the 2000s at the latest, the relevance of socially shared values in the formation of climate-related problem constructions has also been discussed (see also Persson *et al.* 2015; Price *et al.* 2014; Adger *et al.* 2013: 113; Heath and Gifford 2006; Leiserowitz 2006). Here, concrete values from the cultural theory of risk formed part of their explanatory methods (Price *et al.* 2014: 12ff.; Leiserowitz 2006: 57ff.). The authors of this theoretical tradition assume that actors share concepts of desirable forms of social organisation (hierarchical vs. egalitarian; individualist vs. communitarian) (see in more detail Price *et al.* 2014: 10; Kahan *et al.* 2010: 2ff.; Leiserowitz 2006: 49). They further suggest that these arise together with shared concepts of the fragility of nature (tolerant, ephemeral, capricious, benign nature; ibid.) and thus systematically influence risk perception, though the coinciding appearance of values, conceptions of nature, and desirable forms of social organisation have not been tested and no plausible basis for their mutual emergence is provided (see also Krömker 2004: 110f.). It is also questionable whether the use of a schema oriented towards smaller groups can be justified for a complex contemporary (global) society (ibid.).

In addition there are studies on values and climate-related problem construction that draw upon the above-mentioned environmental theoretical traditions. Heath and Gifford (2006: 52), for example, differentiate between so-called anthropocentric values (the conservation of nature for the benefit of humans) and ecocentric values (the conservation of nature for the benefit of the biosphere). They observe that holding the ideal of conservation for the benefit of the biosphere in itself increases the degree of agreement with the statement that climate change is occurring, that it is anthropogenic, and that it will be associated with negative consequences (ibid.: 59). Since these values are similarly presented on a regular basis as explanations for climate-mitigation and adaptation preferences, the development of these ideas will be dealt with further below.

EXCURSUS: EARLY ENVIRONMENTAL VALUE THEORIES FROM ALTRUISM TO SELF-ORIENTATION

Early environmental-sociology research on values and environmental behaviour followed the idea that environmental-conservation values were congruent with altruistic ideals (see in more detail Dietz *et al.* 2005: 343f.). In the 1970s the debate began to be influenced by the issue of whether environmental consciousness was more a vestige of a 'land ethic' (self-transcendence for the benefit of all life forms in the biosphere) or of the 'golden rule' (self-transcendence for the benefit of other humans) (Howell 2013: 282; Dunlap and van Liere 1977; Heberlein 1977). This clearly reflects the dominance of the central concern of this period, namely the discourse around ecological sufficiency, as it is assumes that environmental protection efforts necessitate constraints.

These notions are reflected in the *prosperity hypothesis* formed at the time. Dunlap *et al.* (1983), following Maslow's hierarchy of needs, asserted that environmental orientations motivated by 'higher-order values' (ibid.: 145) come into effect when basic material needs such as the acquisition of food and material security have been fulfilled (see also Dietz *et al.* 2005: 360). The *post-materialism hypothesis* of Ronald Inglehart also follows this discursive framework. Inglehart views nascent environmental orientations as a feature of a comprehensive change of values from materialism to postmaterialism, which he later additionally regarded as typical of the transition from industrial to post-industrial service societies (Inglehart 2006: 117, 1995: 68). From that point of view, with increasing economic development, the willingness 'to make any sacrifice' for the environment (Inglehart 1995: 58) also arose. In place of material efforts, there was an increasing shift towards improving 'environmental quality' and 'quality of life' (ibid.: 61).

These ideas were not uncontested, however. First, the assumption of a linear modernisation theory bound to the hypothesis of prosperity and postmaterialism needs to be critically assessed, since from the outset it makes generalisations about the historical development of western societies in the twentieth century (see also Rössel 2006: 729; Haller 2002: 142). In the 1990s Dunlap and Mertig argued that environmental conservation efforts had increasingly been observed even in the absence of rising standards of living, and thus need not only be the outcome of a condition of luxury (Dunlap and Mertig 1997: 24). This demonstrated that the environmental values spectrum needs to be differentiated still further.

For the western hemisphere, Huber also claims a 'change of values in environmental consciousness' (Huber 2011: 100), characterising developments into the 1990s. As already discussed in Section 4.2, in the 1980s there arose ecological modernisation discourses in which ecology and economy came to be understood less and less as opposites. In the 1990s, these perspectives flowed into value-related environmental sociology. Stern and Dietz (1994), for example, suggested including egoistic values (Stern and Dietz 1994: 69) as environmental values in research, alongside the altruism orientation (human-oriented vs. biosphere-oriented values) which has already been mentioned. The authors assume that these also lead to environmental protection activities when personal gains can be expected (see also Dietz *et al.* 2005: 344). Their considerations have shown up later in climate-related studies, and will therefore be looked at more closely now.

ON THE INFLUENCE OF VALUES ON CLIMATE-MITIGATION PREFERENCES

Even the available studies dealing with the interaction between values and climate-mitigation orientations make multiple references to the value constructions already discussed.

For example, Andreas Nilsson and colleagues (Nilsson and Biel 2008; Nilsson *et al.* 2004) have investigated the influence of altruistic and egoistic orientations on the acceptance of politically initiated climate-mitigation measures. They differentiate between subsidisation, prohibition, taxation, and information campaigns, surveying employees working in both the public and private sectors in Sweden, and using a selection of items from the values model developed in the early 1990s by Shalom Schwartz (1992), which enables the surveying of both altruistic and egoistic orientations. The authors test for the influence of four values, namely the egoistic orientations power and achievement, alongside environmental and human altruism. They observe that universal-altruism orientation significantly increases the acceptance of all named measures (Nilsson *et al.* 2004: 273), while human-altruism, power, and performance orientations provide little explanation for preferential differences (ibid.; also Nilsson and Biel 2008).

Howell (2013) surveys members of two British activist groups, the 'Carbon Rationing Action Groups', who promote lifestyles with low greenhouse gas emissions (ibid.: 283). She also borrows from Shalom Schwartz, and comes to the surprising conclusion that the groups' members demonstrate an essentially more human-altruistic value orientation than an environment-altruistic one.

Williams and Schaefer (2012) carry out qualitative surveys with managers of small and medium-sized British companies involved in environmental and climate preservation, questioning the reasons for their engagement. Their results indicate that both self-orientation and environmental altruism are important motivational aspects for engagement in environmental and climate preservation activities.

In addition, there are studies drawing on the Cultural Theory of Risk (see for example Price *et al.* 2014; Leiserowitz 2006; Pendergraft 1998). In a survey of the US population, Leiserowitz observes that actors with an egalitarian orientation accept national climate-mitigation policies more willingly, while individualistic and hierarchical orientations are accompanied by increasing rejection of the same (Leiserowitz 2006: 59). In an online Australian household-panel survey, Price *et al.* (2014: 15) investigate the influence of these values on the implementation of household-related climate-mitigation measures, such as saving energy and water, or buying environmentally friendly products. It was observed that those surveyed hardly differentiated between various climate-mitigation strategies. Using factor analysis, all the measures were loaded onto one factor and were summarised into an index of 'carbon-relevant behaviour' (ibid.). The results can be interpreted to suggest that sufficiency strategies form the most dominant category, since all the measures considered can be connected to constraints for the interviewees.

In summary, research approaches that examine the influence of values on climate-mitigation preferences often draw upon the environmental and risk-oriented theories outlined above. In environmental sociology, the relevance of altruistic and egoistic values has been discussed as providing explanations for the acceptance of climate-mitigation measures. Risk-oriented cultural theory has examined idealised forms of social organisation, as well as the associated conceptions of nature's fragility, as an influencing factor. It is noticeable that the selection of relevant climate-mitigation measures investigated rarely differentiates between different basic strategies, such as sufficiency, efficiency, or consistency. In addition, the consideration of sufficiency measures appears to have dominated until now. Before going further into the concrete implications of these particular studies, a number of other relevant studies on climate-adaptation should be considered.

ON THE INFLUENCE OF VALUES ON CLIMATE-ADAPTATION PREFERENCES

Only recently has there been an increased effort to investigate the degree to which values can explain preferences for or against climate-adaptation measures (see for example Adger *et al.* 2013; Landauer *et al.* 2013; Leonard *et al.* 2013; Wolf *et al.* 2013; O'Brien and Wolf 2010; O'Brien 2009). It should be noted that adaptive options are conceivable not only for the various impacts of climate change, such as flooding, sea-level rise, heatwaves, and storms, but also for the consequences of each of these individually. The research field thus offers diverse possibilities for the discussion and investigation of relevant value constructions. The current status of published works on the relevance of values by no means matches this level of complexity, however.

As important pioneers, O'Brien and Wolf (2010) direct themselves primarily towards the IPCC's scientific debates on adaptation, and urge a 'values-based approach to vulnerability and adaptation to climate change' (ibid.). They call for more consideration in future of individual and collective values when determining vulnerabilities and local adaptation measures as, they argue, decisions for or against such measures rest much more on these values than supposedly objectively measurable criteria. In implementing adaptation measures, it is thus important to know 'which types of adaptation pathways are perceived as most desirable [...] by individuals and communities' (ibid.).

O'Brien and Wolf have discussed concrete value constructions that might explain decisions for or against specific local 'adaptation pathways' (O'Brien and Wolf 2010). O'Brien also draws on the above-mentioned theoretical traditions in environmental sociology and, in a number of articles, assumes Schwartz's value schema to be of relevance, among which are counted values such as self-direction, self-transcendence, conservation, security, and openness to change.

While these authors have not investigated the possible observable consequences of the values named on the choice of concrete climate-adaptation measures (O'Brien and Wolf 2010; O'Brien 2009: 167), they do refer to Ronald Inglehart in order to discuss typical values and their relevance to adaptation to changed snow conditions in Norway. They thus differentiate between 'traditional values' ('emphasis on rural areas, nature, and the family', O'Brien 2009: 171ff.), 'modern values' ('individualism, economic development, material wealth, technology, and scientific progress', ibid.) and 'postmodern values' ('pluralistic values', 'self-expression and self-realisation', 'well-being', ibid.), any of which, according to their locally dominant form, could lead to different preferences. Until now, however, there have been no studies that have empirically tested their theories.

In a jointly authored work, Wolf presents the empirical results of research carried out in an anthropological study undertaken in Canadian Inuit villages and investigating value orientations and measure preferences (Wolf *et al.* 2013). Here it becomes clear that to understand the various adaptation efforts, much more must be taken into account than environmental values alone. Thus the authors identify 'security', 'tradition', 'freedom', and 'unity' as being of relevance for various climate-adaptation options (ibid.: 556).

The relevance of different value orientations is also discussed in other anthropologically oriented climate-change studies. In these it is noticeable that several of the values described are often similar. Leonard *et al.* (2013), for example, assert that 'preserving the status quo' (ibid.: 630) and 'improving quality of life and well being' (ibid.) play an important role in adaptation to flood risk in Australia. Here the similarity to the 'tradition orientation' values just mentioned, as well as to more self-orientation values such as 'well-being' are conspicuous. Kuruppu, on the other hand, discusses the relevance of altruistic orientation for adaptation to changed water availability in Kiribati, an island nation in the central Pacific Ocean (Kuruppu 2009: 805f.).

Needless to say, when dealing with comparative studies, the degree to which the values used are commensurable must be more closely examined. It cannot, however, be disputed that the descriptions repeat themselves, and with them the discussed repertoire of value orientations that inhibit or promote climate adaptation. The relevance of these should be empirically investigated. To date, however, only a small number of studies can be found that systematically examine the connections between values and adaptation preferences through quantitative research. This task will therefore receive attention within the scope of this study.

In regard to climate adaptation, aspects such as security, conservation, altruism, and self-orientation have been particularly discussed. Interestingly, these values share a close similarity to Schwartz's previously mentioned spectrum of values. The environmental sociological repertoire discussed above can thus be drawn upon to extend the values spectrum used. The discussion below will examine Schwartz's approach in more detail, as well as the consequences for this study that can be derived from it.

5.2.1.3 The research approach of this study: value orientations according to Shalom Schwartz

The previous section demonstrated that Shalom Schwartz's values model aligns with many of the values discussed within environmental sociology, and that the model has also already been discussed as an important explanatory approach to various climate-related problem constructions, as well as climate-mitigation and adaptation preferences. Thus Schwartz's conception of values will be used in this study as an explanatory approach for affiliation to various climate cultures along Europe's coasts.

Schwartz developed his values model in conjunction with earlier studies by Rokeach, Hofstede, and Kluckhohn (Schwartz 1992: 1ff., 2012), and it has since become regarded as 'the most commonly used measure of values' (Dietz *et al.* 2005: 347). His typology includes general motivations and notions with which people, in his opinion, are regularly confronted worldwide (Schwartz 1999: 26, 1999: 4). He thus aspires to formulate an extensive, universal typology of values, potentially enabling a comparison of all human societies (Schwartz 1992: 4).[2]

Schwartz developed several variants of his 'Theory of Basic Values' (Schwartz 2012: 1), whose most commonly used schema consists of ten values[3] representing various motivations (see also Mohler and Wohn 2005: 2). He found this *essential schema* in more than 90% of his globally conducted investigations, which he saw as evidence for its broad validity (Mohler and Wohn 2005: 1; Schwartz 1992: 37f.). Schwartz furthermore assumes that the ten value types relate to each other interactively and therefore form a *structural schema*. His reasoning is that the motivational aims striven for, together with their consequences, can be regarded as being of greater or lesser similarity to one another. He provides a circular diagram, represented in Figure 5.2. He bases this assumption on the idea that structural schemata result from the universal requirements of human existence, such as the 'needs of individuals as biological organisms, requisites of coordinated social interaction, and survival and welfare needs of groups' (Schwartz 1992: 4; see also Roose 2010a: 16; O'Brien 2009: 167).

According to this schema, an additional four values of a higher order (standard types) can be construed, once again relating to one another in two dimensions (Mohler and Wohn 2005: 5; Schwartz 2003: 270). The first postulated dimension is formed by the standard types *self-transcendence* and *self-enhancement*. 'Self-transcendence' is regarded as the superordinate category to the basic values of benevolence and universalism, while 'self-enhancement' lies behind the values of power, personal success, and hedonism. Here it becomes obvious that this schema seems virtually congruent with the above-mentioned values used in environmental sociology, namely altruism (universalism with regard to the biosphere, and benevolence with regard to humans) and ego-orientation (here, self-enhancement).

The second dimension is represented by the standard types *openness to change* and *conservation*. 'Openness to change' lies behind the values of stimulation and

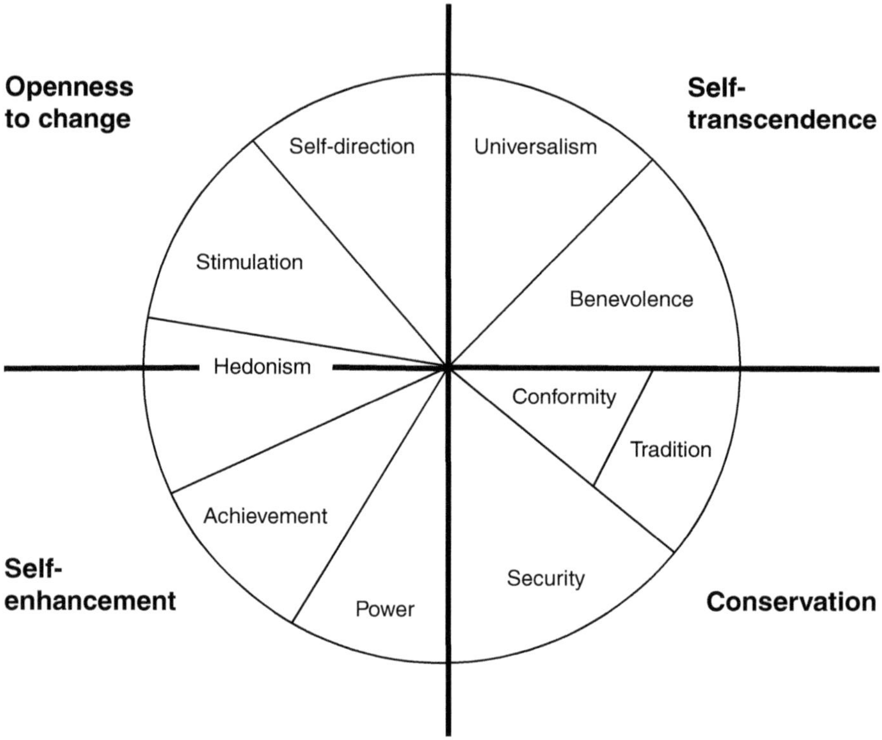

Figure 5.2 Schwartz value model: the ten basic/four standard types.

Source: Schwartz 2003: 270.

self-direction. Security, conformity, and tradition, on the other hand, are formed by the standard type 'conservation'.

Current environmental-sociology studies that relate to climate change and make use of Schwartz's values model have so far been selective in their use, for example choosing individual standard types and their associated items for investigation with regard to climate-related problem perceptions or mitigation preferences (see for example Howell 2013; Nilsson and Biel 2008; Nilsson *et al.* 2004). These authors have thus concentrated primarily, in the classic tradition of environmental sociology, on the relevance of the self-transcendence/self-enhancement dimension, regarding the underlying value types as separate (for example Howell 2013: universalism, benevolence; Nilsson and Biel 2008: universalism, benevolence, power, achievement; Nilsson *et al.* 2004). In the adaptation discourse there appear in addition individual values from the second dimension (openness to change, conservation, tradition, conformity, security), although to date their relevance has not received extensive systematic investigation.

The question thus poses itself, why shouldn't these last-named values also have an influence on the genesis of climate-mitigation preferences or problem perceptions? It is conceivable, for example, that when designing concrete measures, security orientation also plays a role in their acceptance (such as in the installation of CCS technology). In addition, spatial development measures often involve the redesign of concrete places and regions (see Section 4.1). Thus aspects such as conservation and openness to change should be of significance both for climate-mitigation and adaptation measures. For these reasons, I propose further study of a single values model for all three areas (opportunities and problem constructions, climate mitigation, and climate adaptation), and will examine each area to determine the extent to which these values contribute to different perceptions or preferences.

5.2.1.4 Derivation of hypotheses

To be able to draw upon the values discussed in relation to environment and climate-related debates, the influence of individual value types or standard types will also be investigated in this study. Schwartz's values schema will, however, be taken into consideration. Where a differentiation appears necessary, individual value types will be used; where this does not appear necessary, standard types will be used. The following general hypothesis can therefore be formulated:

- **H_{Value}**: Shared value orientations have an influence on climate-culture orientations, in particular on (a) problem and opportunity constructions, (b) climate-mitigation preferences, and (c) climate-adaptation preferences.

On the basis of the previously presented discourses on the influence of concrete value orientations, the following partial hypotheses can additionally be derived:

- **$H_{Value/Universalism}$**: Environmental-altruism orientation has an influence on climate-culture affiliation (Self-transcendence I).
- **$H_{Value/Benevolence}$**: Human-altruism orientation has an influence on climate-culture affiliation (Self-transcendence II).
- **$H_{Value/Self\text{-}enhancement}$**: Ego-orientation has an influence on climate-culture affiliation.
- **$H_{value/Openness\ to\ Change}$**: Openness to change has an influence on climate-culture affiliation.
- **$H_{Value/Security}$**: Security orientation has an influence on climate-culture affiliation (Conservation I).
- **$H_{Value/Tradition\_Conformity}$**: Tradition and conformity orientations have an influence on climate-culture affiliation (Conservation II).

The hypotheses can be formulated in more detail for each of (1) chance and problem constructions, (2) climate-mitigation preferences, and (3) climate-adaptation preference. In order not to increase the complexity of presentation unnecessarily, the possible implications for the genesis of problem constructions and measure preferences needn't always be described to the last detail. This would lead to redundancies that are not expedient here. Instead, the core assumptions will be summarised.

It has been observed in the studies discussed that environmental protection ideals can accompany higher sensitivity to ecological problems. Even in climate mitigation, this increases the acceptance of sufficiency and ecological modernisation measures. One can conclude that a *universalism orientation* generally increases the acceptance for ecologically oriented measures, i.e. measures aligned towards a balancing of humans and nature. To date there are hardly any data relating to adaptation measures, but one might conclude that internalisation of a '*land ethic*' might be associated with similar consequences here. In summary, the following hypotheses can be formulated:

Universalism (Environmental altruism): The higher the orientation towards universalism:

- the greater the probability of sharing a high problem perception (because high estimation of environmental protection leads to openness to climate issues and increases the probability of internalising problem constructions); and
- the greater the probability of a preference for ecologically oriented measures for climate mitigation and adaptation, e.g. greening and desealing of areas.

Additionally, the long-standing environmental-sociology discourse on the '*golden rule*', as well as recent results from Howell (2013), indicate that *human-altruism orientation* plays an important role in climate-mitigation measures. One can conclude that a high manifestation of this orientation will result in preferences for adaptation measures that are aligned with the well-being of others. Giving the following hypothesis:

Benevolence (Human altruism): The higher the orientation towards benevolence:

- the lower the probability of a preference for measures with a high risk for humans or with high associated constraints/costs for other people or communities; and
- the greater the probability of a preference for measures that benefit other people.

It has become apparent that, since the appearance of discourses around ecological modernisation, ego-orientations might also explain ecological and climate-related endeavours (Williams and Schaefer 2012). This provides the following hypothesis:

Self-enhancement: The higher the orientation towards self-enhancement:

- the lower the acceptance of measures that present high costs; and

- the greater the acceptance of measures associated with individual or collective profit or which present possibilities for development.

In more recent studies on climate adaptation, constructions such as 'security', 'tradition', or 'conservation' have been discussed (Leonard *et al.* 2013; Wolf *et al.* 2013). Values from the standard types *openness to change* and *conservation*, which to date have hardly been utilised in studies, can thus now be drawn upon (Schwartz 2003: 270). Of the diverse possible challenges with which human societies might be confronted, climate *change* poses a relatively novel problem construction. The implementation of spatial development itself is often associated with innovation. *Openness to change* may therefore contribute explanations for both vulnerability and resilience constructions:

Openness to change: The higher the openness to change:

- the greater the perception of climate change as an unprecedented, human-made problem;
- the greater the acceptance of measures presented as new or innovative (e.g. technological sinks, forms of sustainable energy forms, accommodation through integration); and
- the greater the acceptance of measures associated with large-scale change.

Conservation orientation might also be of relevance, given that climate-adaptation measures were originally and above-all intended to ensure against natural disasters. Climate-mitigation measures can also be associated with the task of providing security, such as in the implementation of sustainable energies. In both cases, *security orientation* might therefore play an important role in the preference for measures. The following hypothesis can therefore be formulated:

Security: The higher the orientation towards security:

- the more measures are preferred that are associated with low risks or which have proven reliability (e.g. protection strategies such as traditional coastal protection, technological adaptation against storms, etc.).

In the implementation of measures, long-standing traditions can in some cases be drawn upon, such as in dyke-building practices for coastal protection. Today new strategies are also being developed (see Chapter 4). As mentioned, climate change presents a relatively novel problem construction, and tradition and conformity orientations might play a role in whether and the extent to which certain practices are preferred. Thus the following hypothesis can be formulated:

Conformity/Tradition: The higher the orientation towards conformity and tradition:

- the lower the belief in climate change and the lower the perception of the problem;
- the more that measures with proven reliability are preferred (e.g. dyke-building as a protection strategy); and
- the more measures are preferred that preserve the status quo of places and regions and result in the least possible change (e.g. protection strategies).

5.2.2 Beliefs: ecological worldviews

In addition to value orientations, diverse individual and collective beliefs have regularly been discussed as the basis for different perceptions of and ways of handling climate change (Hulme 2016; see also Hoffman 2015; Hulme 2009). In climate-related discussions, the relevance of ecological worldviews has come into particular focus, and in some cases their significance for environmental and climate-related action has been observed through integrated models (e.g. Value-Belief-Norm (VBN) theory; see Steg *et al.* 2005; Stern 2000; Stern *et al.* 1999). Before discussing the relevance of ecological worldviews to ways of handling climate change, the central terminology will be briefly defined.

5.2.2.1 On the terms 'belief' and 'ecological worldview'

Beliefs should be understood in this study as individual or collectively shared assumptions about the world. As already discussed in Section 5.2.1, beliefs can be described as notions related to specific aspects of reality, such as that the temperature of gases can be influenced by light, or that climatic changes are taking a certain course (Dietz *et al.* 2005: 346). Some beliefs can also emerge grouped together, forming 'a coherent set of [...] beliefs' (Dunlap *et al.* 2000: 428). These groupings of presumably more coherent knowledge constructions are also termed worldviews (ibid.).

Following Luckmann, the occurrence of varying worldviews or groupings of knowledge can be explained by processes of social confirmation and individual sedimentation: 'Once a worldview becomes socially established, it presents a compelling system of interpretation to the individual, to which he or she inwardly turns' (Luckmann 1999: 23). Thus knowledge internalised in the past allows only a certain amount of scope for future interpretations and internalisations of new knowledge. This also explains why different worldviews shared socially as background knowledge can ultimately lead to varying climate cultures. Reckwitz describes how internalised worldviews produce tangible action:

> The worldview delivers a cognitive foundation to an ethical evaluation system. If certain things 'are the case', then a certain action is also required; if other interpretations of the world exist, other courses of action also seem 'naturally' appropriate.
>
> (Reckwitz 2000: 456)

In the current study, the relationships that exist between specific worldviews and climate knowledge should be revealed. Above all, the role of ecological worldviews (Dunlap *et al.* 2000: 427) will be discussed, by which is understood the collection of beliefs pertaining to the spectrum lying between humans and nature. This study will speak of ecological worldviews so as, from the outset and also linguistically, not to infer a binary construction between humans and nature, a construction that is again being questioned by many authors (Descola 2013; Christmann and Ibert 2012; Latour 1996, and further remarks). Following Krömker, ecological worldviews should be 'understood as concepts that relate, for example, to the function or nature of ascribed attributes of humans and nature, or to the relationship between them' (Krömker 2004: 82).

5.2.2.2 Central concepts: New Ecological Paradigm (NEP) and Cultural Theory of Risk

In the social-science literature there are two frequently named, and above all qualitatively oriented theoretical streams that take ecological worldviews as their subject, namely the New Ecological Paradigm (NEP) and the previously mentioned Cultural Theory of Risk (see also Krömker 2004: 116ff.). In addition, there is a wealth of social-science studies, in particular from anthropology, describing the diverse and complex beliefs pertaining to the relationship of humans and nature. Elements of all three streams likewise appear in social-science discourse about various climate actions. Thus each will be examined briefly here.

NEW ECOLOGICAL PARADIGM (NEP)

The New Ecological Paradigm has appeared in two forms since the 1970s, on the one hand as a socially shared ecological worldview with increasing prevalence in western societies at the time, and on the other as a sociological research perspective closely tied to environmental sociology (Pellow and Brehm 2013: 230f.; Dunlap and Van Liere 2008 [1978]; Dunlap *et al.* 2000: 425ff.; Catton and Dunlap 1978: 41ff.). As a purportedly new sociological research perspective, Dunlap and his colleagues saw at the time the need for a shift from the Human Exceptionalism Paradigm (HEP) (Catton and Dunlap 1978: 42) to the New Ecological Paradigm (NEP) (ibid.). In what was deemed the older research tradition, the inclusion of the physical environment in the analysis of social action had been dismissed, in accordance with the Durkheimian decree of 'explaining the social only through the social' (see also Krömker 2004: 116).

According to the authors, the HEP perspective contained an *anthropocentric worldview* characteristic of the modern era. This demonstrated itself through the following assumptions: humans have a special position above all other creatures on Earth; human culture develops more quickly than biological nature, thus humans will always be capable of adapting to varying social and nature-dependent conditions; human development is therefore hardly restricted to physical constraints (Catton and Dunlap 1978: 42f.).

By contrast, in the New Ecological Paradigm this perspective is abandoned in favour of an *ecocentric worldview*, and the human species is regarded as one among many. The binary opposition between nature and culture is increasingly relinquished, and human life and social action are viewed as interacting with other species and with the material environment (e.g. food intake or the consumption of other resources). Involvement in ecological processes can therefore also have diverse unintended consequences. In addition, the world appears limited, and therefore there are economic and ecological limits (Catton and Dunlap 1978: 45).

In relation to the new research perspective, the authors argue for the inclusion of, for instance, physical and material entities in social-science analyses. In research on inequality, for example, the availability of nutrition and other resources should be considered (Catton and Dunlap 1978: 46). Since then it has been critically noted that similar perspectives had already appeared in sociological studies of the time, which had however lain outside the authors' field of vision (Pellow and Brehm 2013: 230f.).

Nonetheless, there can be little doubt that Catton and Dunlap were able to empirically observe the spread of the New Ecological Paradigm as an ecological worldview among members of the population in western countries (see also Krömker 2004: 106). Even the idea that the New Ecological Paradigm represented a consistent ecological worldview, encompassing several groups of beliefs, was based upon this observation. In order to capture all relevant aspects of the NEP, in 1978 Dunlap and his colleagues introduced a Likert scale, maintained and subtly revised to the present day (Dunlap and Van Liere 2008 [1978]; Dunlap *et al.* 2000). Using this scale, all manifestations along the axis anthropocentrism (HEP) vs. ecocentrism (NEP) can be measured. The following beliefs have thus been investigated: the occurrence of ecological limits (Dunlap *et al.* 2000: 438); the stability of nature in the face of human intervention (the 'balance of nature'); the rejection of the HEP belief that humans are superordinate to nature and thus possess the ethical right to undertake comprehensive interventions ('human domination'); as well as the Malthusian or Meadovian assumption that the continuation of current developments will lead to ecological catastrophe (ibid.).

ECOLOGICAL WORLDVIEWS OF THE CULTURAL THEORY OF RISK

The ecological worldviews presented by cultural theory are also regularly discussed in current research on climate-related action. As previously mentioned in connection with values, cultural theory includes both values and ecological worldviews as explanatory variables for ways of dealing with risk. The empirical implementation of cultural theory on the whole is problematic, as values and worldviews are theorised in close connection with one another without empirically validating their coupling (see also Krömker 2004: 108). There are, however, authors who have taken an isolated look at the ecological worldviews contained therein (see for example Krömker 2004; Grunenberg and Kuckartz 2003).

The theory takes its starting point from four different, so-called 'myths of nature' (Thompson 2003: 220). These are, first, *benign nature*: nature will continue to renew itself, and human intervention does hardly any damage (the 'individualist' value type). Second, *tolerant nature*: nature is up to a certain degree tolerant (the 'hierarchical' value type). Third, *ephemeral nature*: nature falls swiftly out of balance (the 'egalitarian' value type). Fourth, *capricious nature*: humans can hardly predict what will happen when interventions in nature take place (the 'fatalist' value type).

The individual nature myths of cultural theory thus focus on possible notions of what occurs when humans intervene in nature. In the process, it assumed that nature can be regarded as being stable to varying degrees. In this isolated structure, however, similar themes to those of the NEP are probed, namely the ascribed robustness of nature and the predictability of the consequences of human intervention (see also Krömker 2004: 111).

FURTHER PERSPECTIVES: SPIRITUAL NATURE, AND THE UNITY OF HUMANS AND NATURE

In addition, the literature discusses other worldviews that appear in connection with climate change, and which should therefore be highlighted here. In anthropology, Krömker names *spiritual nature* (Krömker 2004: 118) as a category that regularly emerges (see also Schipper 2010). Within this can be included notions that view nature in transcendental terms. The spectrum here is very broad and spans from monotheistic ideas of a divinely created natural world to animistic concepts in which nature itself is ensouled. Closely connected to this, and briefly hinted at, are the numerous variants on *holistic views of humans and nature*, which have been widely discussed beyond the New Ecological Paradigm, for instance when in a number of world regions, such as Australia, animistic or totemic ideas collide with supposedly western spatial development concepts about climate adaptation (see also the further discussion on climate adaptation).

5.2.2.3 The state of research: ecological worldviews as explanations for climate-related action

The ecological worldviews identified are regularly invoked as explanations for the variation in climate-related responses. In the following, the relevant studies will be examined more closely.

ECOLOGICAL WORLDVIEWS AS EXPLANATIONS FOR VULNERABILITY CONSTRUCTIONS

Krömker studies the extent to which ecological worldviews can explain different opportunity and problem constructions relating to climate change among students in Peru, India, the USA, and Germany (Krömker 2005: 146f, 2004). In the process she uses approaches from all three of the above-mentioned theoretical traditions. Krömker demonstrates that worldwide, at least among young students, country of origin plays a much smaller role than globally shared background knowledge (appearing here in the form of ecological worldview) as a differentiating factor for internalised climate knowledge. The study therefore points to the existence of transnational climate cultures and orders of knowledge connected to the relational concept of space.

In addition, there are studies that focus on nature myths from cultural theory as explanations for varying climate-change problem constructions, and which find significant differences (Price *et al.* 2014; Grunenberg and Kuckartz 2003). In summary, studies suggest that constructions which see nature as less stable (and thus less elastic) lead to higher problem perceptions, and vice versa.

ECOLOGICAL WORLDVIEWS AS EXPLANATIONS FOR CLIMATE-MITIGATION PREFERENCES

Krömer enquires into the acceptance of climate mitigation in general, and of eco-taxes in particular (Krömker 2005: 164). Actors with protection-oriented conceptions of nature (see above) here clearly show a higher acceptance for the strategies assessed (ibid.: 164). Regrettably, she does not investigate any other climate-mitigation strategies, leaving the question open as to whether protection-oriented conceptions of nature also, for example, lead to an increase in the acceptance of technological measures (e.g. renewable energies or technological sinks). Here, contradictions might emerge between compromising nature (e.g. using large areas for solar energy parks) and protecting the climate.

To date, cultural theory's conceptions of nature have been used in particular to explain a preference for sufficiency strategies. Price *et al.* (2014: 18), for example, observe that actors sharing a belief in the limited stability of nature have higher acceptance rates for household-related climate-mitigation measures connected with constraints.

Sufficiency-oriented measures have also been the principal focus when probing NEP orientations to explain climate-mitigation measures. In subtropical Sydney, Deuble and de Dear (2012) come to the conclusion that NEP orientations accompany a higher acceptance of non-ideal indoor summer temperature conditions, as for example in so-called 'green buildings' (ibid.: 26), in which air-conditioning is avoided for the benefit of climate mitigation. Broomell *et al.* (2015) undertake a non-representative, comparative poll in 24 countries. For all measures investigated and regardless of which country the sample originated in, the authors could conclude: 'Environmental worldviews and belief in GW [Global Warming] are stable and consistent predictors' (ibid. 2015: 71). The spectrum of available measures must now, however, be regarded as essentially more complex (see Chapter 4). The issue remains open here whether NEP worldviews are able to explain only sufficiency strategies or also preferences for other strategies.

The study by Hope and Jones (2014) also provides some initial vague indications. Using focus group interviews with a total of 20 people, they investigate the differences between religious groups (Muslim, Christian, and secular) with regard to NEP beliefs and attitudes towards CCS technology. They conclude that high NEP orientation need not necessarily lead to higher CCS acceptance. Since the group sizes were extremely small, however, and the authors hardly differentiated between the individual and group levels in their presentation of the results, one should avoid any (ecologically false) conclusions.

In contrast, initial conclusions about the acceptance of renewable energy are available. In a survey of more than 1000 households in the US states of Oregon and Washington, Steel *et al.* (2015) investigate the influence of NEP orientation on the acceptance of political affirmative action for bioenergy, wind, geothermal, solar, and hydro energy technologies. The authors observe a significant positive influence from NEP orientation on the acceptance of virtually all measures, the exceptions being hydropower and bioenergy (ibid.: 642). The precise reasons behind this can only be guessed at. One possible hypothesis is that NEP orientation opposes any measures that require high levels of intervention into supposedly natural processes.

ECOLOGICAL WORLDVIEWS AS EXPLANATIONS FOR CLIMATE-ADAPTATION PREFERENCES

Also connected to climate-adaptation measures are aspects of the above-mentioned anthropologically discussed worldviews. Thus various anthropological studies have emphasised the significance of different conceptualisations of the relationship between humans and nature. These ecological worldviews are associated in turn with various temporal and territorial constraints on the use of land, and might therefore have implications for climate-adaptation processes (Leonard *et al.* 2013: 630). Heyd and Brooks (2009) also assert that adaptation strategies might depend on the extent 'to which societies see themselves as separate from or part of the wider physical or "natural" environment' (Heyd and Brooks 2009: 269). The western notion that humans inhabit a cultural sphere separate from nature accordingly influences current attitudes towards ways of dealing with the natural environment (ibid.: 273). Unfortunately the authors do not examine their hypotheses empirically.

The influence of conceptions of nature from cultural theory has also since been used to explain preferences for various climate-adaptation strategies (Hajjar and Kozak 2015).

Jopp *et al.* (2015) explore the influence of NEP orientation on the acceptance of adaptation measures among surfing tourists along the Great Ocean Road in Australia. The spectrum of the measures investigated here spans from early warning systems, protection measures, artificial reefs, and retreat from the sea to fire breaks. Here, NEP orientation has hardly any explanatory power for the varying acceptance of the adaptation measures discussed, with the exception of retreat measures: giving up sections of the beach for the benefit of the ocean is here accepted significantly more frequently by actors displaying a high NEP orientation (ibid.: 315). The conclusion is hardly surprising when one calls to mind that the ecocentric worldview of the NEP places nature, not humans, at centre stage. On the whole, though, the study provides no clear conclusions, given that at the same time NEP orientation demonstrated no significant influence on the acceptance of retreat in general (ibid.). Unfortunately, the authors did not investigate accommodation strategies as well, and so it remains unclear what role NEP orientation plays here.

Taken together, it is clear that a wide range of hypotheses about the relevance of various conceptions of nature to adaptation measures have been discussed. To date, though, there have been very few empirical studies.

5.2.2.4 Derivation of hypotheses

Given the exposition above, one can conclude that there is strong evidence that the New Ecological Paradigm is, at least in quantitatively oriented culture-related climate research, the most explored worldview alongside Cultural Theory (see also Price *et al.* 2014: 9). It is also apparent that the themes explored in the NEP (ecological limits, the balance of nature, human dominance over nature, ecological catastrophe) cover the broad thematic diversity of possible central beliefs, including those from other theoretical traditions, and in the process consolidate them into the simple binary schema of anthropocentrism (HEP) vs. ecocentrism (NEP). In addition, it is possible to test regularly whether the NEP continues to deal with a consistent collection of beliefs, and therefore with an ecological worldview (see also Section 7.2 and the consistency tests in the German version of this book, Heimann 2017).

The relevance of the NEP to climate-related action has not yet been fully researched in regard to several aspects. These aspects will be summarised again below. This study will use the NEP to investigate ecocentric and anthropocentric orientations and their role for climate-related action more closely. Now, therefore, the general hypothesis on the relevance of ecological worldviews will be formulated:

- $\mathbf{H}_{\mathrm{EcoWorldview}}$: Ecological worldviews have, as background knowledge, an influence on climate-culture orientation, in particular on socially shared (a) problem and chance constructions, (b) climate-mitigation preferences, and (c) climate-adaptation preferences.

Concerning climate-change-related problem constructions, the above-mentioned research results indicate that NEP and ecocentric orientation increase the belief that climate change is occurring, and that they increase its problem perception and the belief that it is human-made. The general hypothesis regarding problem perception therefore follows:

NEP and problem constructions: The stronger the internalisation of the New Ecological Paradigm (ecocentric worldview) at the expense of the Human Exceptionalism Paradigm (anthropocentric worldview):

- the stronger the belief that climate change is occurring;
- the more likely climate change is perceived as a problem; and
- the greater the belief that climate change is human-made.

If ecocentric beliefs contribute to the selection of only those measures corresponding to them, then the following general hypothesis can be formulated with regard to measures:

NEP and measures in general: The stronger the ecocentric outlook (NEP):

- the higher the probability of a preference for ecocentric measures; and
- the lower the probability of a preference for anthropocentric measures.

These general hypotheses can be further specified. The studies on climate-mitigation measures have shown that NEP orientation can increase the acceptance of climate mitigation in general, as well as for sufficiency measures. It remains unclear what role HEP and NEP orientations have if other strategies are being considered for implementation. The conclusions of Steel *et al.* (2015) and Hope and Jones (2014) indicated that NEP orientation can also lead to the rejection of climate-mitigation measures, such as when humans are placed too centrally in the endeavour or when measures are at the expense of nature and the environment. The central hypothesis on the relevance of NEP orientation for climate-mitigation preferences thus follows:

NEP and climate-mitigation measures: The stronger the internalisation of the New Ecological Paradigm (ecocentric worldview) at the expense of the Human Exceptionalism Paradigm (anthropocentric worldview):

- the more likely climate-mitigation measures will be preferred that involve sufficiency strategies; and
- the less climate-mitigation strategies will be preferred that are associated with the harming of nature and the environment.

Anthropocentric and ecocentric orientations also play a role in efforts by people to adapt to changed environmental conditions, as here it is regularly the case that one must find a balance between human dominance (HEP) and the presumably ecological inclusion of all living things (NEP). The conclusions of Jopp *et al.* (2015) discussed above presented the relevance of the NEP for adaptation measures in which the subordination of humans to other forms of life is practised, for example in relinquishing sections of beaches as part of a retreat strategy. On the other hand, a high HEP or low NEP orientation contributes to action strategies in which humans are prioritised over other species. The issue is thus regularly raised whether the focus of an adaptation measure is on 'struggling against nature' (HEP) or 'living with nature' (NEP). In summary, the following hypotheses present themselves:

NEP and climate-adaptation measures: The stronger the internalisation of the New Ecological Paradigm (ecocentric worldview):

- the more likely climate-adaptation measures will be preferred that do not prioritise humans above other living beings (e.g. retreat strategies, integration strategies; keyword: 'living with water').

HEP and climate-adaptation measures: The stronger the internalisation of the Human Exceptionalism Paradigm:

- the more likely climate-adaptation measures will be preferred that maintain a binary separation between humans and nature (e.g. protection strategies; keyword: 'struggling against the water').

5.2.3 *Place identification*

The idea that identification with places and regions[4] might also be highly relevant to climate-change action was discussed early on by the authors of the IPCC (IPCC 1990: 154). Additionally, in 1991 the journal *Global Environmental Change* published a first thorough thesis (Feitelson 1991). The early social-science discourse appears to have come to a halt, however, or even, as has recently been complained, it 'remains overlooked' (Devine-Wright 2013: 61). Only since the middle of the 2000s have increasing efforts been made to close existing gaps in the research (Adger *et al.* 2011). The following remarks are hence also meant to contribute towards this.

5.2.3.1 *On the concept of 'place identification'*

It has long been considered in social science 'that people identify themselves with a place in certain situations' (Treinen 1965: 73). The decades-long concern with the relevance for action of place identification and the concepts applied goes far beyond the relatively few social-science studies into climate. Since then there has been a diverse repertoire of fields that have been able to make contributions, among them such varied social-science disciplines as sociology, environmental psychology, and human geography. Such abundance goes beyond the scope of this study to discuss comprehensively, but this body of work will nevertheless be broadly outlined (see also Petzold 2013: 85ff.; Richter 2013: 12ff.).[5]

Place identification can be defined as *the extent to which actors feel themselves connected to a place or region* (see also Petzold 2013: 87; Weichhart *et al.* 2006: 21; Mühler and Opp 2004: 21). Gerhards (2000) asserts that there are at least three different determinants, namely (1) the *actor* 'who identifies themself with something in the world' (ibid.: 116), (2) a place or a region which supplies the '*object* of identification' (ibid.), and (3) 'a specific *relation* between the subject and the object [...] that can be described as identification' (ibid., emphases added). For each of the three determinants, the literature provides various approaches as to how precisely each should be understood and how they might be investigated.[6]

In approaches following the socio-cultural perspective, materiality appears of less significance for spatial connectedness (see also Richter 2013: 13f.). Lalli states that places emerge as a 'social construction derived from social communication' (Lalli 1992: 292). He argues that one need not necessarily be dealing with opposites (physical vs. social), but rather that both perspectives could be viewed as complementing each other. Both physical experience of the physical environment and communicative exchange can thus contribute to the genesis of place identification. A convergence of both lines of development thus appears, so that the strict binary distinction between the social and the material need no longer be applied.

Given all of the above, I conclude that places and regions, as the objects of identification, need not necessarily be constructed as container spaces, but may rather be interpreted by actors in diverse possible ways. Hence Mühler and Opp (2004: 12) remind us, for example, that the ascribed urban and regional areas explored in surveys need not be identical to the official administrative area. Massey goes one step further and emphasises that place or locality, as carriers of meaning, do not have to be considered as 'opposite to global space, or a given, delimited unity' (Massey 2006: 28), but rather as 'a continuous project whose identity is formed in relationships and practices' (ibid.).

It should also be noted, however, that places are often regarded at different *scales* (Lalli 1992: 291). Place identification can therefore be comprehended at varying scales, ranging from a single house to a community, a nation, and all the way to the entire globe (see also Feitelson 1991: 403). Differences in the social relevance of action can exist here, depending on whether actors identify with Europe, their country, or their community (ibid.).

Several authors go a step further and draw upon the term *identity* (see also Petzold 2013; Richter 2013; Lalli 1992; Graumann 1983; Proshansky *et al.* 1983; Proshansky 1978). Scannell and Gifford emphasise: 'When place is incorporated into one's self-definition, it is referred to as place identity' (Scannell and Gifford 2013: 66). It would go beyond the scope of this study to look in detail at the extensive work completed on the concept of identity. What remains important to keep in mind, is that place identification has been discussed as a sub-category of a thoroughly more complex concept of individual or collectively shared understandings of the Self and the Other.[7]

5.2.3.2 *Substantive dimensions of place identification: the approach of Lalli*

The issue thus arises as to which substantive dimensions can be associated with place identification. Various aspects have been discussed and empirically investigated in the literature, such as connections to places and regions in general (Mühler and Opp 2004), familiarity with places and regions and memories of the past (Fried 2000), planned future bindings with places (ibid.; Marshall *et al.* 2014: 589), or group ties formed through socially shared knowledge of places (Low 1992; Treinen 1965).

Lalli (1992) has been able to show that many aspects often coincide, and reduces them to five aspects that regularly emerge as content-related dimensions or personal place identification (ibid.: 294ff.). He develops an associated scale for the measurement of these dimensions, enabling their consistency and common appearance to be tested empirically (ibid.).

The first dimension relates to the *perception of external evaluation* Lalli 1992; see also Weichhart *et al.* 2006: 56). This records the degree to which actors ascribe a place a positive or negative image (ibid.). The second dimension, *continuity with personal past* (Lalli 1992: 294) relates to the actors' local past and the relevance of a place for one's own biography. The third dimension includes the *general attachment* to a place, measuring the degree to which actors feel at home in a place. Lalli names the *perception of familiarity* as a fourth dimension. This relates to the feelings of belonging one attaches to a place. The last dimension is *commitment* to a place, i.e. whether actors' view their future as bound to a specific place.

Lalli himself applies his approach explicitly to urban identification processes (Lalli 1992: 295), and therefore names it an 'Urban-identity Scale' (ibid.: 294; see also Section 7.2). He does not, however, exclude the possibility that his theoretical considerations might be applicable at other scales and comes to the conclusion that 'more work needs to be done' (ibid.), an invitation taken up by this study in its explicit application at other scales (see Chapter 7ff.).

5.2.3.3 *The state of research: place identification as an explanation for climate-related action*

In the meantime there have been various studies concerned with the complex interaction between place identification and climate-related problem perception, as well as measure constructions.

PLACE IDENTIFICATION AS AN EXPLANATION FOR VARYING VULNERABILITY CONSTRUCTIONS

The question whether a clear pattern can be observed between place identification and climate-related vulnerability constructions has remained broadly open. One can find in the literature both limiting as well as augmenting impacts.

Dominicis *et al.* (2015) take up the issue of what role connection to place plays with regard to the perception and management of flood risk. They discuss two possible mechanisms. On the one hand, actors with high identification might have developed comprehensive connections with and positively assessed knowledge constructions of a place. Such places will then appear in such a positive light, that a threat will be seen as unlikely (ibid.: 68). This consideration will be labeled here as the *halo-effect hypothesis*. Armas (2006: 1232) has already been able to observe this effect among inhabitants of the city of Bucharest (see also Dominicis 2015: 68).

On the other hand, the opposite effects are conceivable, namely those due to which actors with a strong attachment to a place are also particularly sensitive to threats to it, such as when actors appreciate a place and have an interest in its continued existence (see also Dominicis *et al.* 2015: 68). Actors with a strong place identification could then consistently demonstrate a higher problem perception relating to a place than actors with a weaker identification. These considerations will be termed the *problem-sensitisation hypothesis*. In their own empirical investigations in two Italian cities, however, Dominicis *et al.* have been unable to ascertain the direct influence of place identification on problem perceptions of flood risk (ibid.: 73).

The work of Devine-Wright (2009: 433) also indicates various possible directions in which place identification might have an effect. He works on the assumption that people with stronger place identification also develop a higher attentiveness for information with local relevance, as 'strongly attached individuals would be expected to take an interest in what's going on' (ibid.). Those with stronger identification would accordingly internalise both positive and negative information to a greater extent than others with lower attachment. As climate change can be discussed in very different ways (Christmann *et al.* 2012), it would at the same time be necessary to consider what climate-related information and content the actors are confronted with. This indicates that here we are dealing with complex interactions offering extensive research options for future study, but which owing to limited time and financial resources can only be investigated to a small degree in the present study.

Summarised once again, it remains largely unclear to what extent place identification can explain varying climate-related problem constructions. Both negative and positive correlations appear to be conceivable (e.g. the halo effect and problem sensitisation, respectively).

PLACE IDENTIFICATION AS AN EXPLANATION FOR DIFFERING CLIMATE-MITIGATION PREFERENCES

The influence on spatial planning measures in general. There has long been discussion in the literature about the character of places and regions being altered not only by climate change, but also by the measures themselves (Devine-Wright 2013: 64; IPCC 1990: 154). If measures imply changes to places and regions, then individual or collective attachment to them could influence the acceptance of climate-related measures. Scannell and Gifford assert that 'individuals with strong place attachment may serve as a promising group to whom adaptation and mitigation strategies could be promoted' (Scannell and Gifford 2013: 78f.). In the meantime, interactions between place identification and general measure preferences for climate-mitigation, but also for climate-adaptation (see below), might be observed.

Schweizer *et al.* (2013: 58) measure a positive influence on the readiness to make changes in behaviour for the purpose of climate mitigation. Scannell and Gifford (2013: 75) observe a positive influence on climate-mitigation engagement in general, although they leave it largely unclear about which concrete measures they investigate. As already discussed, it must be assumed that the spectrum of possible measures for responding to climate change requires more differentiated consideration, and it must therefore be asked which concrete measures are of relevance to which explanations (see Chapter 4). The question that particularly arises, therefore, is whether interactions can be observed between place identification and specific climate-mitigation measures. Initial results are already available on this.

Acceptance of concrete strategies. Place identification might only in certain cases have a positive influence on the acceptance of climate-mitigation measures. An idea exists in the literature that can here be termed *development theory*. Devine-Wright (2009) discusses place identification as an influence on the acceptance of the redesign of places for the production of renewable energy. He postulates two possible directions for the effects, which might be dependent on whether the redesign is associated with positive or negative development potential for the place in question (ibid.: 434). If a measure is perceived as inhibiting the development of a certain place, then with increasing place identification there will be increasing rejection of the measure. If, in contrast, the measure can be ascribed a positive contribution to a location's development, then with increasing identification would also be seen increasing acceptance (see also Devine-Wright 2013: 66).

Relevance of the scales considered. It must furthermore be noted that there are authors who discuss the differences in relation to the scale of the places and regions being considered. Vorkinn and Riese (2001: 260), for instance, differentiate between place identification with a Norwegian community in general and with specific areas that would be affected by redesign with the building of a hydropower station. They ascertain that high identification with the community accompanies high acceptance of the building measures, whereas identification with the specific area increases the probability of rejection (ibid.: 259). Feitelson (1991) also suggested early on that place identification can, at different scales, have varying significance for the acceptance of climate-mitigation measures. Strong identification, for example with the entire world as a so-called global place ('sharing the globe', Feitelson 1991: 369), could correspondingly strengthen the acceptance of local climate-mitigation measures, despite a high local cost, if these were to improve the condition of the entire globe (see also Devine-Wright 2013: 61). Following these consideration, interactions at very different scales are possible, from local and regional up to the global level (Feitelson 1991: 403). There is thus also much possibility here for investigation in future studies.

PLACE IDENTIFICATION AS AN EXPLANATION FOR DIFFERING CLIMATE-ADAPTATION PREFERENCES

In the climate-adaptation literature too there are authors who assume that place identification can lead to preferences for certain climate-adaptation measures, as well as those who argue that this can, in certain situations, diminish acceptance.

Pivotal to the debate is the previously discussed development theory, according to which place identification inhibits or supports acceptance if measures are associated with, respectively, negative or positive potential for the development of a place or region (Amundsen 2015; Devine-Wright 2009: 434). Amundsen, for instance, observes in Norwegian coastal municipalities that stronger place identification accompanies strong support for specific development strategies (Amundsen 2015: 268).

How does it look for the acceptance of different adaptation options, though? *Retreat strategies*, for instance, lead to the relinquishing of territories and accordingly reduce the possibility of their development. If one is to remain true to the development theory, then increasing place identification should see decreasing acceptance for retreat strategies. On the other hand, increasing identification could also lead to an increase in acceptance if measures are ascribed positive development potential, as would be the case, for example, if the technological adaptation of buildings against storms were undertaken.

Adger *et al.* (2011) have contended that in making climate-relevant political decisions, the criteria applied are often too functional and economic, and that the 'symbolic and psychological aspects' are too seldom taken into account, such as the retention of artefacts that might be considered part of the cultural heritage of specific actor groups (ibid.: 2). Devine-Wright (2013: 65) remarks that these procedures have already led to resistance against measures. This appears, for instance, in Great Britain in the emergence of nationwide action groups such as the 'National Voice for Coastal Communities' (ibid.) who, owing to strong local attachments, call for protection strategies. Scannell and Gifford also argue that attachment to place can increase the willingness for protection: 'We postulate that strong connectedness to place is important to climate change attitudes and behaviors, because it can sometimes engender place-protective intentions' (Scannell and Gifford 2013: 66). Following these considerations leads us to deduce that with increasing place identification, the preference for protection measures that maintain the status quo of a place (such as raising dykes and other technological security measures) must also increase.

In total, the discourse indicates a varying relevance of place identification for different options. Thus, for adaptation measures, it is also necessary to investigate the explanatory force of place identification for clearly shared preferences for specific measures.

5.2.3.4 *Derivation of hypotheses*

Consideration of the state of research shows that place identification has already been discussed for all the considered aspects of climate-culture orientations. Many issues remain, however, for this study to test empirically.

First, a hypothesis will be formulated that takes up the explanatory force of all three observed levels of climate-culture orientation:

- $\mathbf{H}_{\text{Ident}}$: Place identification has an influence on climate-culture orientations, namely on socially shared (a) problem and opportunity constructions, (b) climate-mitigation preferences, and (c) climate-adaptation preferences.

With regard to the various problem constructions, it has been seen that the effects which place identification has can take several possible directions. According to the *halo-effect hypothesis* discussed above, it is assumed that with increasing place identification comes a decrease in problem perception. The opposite case is true according to the *problem sensitisation hypothesis*. Thus the following hypothesis can be formulated for verification:

Place identification and problem constructions: The stronger the place identification:

- the less climate change is perceived as a local problem (halo effect); and
- the more climate change is perceived as a local problem (problem sensitisation).

Furthermore, it can be seen that in connection with climate-mitigation measures, it has been suggested that increasing place identification will increase the acceptance of any measure that also supports the potential for development of a place or region, and vice versa. This gives the following hypothesis:

Place identification and climate-mitigation measures: The stronger the place identification:

- the higher the acceptance of climate-mitigation measures to which positive development potential can be ascribed;[8] and
- the lower the acceptance of measures that would be obstacles to the positive development of a place.

Development theory makes an appearance in climate-adaptation studies too. It is thus possible to derive hypotheses with regard to the following measures:

Place identification and climate-adaptation measures: The stronger the place identification:

- the more climate-adaptation measures will be preferred to which positive development potential can be ascribed, in particular
  - there will be increased acceptance of protection strategies that guard a place or region's development and
  - there will be increased acceptance of integration strategies that maintain a place or region in the long term
- the more climate-adaptation measures will be rejected to which negative development potential can be ascribed, in particular
  - the acceptance of retreat strategies will decrease, as these cause a place or region to degenerate.

5.2.4 Integration: the VBI model for explaining climate-culture differences

In this section, three existing principal forms of background knowledge have been given as an explanation for different climate-culture orientations, namely values, beliefs (specifically, ecological worldviews), and identification with places/regions. As explanations for different forms of climate-related action, these approaches have largely been utilised separately from one another. In the available literature, only values and beliefs have been observed in combination (such as in the VBN approach used by Stern and colleagues to explain environmental protection action in general; see Stern 2000; Stern *et al.* 1999).

It would therefore seem sensible to integrate values, beliefs, and identification with a place or region in a general approach towards studying climate-related action, as they all have been discussed as relevant for explaining climate-change-related behaviour.

The central idea, based on Douglas and Wildavsky's formulation (1982: 8) mentioned at the start, can thus be expanded like this: '*Common values, beliefs, and identities lead to common fears and practices against climate change*'. The integrated approach will therefore be termed the VBI model (**V**alues, **B**eliefs, and place **I**dentification). The applicability of the VBI model for explaining climate-culture orientations will be tested in the empirical part of this study.

Finally, it should be noted that, as is usual in the social sciences, one of the basic problems consists of the fact that there are always more potential causal directions possible. Thus, for example, internalised knowledge about climate change could also lead in the long term to changes in values, beliefs, and identification. The empirical testing of the direction of causation is always based upon fundamental theoretical decisions that require a reduction of complexity and which are particularly necessary for the analysis of cross-sectional data (see also Roose 2010b: 33). The present study works on the basic assumption that background knowledge can be treated as an explanatory factor, an assumption founded on the theoretical traditions previously presented – environmental sociology (Section 5.2.1), cultural theory (ibid.), and sociology of knowledge (Sections 2.1 and 5.2.2) – in which, as described, the sedimentation of knowledge and associated selection processes in the internalisation of knowledge provide a rigorous theoretical foundation.

5.3 Societal fields and place/region

Alongside background knowledge, there are further social factors that are suggested in the literature as explanations for different climate-culture orientations. This study takes an interest in the relevance of such social factors, in particular field affiliation and the significance of place constructions. In addition, social demographic factors will act as control variables. The concept of space has already been considered in depth in Chapter 2, thus at this point discussion will be restricted to central aspects in the explanation of climate-culture differences.

5.3.1 Field affiliation: administration, politics, the economy, and civil society

The notion that societies based on the division of labour tend to form systematically differentiated stocks of knowledge and (sub-)group-specific cultural patterns can be considered a basic tenet of sociology. Max Weber already assumed the existence of various 'value systems' (Weber 2010: 531), such as the economy and politics. At the height of structural functionalism, Parsons distinguished between various partial systems such as politics, the economy, or art and culture, laying the foundations for later system-theory approaches such as that of Niklas Luhmann, in which individual actors become increasingly unimportant (Schroer 2006: 141; Reckwitz 2000: 81; Parsons 1961: 38). From increasingly action-theoretical perspectives too, there is evidence that relational culture-space boundaries are able to cut across and around communities. Berger and Luckmann (1966) differentiate, for example, between primary and secondary socialisation[9] in knowledge acquisition and emphasise that they know of no social formations 'in which there is no distribution of labour, and thus no social distribution of knowledge' (ibid.). It seems all the more astonishing then, that comparative-cultural research has long held on to places and regions such as the nation state as its units of comparison (Rippl and Seipel 2008: 21). From all these consideration it could ultimately be concluded that cultural boundaries can also exist between functionally differentiated groups.

Some initial considerations have been discussed and observed in the climate-related literature, according to which socially shared vulnerability and resilience constructions systematically vary between functionally differentiated groups. Otto-Banaszak *et al.* (2011) report within Europe on systematic differences in climate-related problem perception and adaptation-measure preferences among administrators, politicians, academics, journalists, and civil society actors. Nilsson *et al.* (2004) observe differences in value orientations between those in the public and private sectors.

Accordingly, it must be possible to trace differences in actors' (climate) knowledge and volition back to their functionally differentiated group affiliation. The question arises to what extent group affiliation tied to functional differentiation also plays a role in discriminating between cultural spaces of shared knowledge. This study will follow differentiation theory's governance classification, which discriminates between the fields of politics, administration, the economy, and civil society (see also Section 7.1, as well as Frommer 2010: 66; Zürn 2008: 559). This will allow for empirical testing of the extent to which field affiliation provides alternative explanations to local/regional affiliations for differences in climate-culture preferences. The following hypothesis can thus be formulated:

- $\mathbf{H}_{\text{Field}}$: Field affiliation (administration, politics, economy, NGO) has an influence on climate-culture orientations.

5.3.2 *Local/regional affiliation*

In this study it has already been comprehensively discussed that, owing to recent social-science discourse returning to altered relationships between society and territory, cultures will be considered along a relational concept of space, as aspects of vulnerability and resilience constructions shared socially among actors. This approach makes it possible to make an a priori social-theoretical and methodological separation between culture and place/region. It does not, however, exclude the continuing relevance of places and regions and should serve much more to bring their precise relationship into focus. In social-science climate research there are various positions on the relevance of place and regional affiliations as discriminants of climate culture. These will be briefly sketched out here. Drawing upon this, hypotheses for the empirical work can be derived.

Several authors adhere to *cultural convergence and homogenisation theories* (see Section 2.2) and view the ways of dealing with climate change as a paradigmatic example for the emergence of a 'world society' (Weiss and Burke 2011: 1057; see also Erhardt-Martinez *et al.* 2015: 218f.). Associated with this is the notion of the declining significance of places and regions for climate-culture formations. Erhardt-Martinez *et al.* (2015) draw closely on the neo-institutionalist theory introduced by Meyer (2007), according to which the rise of supranational (or often global) institutions created to overcome commonly perceived climate-related challenges can also lead to the worldwide dissemination of similar stocks of knowledge, for example as a 'common set of norms, values, and cultural rules' (Erhardt-Martinez *et al.* 2015: 218). In a similar way, Weiss and Burke (2011) state that the common treatment and recognition of the supposedly global problem of climate change is not restricted to creating only transnational institutions such as the IPCC or UN or EU climate-mitigation programmes, but also common interests, values, and ideas, which in an ideal case might lead to a 'movement from an international society of "rivals" towards a world society of "friends"' (ibid.: 1070). It can be seen from these considerations that the categorisation of actors according to territorial affiliations such as national affiliation has little explanatory force for socially differentiated vulnerability and resilience constructions.

On the other hand, there are authors who are proponents of *local and regional differentiation theory* and who continue to emphasise *local/regional* particularities in dealing with climate change, giving more weight to *social conditions* than to observable differences in climate and physical geography (see among others Adger *et al.* 2013, 2011; Landauer *et al.* 2013; Christmann *et al.* 2012; Crate 2011; Biesbroek *et al.* 2010; O'Brien 2009; Brooks *et al.* 2005; Bulkeley and Kern 2004). Meanwhile, in the field of anthropology there is now a new strand of research called 'place-based community research' (Crate 2011: 179), in which differences in the ways small, supposedly 'locally situated' communities deal with climate change are closely examined.

Places and regions have been compared at quite different scales. Several authors observe differences in the constructions of climate-related problems and opportunities at the *urban and community levels* (Martinez *et al.* 2014; Christmann *et al.* 2012), while Knoblauch and Stelljes (2014) discuss differences in climate-change problem constructions at the *regional level*, comparing German federal states. In addition, studies can be found which postulate a continuance of *national differences* with regard to the implementation of climate-mitigation and adaptation measures (Greiving and Fleischauer 2012; Biesbroek *et al.* 2010; Abildgaard 2000). It is important to highlight the work of Karen O'Brien (2009: 171) here, who argues that nationally determined perceptions and values can have an impact on the different ways of dealing with climate change. Abildgaard (2000) describes national peculiarities in Danish climate-mitigation practices of the 1990s. Meanwhile, Biesbroek *et al.* (2010: 445) describe how not only differing national adaptation strategies are being implemented, but also varying national information strategies, which can themselves lead to differences in nationally shared stocks of knowledge about adaptation.

In addition, *transnational commonalities and differences* have been observed (see for example Otto-Banaszak *et al.* 2011). Mez (1997) views Denmark and Germany at the end of the 1990s as supposed trendsetters for climate-mitigation measures. Otto-Banaszak *et al.* (2011: 218) postulate systematic differences regarding adaptation preferences between the old EU15 states and those that joined from eastern Europe after 2004. Considering possible reasons, Gerhards and Lengfeld (2008) conclude that in several EU15 states a fundamentally higher environmental consciousness dominates than in many eastern European countries.

The previous considerations make it clear that not only historical borders at various scales, but other factors can also contribute to the explanation of climate-culture differences. As suggested in Chapter 2, these can each be explored on an aspectual basis.

The central criticism against 'methodological nationalism' is that comparative work grants a priori too high a status to national borders (see Section 2.2). After the further discussion in this section, however, I conclude that it remains an open question what their explanatory power for social phenomena is today. In order not to exceed the remit of this study, the discussion here will be limited to the current relevance of national borders.

The hypothesis follows:

- $\mathbf{H}_{\text{Place}}$: National affiliation has an influence on climate-culture orientation.

Differentiation theory will therefore be discarded in favour of *homogenisation theory* if this helps to ascertain any significant influence of national affiliation on vulnerability or resilience constructions, especially taking into account the control variables (see Chapter 9ff.). Because an aspectual approach is being taken, it should be emphasised that difference and homogeneity can each occur simultaneously under different aspects such as (1) vulnerability constructions, (2) climate-mitigation cultures, or (3) climate-adaptation cultures. Furthermore, it should be verified whether a possibly *restricted transnational homogeneity* exists, for instance when, as described by Otto-Banaszak *et al.* (2011) or Gerhards and Lengfeld (2008), commonalities between several countries exist that are significantly delimitable with respect to other countries.

To take a closer look at the possible local and regional relevance for differentiating internalised climate-culture knowledge constructions, Chapter 6 will deal with the historical constructions of the climate-political frameworks being investigated and thus outline the development of the global, European, and national levels.

5.3.3 *Control variables: socio-demographic factors*

Several socio-demographic factors appear in the literature to which differentiating effects can be ascribed regarding climate culture, as well as several of the constructions of background knowledge discussed here.

Age. Many authors report the significant influence of age on climate-related vulnerability and resilience constructions (Steel *et al.* 2015; Kuckartz 2009; Weber 2008). Weber observed among the German population a decreasing 'sense of threat' (Weber 2008: 184) with increasing age. Contrastingly, Safi *et al.* (2012: 1052) find no age-related differences in rural Nevada regarding either climate-change problem perceptions or measure preferences. In Weber's study (2008: 188), the acceptance for climate-mitigation as well as adaptation measures falls with increasing age. In two states of the USA, Steel *et al.* (2015: 642) observe with increasing age a higher rate of rejection.

In addition, a potential interaction with other variables has been discussed. Gerhards and Lengfeld (2008) observe the significant influence of age on environmental consciousness (ibid.: 9). They trace this back to cohort effects, assuming that the generations born before 1960 were raised without the presence of ecological discourse, and accordingly could hardly develop an environmental consciousness. Levey and Silver (2006: 679) observe in Japan and the USA that traditionalist orientation increases with age, i.e. conservatism increases incrementally. It is conspicuous that, on the whole, the degree to which these observations are related to the effects of a particular cohort, period, or age group is rather seldom investigated (see for example Weber 2008: 183; Franzen and Meyer 2004: 122). Therefore, given the variety of possible influences which the variable *Age* can have, it will be used as a control variable in this study.

Gender. Differences ascribed to gender have also been described in the literature. Kuckartz (2009: 5) has observed a significantly larger consciousness of climate problems among men. Many studies, in contrast, report a higher effect among women (see for example Crona *et al.* 2013: 527; McCright 2010: 78; Weber 2008: 184). Resilience constructions also vary by gender according to some authors. Weber (2008: 188), for instance, observes a greater willingness to implement climate-adaptation measures among men, although she finds no difference with regard to climate mitigation. On the other hand, Steel *et al.* (2015: 642) can make out no significant difference with regard to alternative energy. Scannell and Gifford (2013: 75) describe a higher rate of climate-measure engagement among women than men in several Norwegian communities. In principle, however, from a social-constructivist perspective one should warn against essentialist considerations or inappropriate biologisms. It is nevertheless unquestionable that group-specific socialisation processes, ascribed roles, and other factors can lead to gender-specific differences. The studies considered therefore suggest that gender-specific influences should be controlled for.

Education. Since the beginning of research in environmental sociology it has regularly been observed that higher levels of education increase environmental consciousness (Gerhards and Lengfeld 2008: 9; Franzen and Meyer 2004: 129; Dunlap *et al.* 2000: 429; Dunlap and Catton 1979: 249). Different forms of climate knowledge all regularly appear to follow this logic. Thus authors describe how higher levels of education can increase climate-related problem perception as well as measure preferences (Chen *et al.* 2015: 372; Steel *et al.* 2015: 642; Crona *et al.* 2013: 528; McCright 2010: 77; Kuckartz 2009: 7). Investigations into climate-culture differences must therefore also closely observe the influence of educational background. In this study, however, this procedure will be disregarded, as all actors considered already present a high level of education (see Table 7.3 Socio-demographic composition of data in Section 7.3).

5.4 Interim conclusion

This chapter has considered various forms of background knowledge that may explain climate-culture differences.

Let us once again consider the vulnerability and resilience constructions as discussed in the literature. Regarding vulnerability constructions, to date numerous attempts have been made to explain variance in problem perceptions. In contrast, approaches that deal with measure preferences regularly restrict themselves to relatively abstract categories (e.g. climate mitigation or adaptation in general). For climate-mitigation measures, perspectives that associate climate-mitigation with sufficiency strategies, or which are limited to only one kind of strategy (e.g. renewable energy or technological sinks), still dominate. With the adaptation measures observed it was also noticeable that there are very few studies which focus on systematic differences with regard to the various possible strategies. On the whole, it still remains to be ascertained which *different strategy preferences* are shared by actors and what *explanatory factors* contribute to differences.

In order to close these gaps in the research, first the concept of *orders of knowledge* was introduced in this chapter. Orders of knowledge imply here regularly observable characteristics and patterns of individual or collectively shared knowledge constructions. In the further course of this study, existing orders of climate knowledge will be investigated. The first question to consider will be the ways in which actors share vulnerability and resilience constructions, if at all.

Connected to this is the issue of how the observed climate-culture differences can be explained. Orders of knowledge imply *interactions between forms of background knowledge and climate knowledge*. The possible explanatory relevance of various forms of background knowledge has thus been discussed here, namely value orientations, beliefs, and place identification.

Existing orders of knowledge can systematically vary according to *further social factors*. In this study, the relevance of field affiliation and national affiliation will be examined.

Notes

1 Authors such as Dunlap, van Liere, Heberlein, or Cotgrove first introduced the concept of values to debates about new forms of environmental awareness (Dunlap and Van Liere 2008 [1978], 1977; Cotgrove and Duff 1981; Heberlein 1977). The first Environmental Sociology section of the American Sociological Association (ASA) was established in 1976 (Dunlap 2000: 21).

2 Briefly at this point, I should refer to the comprehensive epistemological and methodological debates that continue to the present, and whose extreme positions range from an assumption of global, universal comparability to the impossibility of cultural comparison (see in more detail Rippl and Seipel 2008: 25ff.; Cappai 2005).

3 For a detailed description of each value, see Schwartz 2012, 2003, 1992; Roose 2010a; Mohler and Wohn 2005.

4 As will be made clear in the following, identification can also be made with large-scale *regions*, such as countries or even continents. For simplicity, this study will refer to *place identification*, even where it is more precisely regions that are being spoken of.

5 Weichhart *et al.* (2006: 27) offer an initial overview. Deviating from several authors, the following will use the concept of place identification (see also Petzold 2013: 85). The associated phenomenon appears in the literature as, among others, 'place identity' (Proshansky *et al.* 1983), 'attachment to place' (Feitelson 1991), 'place attachment' (Scannell and Gifford 2013), 'sense of place' (Cunsolo Willox *et al.* 2012), 'space-related identity' (Weichhart 1990; Weichhart *et al.* 2006), 'local identity' (Richter 2013), or even as 'urban Identity' (Lalli 1992).

6 For a detailed description of these approaches, see for (1) the level of the actor (Richter 2013: 14f.; Belk 1992; Low 1992: 165; Marcus 1992: 87), (2) place or region as the 'object of identification' (Richter 2013: 13f.; Proshansky *et al.* 1983; Treinen 1965: 254), and (3) subject/object relations (Mühler and Opp 2004: 15; Proshansky 1978).

7 Deviating from the authors mentioned and for the sake of simplification, this study will refer consistently to place *identification*, even where others might here speak of *identity*.

8 To be precise, this hypothesis can only be verified if the contribution ascribed to the development of the place or region is also surveyed. This is not possible on the basis of the current data. However, the existing discourse on measures can be drawn upon, according to which ecological modernisation approaches are more development oriented than, for example, sufficiency approaches.

9 Primary socialisation here means 'the first phase in which a person, during their childhood, becomes a member of society' (Berger and Luckmann 1966: 150). Secondary socialisation relates to the formation of 'subworlds' (ibid.: 158), such as in the form of specialised knowledge 'as a result of [the] division of labour' (ibid.: 159).

References

Abildgaard, Jorgen (2000) 'Energiepolitik und Energieplanung in Dänemark'. In: Jänicke, Martin and Jörgens, Helge (ed.): *Umweltplanung im internationalen Vergleich. Strategien der Nachhaltigkeit*. Berlin/Heidelberg: Springer, pp. 39–52.

Adger, W. Neil, Barnett, Jon, Brown, Katrina, Marshall, Nadine and O'Brien, Karen (2013) 'Cultural dimensions of climate change impacts and adaptation'. *Nature Climate Change*, 3 (2): 112–117.

Adger, W. Neil, Barnett, Jon, Chapin III, F.S., and Ellemor, Heidi (2011) 'This Must Be the Place: Underrepresentation of Identity and Meaning in Climate Change Decision-Making'. *Global Environmental Politics* 11: 1–25.

Amundsen, Helene (2015) 'Place Attachment as a Driver of Adaptation in Coastal Communities in Northern Norway'. *Local Environment: The International Journal of Justice and Sustainability* 20 (3): 257–276.

Armas, Iuliana (2006) 'Earthquake Risk Perception in Bucharest, Romania'. *Risk Analysis* 26 (5): 1223–1234.

Belk, Russell W. (1992) 'Attachment to Possessions'. In: Altman, Irwin and Low, Setha M. (ed.): *Place Attachment. Human Behavior and Environment 12*. New York/London: Plenum Press, pp. 37–62.

Berger, Peter L. and Luckmann, Thomas (1966) *The Social Construction of Reality: A Treatise in the Sociology of Knowledge*. Garden City, New York: Doubleday.

Biesbroek, G. Robbert, Rob, J. Swart, Carter, Timothy R., Cowan, Caroline, Heinrichs, Thomas, Mela, Hanna, Morecroft, Michael D., and Rey, Daniela (2010) 'Europe Adapts to Climate Change: Comparing National Adaptation Strategies'. *Global Environmental Change* 20: 440–450.

Brooks, N., Adger, W.N., and Kelly, P.M. (2005) 'The Determinants of Vulnerability and Adaptive Capacity at the National Level and the Implications for Adaptation'. *Global Environmental Change* 15 (2): 151–163.

Broomell, Stephen B., Budescu, David V., and Por, Han-Hui (2015) 'Personal Experience with Climate Change Predicts Intentions to Act'. *Global Environmental Change* 32: 67–73.

Bulkeley, Harriet and Kern, Kristine (2004) 'Local Climate Change Policy in the United Kingdom and Germany'. Discussion Paper No. SP IV 2004–103. Abteilung Zivilgesellschaft, Konflikte und Demokratie des Wissenschaftszentrums Berlin für Sozialforschung.

Cappai, Gabriele (2005) 'Der interkulturelle Vergleich. Herausforderungen und Strategien einer sozialwissenschaftlichen Methode'. In: Srubar, Ilja, Renn, Joachim, and Wenzel, Ulrich (ed.) (2005): *Kulturen vergleichen. Sozial- und kulturwissenschaftliche Grundlagen und Kontroversen*. Wiesbaden: VS Verlag für Sozialwissenschaften, pp. 48–78.

Catton, William R. and Dunlap, Riley E. (1978) 'Environmental Sociology. A New Paradigm'. *The American Sociologist* 13: 41–49.

Chen, Zheng-Ao, Li, Qi, Liu, Lan-Cui, Zhang, Xian, Kuang, Liping, Jia, Li, and Liu, Guizhen (2015) 'A Large National Survey of Public Perceptions of CCS Technology in China'. *Applied Energy* 158: 366–377.

Christmann, Gabriela B. and Ibert, Oliver (2012) 'Vulnerability and Resilience in a Socio-Spatial Perspective'. *Raumforschung und Raumordnung* 70: 259–272.

Christmann, Gabriela B., Heimann, Thorsten, Mahlkow, Nicole and Balgar, Karsten (2012) 'Klimawandel als soziale Konstruktion?' *Zeitschrift für Zukunftsforschung 1* (urn:nbn:de:0009–32–34126). Available at: www.zeitschrift-zukunftsforschung.de/ausgaben/2012/1/3412/Christmann_et_al (accessed 28 October 2013).

Cotgrove, Stephen and Duff, Andrew (1981) 'Environmentalism, Values and Social Change'. *The British Journal of Sociology* 32 (1): 92–110.

Crate, Susan A. (2011) 'Climate and Culture: Anthropology in the Era of Contemporary Climate Change'. *The Annual Review of Anthropology 2011*. 40: 175–194. DOI: 0.1146/annurev.anthro.012809.104925 (accessed 29 October 2013).

Crona, Beatrice, Wutich, Amber, Brewis, Alexandra, and Gartin, Meredith (2013) 'Perceptions of Climate Change: Linking Local and Global Perceptions through a Cultural Knowledge Approach'. *Climatic Change* 119: 519–531.

Cunsolo Willox, Ashlee, Harper, Sherilee L., Ford, James D., Landman, Karen, Houle, Karen, Edge, Victoria L., and The Rigolet Inuit Community Government (2012) '"From this Place and of this Place:" Climate Change, Sense of Place, and Health in Nunatsiavut, Canada'. *Social Science & Medicine* 75: 538–547.

Descola, Philipp (2013) *Beyond Nature and Culture*. Chicago: University of Chicago Press.

Deuble, Max Paul and de Dear, Richard John (2012) 'Green Occupants for Green Buildings: The Missing Link?' *Building and Environment* 56: 21–27.

Devine-Wright, Patrick (2009) 'Rethinking NIMBYism: The Role of Place Attachment and Place Identity in Explaining Place-Protective Action'. *Journal of Community & Applied Social Psychology* 19: 426–441.

Devine-Wright, Patrick (2013) 'Think Global, Act Local? The Relevance of Place Attachments and Place Identities in a Climate Changed World'. *Global Environmental Change* 23: 69–69.

Dietz, Thomas, Fitzgerald, Amy, and Shwom, Rachael (2005) 'Environmental Values'. *Annual Review of Environment and Resources* 30: 335–372.

Dominicis, Stefano De, Fornara, Ferdinando, Cancellieri, Uberta Ganucci, Twigger-Ross, Clare and Bonaiuto, Marino (2015) 'We Are at Risk, and So What? Place Attachment, Environmental Risk Perceptions and Preventive Coping Behaviours'. *Journal of Environmental Psychology* 43: 66–78.

Douglas, Mary and Wildavsky, Aaron (1982) *Risk and Culture. An Essay on the Selection of Technological and Environmental Dangers*. Berkeley/Los Angeles/London: University of California Press.

Dunlap, Riley E. (2000) 'The Evolution of Environmental Sociology. A Brief History and Assessment of the American Experience'. In: Redclift, Michael and Woodgate, Graham (ed.) *The International Handbook of Environmental Sociology*. Cheltenham/Nothampton: Edward Elgar Publishing, pp. 21–39.

Dunlap, Riley E. and Catton, William R. (1979) 'Environmental Sociology'. *Annual Review of Sociology* 5: 243–273.

Dunlap, Riley E. and Mertig, Angela G. (1997) 'Global Environmental Concern: An Anomaly for Postmaterialism'. *Social Science Quarterly* 78 (1): 24–29.

Dunlap, Riley E. and Van Liere, Kent D. (1977) 'Land Ethic or Golden Rule: Comment on "Land Ethic Realized" by Thomas A. Heberlein'. *Journal of Social Issues* 33 (3): 200–207.

Dunlap, Riley E. and Van Liere, Kent D. (2008 [1978]) 'The New Environmental Paradigm'. *Journal of Environmental Education* 40 (1): 19–28.

Dunlap, Riley E., Grieneeks, J. Keith, and Rokeach, Milton (1983) 'Human Values and Pro-environmental Behavior'. In: Conn, W.D. (ed.) *Energy and Material Resources. Attitudes, Values, and Public Policy*, Boulder, CO: Westview, pp. 145–169.

Dunlap Riley E., Van Liere, Kent D., Mertig, Angela G., and Jones, Robert Emmet (2000) 'Measuring Endorsement of the New Ecological Paradigm: A Revised NEP Scale'. *Journal of Social Issues* 56 (3), 425–442.

Erhardt-Martinez, Karen, Rudel, Thomas K., Norgaard, Kari Marie, and Broadbent, Jeffrey (2015) 'Mitigating Climate Change'. In: Dunlap, Riley E. and Brulle, Robert J. (ed.): *Climate Change and Society. Sociological Perspectives*. New York: Oxford University Press, pp. 199–234.

Feitelson, Eran (1991) 'Sharing the Globe. The Role of Attachment to Place'. In: *Global Environmental Change*, December 1991: 396–406.

Franzen, Axel and Meyer, Reto (2004) 'Klimawandel des Umweltbewusstseins? Eine Analyse des ISSP 2000'. *Zeitschrift für Soziologie* 33 (2, April): 119–137.

Fried, Marc (2000) 'Continuities and Discontinuities of Place'. *Journal of Environmental Psychology* 20: 193–205.

Frommer, Birte (2010) *Regionale Anpassungsstrategien an den Klimawandel. Akteure und Prozess*. Darmstadt: IWAR.

Gerhards, Jürgen (2000) 'Regionale Identifikation und Loyalität zu Gruppennormen: Empirische Befunde aus der Umfrageforschung'. In: Metze, Regina, Mühler, Kurt, and Opp, Karl-Dieter (ed.): *Normen und Institutionen: Entstehung und Wirkungen*. Leipziger Soziologische Studien Volume 2. Leipzig: Universitätsverlag, pp. 115–136.

Gerhards, Jürgen and Lengfeld, Holger (2008) 'The Growing Remit of the EU in Environmental and Climate Change Policy and Citizens' Support across the Union'. BSSE Working Paper. – Berliner Studien zur Soziologie Europas No. 11. Available at: http://nbn-resolving.de/urn:nbn:de:0168-ssoar-110516 (accessed 11 October 2015).

Graumann, Carl F. (1983) 'On Multiple Identities'. *International Social Science Journal* 35: 309–321.

Greiving, Stefan and Fleischauer, Mark (2012) 'National Climate Change Adaptation Strategies of European States from a Spatial Planning and Development Perspective'. *European Planning Studies* 20 (1): 27–48.

Grunenberg, Heiko and Kuckartz, Udo (2003) 'Klimawandel, Zukunftsszenarien und Fortschrittsbegriff '. In: Umweltbundesamt (ed.): *Umweltbewusstsein im Wandel. Ergebnisse der UBA-Studie Umweltbewusstsein in Deutschland 2002*. Opladen: Leske & Budrich, pp. 169–187.

Hajjar, Reem and Kozak, Robert A. (2015) 'Exploring Public Perceptions of Forest Adaptation Strategies in Western Canada: Implications for Policy-Makers'. *Forest Policy and Economics* 61: 59–69.

Haller, Max (2002) 'Theory and Method in the Comparative Study of Values. Critique and Alternative to Inglehart'. *European Sociological Review* 18 (2): 139–158.

Heath, Yuko and Gifford, Robert (2006) 'Free-Market Ideology and Environmental Degradation: The Case of Belief in Global Climate Change'. *Environment and Behavior* 38 (1): 48–71.

Heberlein, Thomas A. (1977) 'Norm Activation and Environmental Action: A Rejoinder to R.E. Dunlap and K.D. Van Liere'. *Journal of Social Issues* 33 (3): 207–211.

Heimann, Thorsten (2017): *Klimakulturen und Raum. Umgangsweisen mit Klimawandel an europäischen Küsten*. Wiesbaden: Springer VS.

Heyd, Thomas and Brooks, Nick (2009) 'Exploring Cultural Dimensions of Adaptation to Climate Change'. In: Adger, Neil, Lorenzioni, Irene, and O'Brien, Karen: *Adapting to Climate Change: Thresholds, Values, Governance*, pp. 269–282.

Hitlin, Steven and Piliavin, Jane Allyn (2004) 'Values: Reviving a Dormant Concept'. *Annual Review of Sociology* 30: 359–593.

Hoffman. Andrew J. (2015) *How Culture Shapes the Climate Change Debate*. Stanford: Stanford University Press.

Hope, Aimie L.B. and Jones, Christopher R. (2014) 'The Impact of Religious Faith on Attitudes to Environmental Issues and Carbon Capture Storage (CCS) Technologies: A Mixed Methods Study'. *Technology in Society* 38: 48–59.

Howell, Rachel A. (2013) 'It's Not (Just) "the Environment, Stupid!" Values, Motivations, and Routes to Engagement of People Adopting Lower-Carbon Lifestyles'. *Global Environmental Change* 23: 281–290.

Huber, Joseph (2011) *Allgemeine Umweltsoziologie*. Wiesbaden: Springer VS.

Hulme, Mike (2009) *Why We Disagree about Climate Change*. Cambridge/New York: Cambridge University Press.

Hulme, Mike (2016) *Weathered: Cultures of Climate*. Los Angeles/London/New Delhi/Singapore/Washington DC/Melbourne: Sage.

Inglehart, Ronald (1995) 'Public Support for Environmental Protection: Objective Problems and Subjective Values in 43 Societies'. *PS Political Science and Politics* 28 (1): 57–72.

Inglehart, Ronald (2006) 'Mapping Global Values'. *Comparative Sociology* 5 (2/3): 115–136.

IPCC Intergovernmental Panel on Climate Change (1990) *Climate Change. The IPCC Response Strategies*. Geneva: WMO/UNEP. Available at: www.ipcc.ch/publications_and_data/publications_ipcc_first_assessment_1990_wg3.shtml (accessed 28 May 2014).

IPCC Intergovernmental Panel on Climate Change (1996) *Climate Change 1995. Impacts, Adaptations and Mitigation of Climate Change: Scientific-Technical Analyses*. Cambridge/New York/Melbourne: Cambridge University Press.

IPCC Intergovernmental Panel on Climate Change (ed.) (2007) *Climate Change 2007: Impacts, Adaptation and Vulnerability*. Contribution of Working Group II to the Fourth Assessment Report of the Intergovernmental Panel on Climate Change. Cambridge/New York: Cambridge University Press.

IPCC Intergovernmental Panel on Climate Change (ed.) (2014) *Climate Change 2014: Mitigation of Climate Change*. Contribution of Working Group III to the Fifth Assessment Report of the Intergovernmental Panel on Climate Change. Cambridge/New York: Cambridge University Press.

Jopp, Ryan, Mair, Judith, Delacy, Terry, and Fluker, Martin (2015) 'Climate Change Adaptation: Destination Management and the Green Tourist'. *Tourism Planning & Development* 12 (3): 300–320.

Kahan, Dan M., Jenkins-Smith, Hank and Braman, Donald (2010) 'Cultural Cognition of Scientific Consensus'. *Journal of Risk Research 2010*, iFirst Article: 1–28. Available at: http://papers.ssrn.com/sol3/papers.cfm?abstract_id=1549444 (accessed 3 November 2015).

Kluckhohn, Clyde (1967 [1951]) 'Values and Value-Orientations in the Theory of Action'. In: Parsons, Talcott and Shils, Edward (ed.): *Toward a General Theory of Action*. Cambridge, MA: Harvard University Press, pp. 388–433.

Knoblauch, Doris and Stelljes, Nico (2014) 'Regional Perspectives Concerning Climate Change and Coastal Adaptation. A Comparison between Mecklenburg-Western Pomerania and Schleswig-Holstein'. In: Martinez, Grit, Fröhle, Peter, and Meier, Hans-Joachim (ed.): *Social Dimensions of Climate Change Adaptation in Coastal Regions. Findings from Transdisciplinary Research*. München: Oekom, pp. 45–59.

Krömker, Dörthe (2004) *Naturbilder, Klimaschutz und Kultur*. Weinheim/Basel/Berlin: Beltz Verlag.

Krömker, Dörthe (2005) 'Naturbilder – ein kulturbedingter Faktor im Umgang mit dem Klimawandel'. *Umweltpsychologie* 9 (2): 146–171.

Kuckartz, Udo (2009) 'Europeans' Perceptions of Climate Change and Global Warming: A Micro-Macro Analysis'. Conference Paper presented at ESA2009 – 9th Conference of the European Sociological Association, Lisbon Research Network 12 'Environment and Society'. Available at: www.klimabewusstsein.de/dateien/Lisbon09.pdf (accessed 20 October 2015).

Kuruppu, Natasha (2009) 'Adapting Water Resources to Climate Change in Kiribati: The Importance of Cultural Values and Meanings'. *Environmental Science & Policy* 12: 799–809.

Lalli, Marco (1992) 'Urban-Related Identity: Theory, Measurement, and Empirical Findings'. *Journal of Environmental Psychology* 12: 285–303.

Landauer, Mia, Haider, Wolfgang, and Pröbstl-Haider, Ulrike (2013) 'The Influence of Culture on Climate Change Adaptation Strategies: Preferences of Cross-Country Skiers in Austria and Finland'. *Journal of Travel Research* XX (X): 1–15.

Latour, Bruno (1996) 'On Actor-Network Theory. A Few Clarifications'. In: *Soziale Welt* 47 (4): 396–381.

Leiserowitz, Anthony (2006) 'Climate Change Risk Perception and Policy Preferences: The Role of Affect, Imaginary, and Values'. *Climatic Change* 77: 45–72.

Leonard, Sonia, Parsons, Meg, Olawsky, Knut and Kofod, Frances (2013) 'The Role of Culture and Traditional Knowledge in Climate Change Adaptation: Insights from East Kimberley, Australia'. *Global Environmental Change* 23: 623–632.

Levey, Tania and Silver, Catherine B. (2006) 'Gender and Value Orientations – What's the Difference!? The Case of Japan and the United States'. *Sociological Forum* 21 (4): 659–691.

Low, Setha M. (1992) 'Symbolic Ties That Bind. Place Attachment in the Plaza Environmental Memories'. In: Altman, Irwin and Low, Setha M. (ed.): *Place Attachment. Human Behavior and Environment 12*. New York/London: Plenum Press, pp. 165–185.

Luckmann, Thomas (1999) 'Wirklichkeiten: individuelle Konstitution und gesellschaftliche Konstruktion'. In: Hitzler, Ronald, Reichert, Jo, and Schröer, Norbert (ed.): *Hermeneutische Wissenssoziologie. Standpunkte zur Theorie der Interpretation*. Konstanz: UVK, pp. 17–28.

Marcus, Clare Cooper (1992) 'Environmental Memories'. In: Altman, Irwin and Low, Setha M. (ed.): *Place Attachment. Human Behavior and Environment 12*. New York/London: Plenum Press, pp. 87–112.

Marshall, Nadine Anne, Dowd, Anne-Maree, Fleming, Aysha, Gambley, Clair, Howden, Mark, Jakku, Emma, Larsen, Carl, Marshall, Paul Augustine, Moon, Katie, Park, Sarah, and Thorburn, Peter John (2014) 'Transformational Capacity in Australian Peanut Farmers for better Climate Adaptation'. *Agronomy for Sustainable Development* 34: 583–591.

Martinez, Grit, Frick, Fanny, and Gee, Kira (2014) 'Socioeconomic and Cultural Issues in the Planning, Implementation and Transfer of Adaptation Measures to Climate Change. The Example of Two Communities on the German Baltic Sea Coast'. In: Martinez, Grit, Fröhle, Peter, and Meier, Hans-Joachim (ed.): *Social Dimensions of Climate Change Adaptation in Coastal Regions. Findings from Transdisciplinary Research*. München: Oekom, pp. 203–215.

Massey, Doreen (2006) 'Keine Entlastung für das Lokale'. In: Berking, Helmuth (ed.): *Die Macht des Lokalen in einer Welt ohne Grenzen*. Frankfurt/New York: Campus, pp. 25–31.

McCright, Aaron M. (2010) 'The Effects of Gender on Climate Change Knowledge and Concern in the American Public'. *Population and Environment* 32: 66–87.

Meyer, John W. (2007) 'Globalization. Theory and Trends'. *International Journal of Comparative Sociology* 48 (4): 261–273.

Mez, Lutz (1997) 'Klimaschutzpolitik als CO_2-Minderungspolitik. Dänemark und Deutschland im nationalen Alleingang'. In: Metz, Lutz and Jänicke, Martin (ed.): *Sektorale Umweltpolitik: Analysen im Industrieländervergleich*. Berlin: Edition Sigma, pp. 15–31.

Mohler, Peter P. and Wohn, Katrin W. (2005) 'Persönliche Wertorientierungen im European Social Survey'. In: *ZUMA-Arbeitsbericht* No. 2005/01, 1–20.

Mühler, Kurt and Opp, Karl-Dieter (2004) *Region und Nation. Zu den Ursachen und Wirkungen regionaler und überregionaler Identifikation*. Wiesbaden: VS für Sozialwissenschaften/GWV Fachverlage.

Nilsson, Andreas and Biel, Anders (2008) 'Acceptance of Climate Change Policy Measures: Role Framing and Value Guidance'. *European Environment* 18 (4): 203–215.

Nilsson, Andreas, Borgstede, Chris von, and Biel, Anders (2004) 'Willingness to Accept Climate Change Strategies: The Effect of Values and Norms'. *Journal of Environmental Psychology* 24, 267–277.

O'Brien, Karen (2009) 'Do Values Subjectively Define the Limits to Climate Change Adaptation?'. In: Adger, Neil, Lorenzioni, Irene and O'Brien, Karen: *Adapting to Climate Change: Thresholds, Values, Governance*, pp. 164–181.

O'Brien, Karen and Wolf, Johanna (2010) 'A Values-Based Approach to Vulnerability and Adaptation to Climate Change'. In: *Wires Climate Change 2010*. Available at: http://wires.wiley.com/WileyCDA/WiresArticle/wisId-WCC30.html (accessed 11 October 2015).

Otto-Banaszak, Ilona, Matczak, Piotr, Wesseler, Justus, and Wechsung, Frank (2011) 'Different Perceptions of Adaptation to Climate Change: A Mental Model Approach Applied to the Evidence from Expert Interviews'. *Regional Environmental Change* 11: 217–228.

Parsons, Talcott (1961) 'An Outline of the Social System'. In: Parsons, Talcott, Shils, Edward, Naegele, Kaspar D., and Pitts, Jesse R. (ed.): *Theories of Society. Foundations of Modern Sociological Theory*. New York: The Free Press, pp. 30–79.

Pellow, David N. and Brehm, Hollie Nyseth (2013) 'An Environmental Sociology for the Twenty-First Century'. *Annual Review of Sociology* 39: 229–250.

Pendergraft, Curtis A. (1998) 'Human Dimensions of Climate Change: Cultural Theory and Collective Action'. *Climatic Change* 39: 643–666.

Persson, Johannes, Sahlin, Nils-Eric, and Wallin, Annika (2015) 'Climate Change, Values, and the Cultural Cognition Thesis'. *Environmental Science and Policy* 52: 1–5.

Petzold, Knud (2013) 'Mobilität und Identität. Eine theoretische und empirische Exploration am Beispiel multilokalisierter Akteure'. In: Hömke, Maik (ed.): *Mobilität und Identität*. Wiesbaden: Springer VS, pp. 79–122.

Price, Jennifer C., Walker, Ian A., and Boschetti, Fabio (2014) 'Measuring Cultural Values and Belief about Environment to Identify their Role in Climate Change Responses'. *Journal of Environmental Psychology* 37: 8–20.

Proshansky, Harold M. (1978) 'The City and Self-Identity'. *Environment and Behavior* 10 (2): 147–169.

Proshansky, Harold M., Fabian, Abbe K., and Kaminoff, Robert (1983) 'Place-Identity: Physical World Socialization of the Self'. *Journal of Environmental Psychology* 3: 57–83.

Reckwitz, Andreas (2000) *Die Transformation der Kulturtheorien*. Weilerswist: Velbrück.

Richter, Ralph (2013) 'Ortsbezogene Identität. Die kognitive Repräsentanz von Orten im Zeichen zunehmender Wohnmobilität'. In: Hömke, M. (ed.): *Mobilität und Identität*. Wiesbaden: Springer VS, pp. 11–30.

Rippl, Susanne and Seipel, Christian (2008) *Methoden kulturvergleichender Forschung*. Wiesbaden: VS Verlag für Sozialwissenschaften.

Roose, Jochen (2010a) *Der Index kultureller Ähnlichkeit. Konstruktion und Diskussion*. BSSE Working Paper No. 21, Berlin.

Roose, Jochen (2010b) *Vergesellschaftung an Europas Binnengrenzen. Eine vergleichende Studie zu den Bedingungen sozialer Integration*. Wiesbaden: VS Verlag für Sozialwissen-schaftenGWV Fachverlage.

Rössel, Jörg (2006) 'Ronald Inglehart: Daten auf der Suche nach einer Theorie: Analysen des weltweiten Wertewandels'. In: Moebius, Stephan and Quadflied, Dirk (ed.): *Kultur. Theorien der Gegenwart*. Wiesbaden: VS Verlag für Sozialwissenschaften, pp. 722–733.

Safi, Ahmad Saleh, Smith, William James Jr and Liu, Zhnongwei (2012) 'Rural Nevada and Climate Change: Vulnerability, Beliefs, and Risk Perception'. *Risk Analysis* 32 (6): 1041–1059. DOI: 10.1111/j.1539-6924.2012.01836.x.

Scannell, Leila and Gifford, Robert (2013) 'Personally Relevant Climate Change: The Role of Place Attachment and Local Versus Global Message Framing in Engagement'. *Environment and Behavior* 45 (1): 60–85.

Schäfers, Bernhard (2006) 'Soziales Handeln und seine Grundlagen: Normen, Werte, Sinn'. In: Korte, Hermann and Schäfers, Bernhard (ed.): *Einführung in Hauptbegriffe der Soziologie*. Wiesbaden: VS Verlag für Sozialwissenschaften, pp. 25–43.

Schipper, E. Lisa F. (2010) 'Religion as an Integral Part of Determining and Reducing Climate Change and Disaster Risk: An Agenda for Research'. In: Voss, Martin (ed.): *Der Klimawandel. Sozialwissenschaftliche Perspektiven*. Wiesbaden: VS Verlag für Sozialwissenschaften, pp. 377–393.

Schroer, Markus (2006) *Räume, Orte, Grenzen. Auf dem Weg zu einer Soziologie des Raums*. Frankfurt am Main: Suhrkamp.

Schwartz, Shalom H. (1992) 'Universals in the Content and Structure of Values: Theoretical Advances and Empirical Tests in 20 Countries'. *Advances in Experimental Social Psychology* 25: 1–65.

Schwartz, Shalom H. (1999) 'A Theory of Cultural Values and Some Implications for Work'. *Applied Psychology: An International Review* 48 (1): 23–47.

Schwartz, Shalom H. (2003) 'A Proposal for Measuring Value Orientations across Nations. Chapter 7: Questionnaire Development Package of the European Social Survey'. Available at: www.europeansocialsurvey.org/docs/methodology/core_ess_questionnaire/ESS_core_questionnaire_human_values.pdf (accessed 6 October 2015).

Schwartz, Shalom. H. (2012) 'An Overview of the Schwartz Theory of Basic Values'. *Online Readings in Psychology and Culture*, 2 (1). Available at: http://dx.doi.org/10.9707/2307-0919.1116 (accessed 8 October 2015).

Schweizer, Sarah, Davis, Shawn, and Thompson, Jessica Leigh (2013) 'Changing the Conversation about Climate Change: A Theoretical Framework for Place-Based Climate Change Engagement'. *Environmental Communication* 7 (1): 42–62.

Steel, Brent S., Pierce, John C.. Warner, Rebecca L., and Lovrich, Nicholas P. (2015) 'Environmental Value Considerations in Public Attitudes about Alternative Energy'. *Environmental Management* 55 (3): 634–645.

Steg, Linda, Dreijerink, Lieke and Abrahamse, Wokje (2005) 'Factors Influencing the Acceptability of Energy Policies: A Test of VBN-Theory'. *Journal of Environmental Psychology* 25: 415–425.

Stern, Paul (2000) 'Toward a Coherent Theory of Environmentally Significant Behaviour'. *Journal of Social Issues*, 56 (3): 407–424.

Stern, Paul C. and Dietz, Thomas (1994) 'The Value Basis of Environmental Concern'. *Journal of Social Issues* 50 (3): 65–84.

Stern, Paul, Dietz, Thomas, Abel, Troy, Guagnano, Gregory A., and Kalof, Linda (1999) 'A Value-Belief-Norm Theory of Support for Social Movements: The Case of Environmentalism'. *Human Ecology Review* 6 (2): 81–97.

Thompson, Michael (2003) 'Cultural Theory, Climate Change and Clumsiness'. *Economic and Political Weekly* 38 (48, January): 5107–5112.

Treinen, Heiner (1965) 'Symbolische Ortsbezogenheit. Eine soziologische Untersuchung zum Heimatproblem'. *Kölner Zeitschrift für Soziologie und Sozialpsychologie* 17: 73–97.

Van Deth, Jan W. and Scarbrough, Elinor (1995) 'The Concept of Values'. In: Van Deth, Jan W. and Scarbrough, Elinor (ed.): *The Impact of Values. Beliefs in Government* Volume 4. Oxford/New York: Oxford University Press, pp. 21–47.

Vorkinn, Marit and Riese, Hanne (2001) 'Environmental Concern in a Local Context. The Significance of Place Attachment'. *Environment and Behavior* 33 (2): 249–263.
Weber, Max (2010) 'Zwischenbetrachtung: Theorie der Stufen und Richtungen Religiöser Weltablehnung'. In: Weber, Max: *Religion und Gesellschaft. Gesammelte Aufsätze zur Religionssoziologie*. Frankfurt am Main: Zweitausendeins, pp. 531–559.
Weber, Melanie (2008) *Alltagsbilder des Klimawandels. Zum Klimabewusstsein in Deutschland.* Wiesbaden: VS Verlag für Sozialwissenschaften/GWV Fachverlage.
Weichhart, Peter (1990) *Raumbezogene Identität. Bausteine einer Theorie räumlich-sozialer Kognition und Identifikation.* Stuttgart: F. Steiner.
Weichhart, Peter, Weiske, Christine, and Werlen, Benno (2006) *Place Identity und Images. Das Beispiel Eisenhüttenstadt.* Wien: Wohlschlägl.
Weiss, Thomas G. and Burke, Martin J. (2011) 'Legitimacy, Identity and Climate Change: Moving from International to World Society?' *Third World Quarterly* 32 (6): 1057–1072.
Williams, Sarah and Schaefer, Anja (2012) 'Small and Medium-Sized Enterprises and Sustainability: Managers' Values and Engagement with Environmental and Climate Change Issues'. *Business Strategy and the Environment* 22: 173–186.
Wolf, Johanna, Allice, Illana, and Bell, Trevor (2013) 'Values, Climate Change, and Implications for Adaptation: Evidence from two Communities in Labrador, Canada'. *Global Environmental Change* 23: 548–562.
Zürn, Michael (2008) 'Governance in einer sich wandelnden Welt: eine Zwischenbilanz'. In: Schuppert, Gunnar Folke and Zürn, Michael (ed.): *Governance in einer sich wandelnden Welt. PVS – Politische Vierteljahresschrift, Special Issue*, 41/2008. Wiesbaden: VS Verlag für Sozialwissenschaften, pp. 553–580.

6 On the historical construction of political–administrative climate-change frameworks in the research area

In order to understand the political–administrative context of climate-knowledge internalisation processes, this chapter will take a closer look at its historical construction in the research area. In the discourse on globalisation (see Sections 2.2 and 5.3) there has been mention of a 'deterritorialisation of the political' and the 'end of the nation state' (Schroer 2006: 195). As already discussed in Section 5.3, the global climate discourse could lead to common problem constructions and action strategies, and so also contribute to cultural homogenisation processes. It is therefore necessary to explore more closely the central developments at the global (Section 6.1) and European levels (Section 6.2) over the last decades. This chapter will, in addition, consider the development of national climate politics (Section 6.3).

6.1 Global level

The objective of stabilising global greenhouse gas concentrations through global climate mitigation has been described as a global 'tragedy of the commons' (Wilson and Piper 2010: 45; the term borrows from Hardin 1968). From this point of view, inadequate global regulation of greenhouse gas emissions leads to further increases in atmospheric concentrations. The United Nations sided with this perspective and therefore wanted to create a globally applicable regulatory framework for use with the atmosphere, more specifically for the emission of greenhouse gases. In 1992 the participants of the UN Conference in Rio adopted the United Nations Framework Convention on Climate Change (UNFCCC 2014a, 1992). There, however, compliance dates and scales were not specified, with future negotiations intended to specify more concrete terms.

In December 1997, participants of the third Conference of the Parties (COP3) in Kyoto finally adopted an additional protocol naming concrete climate-mitigation targets for the period 2008–2012 (the 'Kyoto Protocol', see Art. 3, UNFCCC 1998: 3). By this time, industrial nations named as the most responsible for emissions should have decreased their carbon-dioxide emissions by a total of 5%,[1] taking 1990 as a reference year (ibid.). Annex B of the protocol earmarked different proportions for each participating country, in order to offer room for individual negotiation.

The Kyoto Protocol was officially able to come into force in February 2005, once 55 of the then 191 participating nations[2] had ratified it (UNFCCC 2014b). As the protocol was meant to expire by 2012, negotiations for a successor protocol soon began. Shortly before its expiration in December 2012, a last-minute extension of the Kyoto Protocol to 2020 ('Kyoto II') was agreed in Doha, with adapted reduction figures for the participating parties (UNFCCC 2012).

The countries participating in Kyoto II were responsible for less than 15% of worldwide emissions, however (ibid.). Canada, Japan, New Zealand, and Russia had backed out of the protocol, and the USA, the world's greatest emitter, had never ratified it (UNFCCC 2012).[3] By the time of the publication of the IPCC Report of Working Group I in 2013, the amounts of atmospheric greenhouse gases had continued to rise. Global efforts as part of the Framework Convention on Climate Change were unable to halt the increasing global trend and it remains to be seen whether the accord signed for the period after 2020 will be better able to meet its targets (see also Section 4.2).

The outcome of this is that adaptation measures have gained in significance. In the 1992 Framework Convention, participating nations had obligated themselves in Article 4 not only to climate mitigation, but also to developing adaptation measures for altered climate conditions (see Art. 4, UNFCCC 1992: 6). It would take ten years, however, before this achieved increasing worldwide significance. Climate adaptation had been discussed by the IPCC in the 1990s, but rather critically as an opponent to climate mitigation. With the third assessment report from 2001, a new perspective gained traction: the impacts of climate change were already being observed, making adaptation measures an urgent necessity. Both strategies have since been increasingly viewed by the IPCC as mutually complementary (Beck *et al.* 2011: 16ff.; UNFCCC 1992).

In 2001 the participants of COP7 in Marrakesh compiled the first guidelines for the achievement of adaptation programmes for developing nations (UNFCCC 2015a). At the meeting of the parties in 2005, the design of an adaptation fund was put in concrete terms. The Nairobi working programme was also adopted at this point, its aim being to assist nations in improving their understanding of the impacts of climate change (ibid.; Eichhorst and Madry 2013). In 2008 a supervisory body for the funds was established; since 2011 this has been permanently located in Bonn. Climate adaptation has thus for a long time not only taken place locally, but it has also seen an increasing specialisation of institutions at the global level.

At the COP16 in Cancún in 2010, the parties first formulated the target of giving climate adaptation and mitigation equal priority (UNFCCC 2015b; Eichhorst and Madry 2013). The IPCC finally named the 'climate-resilient pathways' to be striven for in the plural: 'Climate-resilient pathways are sustainable-development trajectories that combine adaptation and mitigation to reduce climate change and its impacts' (IPCC 2014: 87). This marked an important turning point in global discourse. The IPCC now emphasised the diversity of possible measures, various combinations of which should be made possible. This indicates that future consideration of the different priorities of actors in the ways of dealing with climate change will gain in significance.

6.2 European level

Historically viewed, the thematic development of the European discourse on climate change resembles its global counterpart, as the European Union itself focused initially on climate-mitigation measures (see Wilson and Piper 2010: 52 and Section 6.1). In global climate negotiations under the United Nations Framework Convention, the EU continues to be an individual actor. As early as the 1990s, it had negotiated a common reduction target for all EU nations (8% by 2012 compared with 1990; see also Section 4.2).[4]

The European Climate Change Programme (ECCP) was launched in 2000, for the years 2000 to 2004, with the awareness that achievement of the targets would require systematic efforts not only for the individual EU member states, but also for the institutions of the European Union as a whole (EC 2000). An important outcome in 2003 was the joint decision to introduce a European Emissions Trading Scheme (EU ETS).

The second European Programme for Climate Change (ECCP II), beginning in 2005, also increasingly established Working Groups to implement climate-adaptation measures (Wilson and Piper 2010: 53; EC 2005). The publication in 2009 of a Green Paper (EC 2007) and its core findings in a White Paper (EC 2009c) was intended to initiate a broad consultation process on climate-change adaptation. It also addressed coastal areas as a particular problem area and proposed cross-border strategies to implement adaptation processes, for example via transnational arrangements for integrated coastal zone management (ICZM) or joint planning for maritime spatial planning (ibid.: 13).

Central to the 2013 European adaptation strategy is the financial and communicative support of measures undertaken by member states, measures to generate knowledge through scientific programmes, and the promotion of specific sectors (EC 2013a: 4ff.). Decisions in the field of spatial development are often delegated to individual member states. The EU Commission has included the issue of climate adaptation in the main funding programmes for the period 2014–2020 (ibid.: 11f.). The promotion of transnational adaptation projects is seen as an important task of the EU (ibid.: 5). Roughly summarised, the EU is less concerned with advocating or rejecting concrete forms of climate adaptation, such as local protection, integration, or retreat strategies relating to flooding (see Section 4.3). More central is the elaboration of funding possibilities and the framework conditions meant to promote decisions by individual member states on the implementation of various measures by spatial development actors.

By contrast, the EU has been able to formulate more extensive targets for climate mitigation, promoting concrete strategies such as sufficiency, efficiency, and consistency (see Chapter 4). A climate and energy package was adopted in 2009, at the heart of which is the so-called '20–20–20 targets' (EC 2013b: 4ff.).

This is a voluntary commitment programme to accelerate three strategies in the period to 2020: first, to reduce greenhouse gas emissions in the EU by 20% compared to 1990; second to increase the share of renewable energies to 20% of the gross energy consumption; and third to improve energy efficiency in EU countries by 20% (ibid.). Within this framework, directives on renewable energy and efficiency have been passed dividing the 20% targets among the member states, setting national targets and sanctions for non-compliance (EC 2012, 2009b).

The EU is currently planning for the period after 2020. At the beginning of 2014, the Commission presented proposals for new targets for this period (EC 2014, 2013b). CO_2 emissions should accordingly be reduced by 40% compared to 1990 levels. More ambitious targets up until 2050 have already been formulated (between 80% and 95%) (EC 2014).

It should be noted that the EU has taken an active international role, pushing forward global commitments and attempting to implement European voluntary commitments. It will later be shown, however, that its outward unity does not reflect an internal one. There are differences in both the implementation and the acceptance of climate strategies between different actors.

6.3 National level

6.3.1 *Germany: 'energy revolution' and 'federal fragments'*

Here we shall take a closer look at the climate-mitigation and adaptation policies of the Federal Republic of Germany.

Climate-mitigation policies in Germany

In 2002, within the framework of Kyoto burden-sharing within the European Union, Germany pledged itself to a 21% reduction in its greenhouse gas emissions by 2012 (compared with 1990) (EC 2002).[5] Together with Denmark, Germany thus set the second-highest target for the relative reduction of emissions among all the countries involved, with only Luxemburg formulating a higher value of 28%. According to the UNFCCC, by the end of 2012 a reduction of 24.8% had been achieved (UNFCCC 2014c: 9), suggesting that Germany had not only met its target, but surpassed it. A further reduction of 14% in comparison to 2005 is envisaged for the current period up to 2020, and in this context the German government has communicated a reduction target of 40% in comparison to 1990 (BMUB 2014b: 7; EC 2009a).

To achieve these targets, measures explicitly related to climate-mitigation have been discussed and implemented in national policy since the 1990s. In 1990 the German government set the climate-political goal that energy-related CO_2 emissions should be reduced by between 25% and 30% by 2005 (Mez 1997: 26). The main focus at the time lay above all on measures for energy saving and efficiency, and the substitution of high-carbon energy sources such as coal and oil for low-carbon alternatives such as gas (ibid.: 26). The first subsidies for individual renewable energy sources such as solar-power plants were also made available at this point (ibid.: 29).

Emissions of the most important Kyoto greenhouse gases have been in decline in Germany since 1990. A peculiarity of Germany should be taken into account when considering these emission levels, however. The Kyoto Protocol makes explicit mention of 'countries that are undergoing the process of transition to a market economy' (UNFCCC 1998: 20). Post-socialist countries were provided special rights in order not to unnecessarily impede the process of transformation, including, for example, Poland (see Section 6.3.4). Germany is not explicitly listed in the Kyoto Protocol as a transitioning nation (ibid.). It should be taken into account, though, that Germany also had to cope with transformation processes following reunification. In the post-socialist states, many industrial operations had to be closed (Fritsch *et al.* 2010). The per capita emissions subsequently sank between 1990 and 1995 by around 13%.[6] This can be ascribed not only to national climate policy, however, but also to a large extent to the transformation processes in place (Kühne 2002: 83ff.). So although not given transitional status in the Kyoto Protocol, the choice of 1990 as its main base year was to Germany's advantage as it nevertheless benefitted from the results of its transformation.

In 2007 an extended, integrated programme for energy and climate mitigation appeared that aimed at new emissions targets and measures by 2020 (Bundesregierung 2007) and provided a basis for post-Kyoto climate negotiations (see Section 6.1). In 2014 it emerged that Germany might not reach its targets by 2020, leading to the adoption of an additional programme, the so-called *Climate-Migitation Action Programme 2020 (Aktionsprogramm Klimaschutz 2020)* (BMUB 2014b). All programmes concentrate on the various central strategies discussed in Chapter 4. On the whole, efficiency and consistency strategies dominate, i.e. strategies for ecological modernisation, above all those which target an increase in the share of renewable energies and improvements in energy efficiency.

Let us first consider renewable energies. By 2012 the German energy market already derived a quarter (26%) of supplies from renewable energy. In the 2013 coalition agreement, additional targets of 40–45% by 2025 and 55–60% by 2035 were specified (CDU/CSU/SPD 2013: 37). The German government has thus aimed at a swift energy-transition process, and in the meantime all involved parties have described these transition targets with recourse to the concept of an 'Energiewende' ('energy revolution'; ibid.: 36), a term from the early debates on ecological modernisation of the 1970s and 80s.

This large consensus between the major parties on a swift energy transition would hardly have been foreseeable before 2011. Until this date there had continually been large differences both within the population and between the political parties regarding the assessment of the composition and development of the German energy mix. The debate in Germany was initially less connected to climate discourse and much more to the use of nuclear energy. As Roose points out, since the 1970s, the dispute had regularly resulted in much greater conflicts between different social groups than climate-related questions (Roose 2012: 91). In 2002 a 'law for the orderly termination of nuclear-energy use' (BMJ 2002) was adopted by the then coalition of the SPD and Greens. This made provision for the first time for a medium-term phasing out of nuclear energy and set maximum power limits on the running time of nuclear reactors (ibid.). In order to once again extend the use of the comparatively cost-effective and virtually climate-neutral nuclear plants, a new coalition of the CDU/CSU and FDP in 2010 initially extended their lifespan (BMJ 2010). This position changed in March 2011, however, when an earthquake and tsunami on the Japanese east coast led to a core meltdown in the nuclear power station near the city of Fukushima (Maubach 2014: 9ff.). The coalition then withdrew the lifespan extension, deciding instead to withdraw from nuclear energy as soon as possible.[7]

To make the production and purchase of renewable energies possible, a law on the sale of electricity to the grid (the so-called 'Stromeinspeisungsgesetz') was adopted in 1991. In 2000 it was transferred to the renewable energy law ('Erneuerbare-Energien-Gesetz'), before undergoing a final revision in 2014 (BMJV 2014). Its most important mechanisms are the supply and purchase requirements as well as fixed feed-in tariffs. Regarding renewable energy in Germany, it should be noted that the most important parties have appeared to agree on the fundamental direction of development since at least 2011.

With regard to *efficiency strategies* in Germany, the current government's energy concept provides for a 20% reduction in primary energy consumption compared to 2008, which equates to a 2.1% annual increase in energy efficiency. Recently, however, it has become apparent that actors involved in Germany have been unable to agree on the uniform implementation of the directive (Schultz 2012). The two-year implementation deadline was not met by 2014, resulting in the EU launching an infringement procedure (Dehmer 2014). This demonstrates that within German political discourse, efficiency strategies ultimately held a less broad consensus among the actors involved than increasing the share of renewable energies.

Climate-adaptation policies in Germany

Just as at the global and European level, the development in Germany of a national adaptation discourse in politics and administration lagged behind the discussions on climate mitigation. It gained momentum in the mid-2000s, and in 2005 a *national vulnerability study* commissioned by the Federal Environment Agency (Umweltbundesamt) was published to determine the existing and projected impacts of climate change on Germany (UBA 2005). In the same year the government for the first time pushed for the establishment of a national climate-adaptation programme. They justified this step with the argument that it was clear climate-mitigation measures alone would 'not lead to a defence against a possible natural, human-influenced climate change' (BMU 2005: 40). In December 2008, the federal cabinet finally adopted the *German Strategy for*

Adaptation to Climate Change (*Deutsche Anpassungsstrategie an den Klimawandel, DAS*) (Bundesregierung 2008). This was not judicially binding and was meant more to deliver a national framework for medium-term, participative adaptation processes (ibid.: 4; BMVBS 2010: 101), allowing for future adaptive actions to be made by the government, while serving to orient other actors, both in politics and administration at the state, regional, and municipal levels, and in the economy and civil society (see also BMVBS 2010: 101).

The adaptation strategy allowed for the development of a national *action plan* for implementing concrete strategies (Bundesregierung 2011). This focuses above all on actions by the federation that would be supplemented by the federal states and municipalities, and divides these actions under four pillars. Under the framework of the first pillar ('supply knowledge, inform, empower'; ibid.: 16) the research resources of the federation are increased to enable a broad institutional basis for the generation of knowledge through climate research. In addition, institutions are created or given support to enable communication and transfer of information among diverse social groups.[8] The second pillar ('the regulatory framework of the federal government'; ibid.: 30) allows for the testing of legal frameworks and regulation, as well as the implementation of support programmes and economic incentives. The third pillar is concerned with 'activities under the direct responsibility of the federation' (ibid.: 34f.). Finally, the fourth pillar ('international responsibility'; ibid.: 36) names the support of international cooperation, such as the implementation of transnational European adaptation projects, or in the area of international research and development.

In Germany's national adaptation process spatial development themes are of high, but not central importance, standing alongside health, tourism, and the financial sector. The strategic implementation thus follows the federal structure of the country. Coordination and information offerings in particular are expedited by the federal government, alongside the judicial framework legislation. Subsidies are also provided that can be participatively accessed by different actors. Certainly, within the framework of the adaptation strategy the government identifies and researches possible measures for spatial development, including the selection of specific implementation options on the ground, such as the issue of retreat, protection, or integration in the handling of water (see Section 4.3), but on the whole it is actors at the levels below which must deal with these issues.

Additionally, different legal frameworks exist across the federal states, regions, and municipalities. In February 2013, the Ministerial Conference of the Federal States agreed a circular resolution recommending the formal incorporation of risk provisions for flood and coastal protection, heat impact, and water scarcity into regional planning in order to stabilise spatial development measures at the national level (MKRO 2013: 1). Several municipalities have drafted concepts for adaptation (see for example Hansestadt Lübeck 2014, Hansestadt Rostock 2013). However, these apply to only a few of the coastal towns on the North Sea and Baltic coasts considered in this study (see Section 7.1).

6.3.2 The Netherlands: cooperation and innovation

Fischer and Foißner (2002: 153) have described how the 'constant threat' of the sea has regularly required consensual spatial planning related decisions in the Netherlands. The example demonstrates that spatial development and the associated handling of water is traditionally of high significance in the Netherlands. This can also be observed in how climate change is dealt with.

Climate-mitigation policies in the Netherlands

In the first EU Kyoto Burden-Sharing arrangement, the Netherlands committed itself to a 6% reduction in greenhouse gas emissions by 2012 compared to 1990 (EC 2002).[9] In doing so, its government representative negotiated a reduction rate of less than the EU average of 8% (ibid.). According to the UNFCCC, a value of 9.5% had been attained by 2012 (UNFCCC 2014c: 9). For the next Kyoto period, until 2020, a target of 16% in comparison to 2005 has been envisaged (EC 2009a).

The efforts of Dutch climate-mitigation policy can be traced back to before the Kyoto negotiations. Pettenger (2007: 57f.) counts the Netherlands among those countries that identified climate change as a serious problem very early on in their environmental policy. This occurred in 1989 within the framework of implementing a system of national environmental planning, in the first so-called *National Environmental Policy Plan (NEPP) (Nationaal Milieubeleidsplan, NMP)*.

With the development of the NEPPs, the Dutch may be considered innovative pioneers in the implementation of national environmental planning instruments, as NEPP elements were later adopted by other countries in a similar form for environmental policy planning processes, for example in the USA and Austria (ibid.). The implementation of the first NEPP occurred within the conceptual context of the 1987 Brundtland Report, at which time the central target was to achieve sustainable development of the Netherlands within one generation (Busch and Jörgens 2014: 244; Luitwieler 2000: 15; Tellegen 1989: 340). This would be accomplished above all through the improvement of resource efficiency as well as through the decoupling of economic growth and damage to the environment. The NEPPs are thus also considered an early form of political programme for ecological modernisation (Wilson and Piper 2010: 90 and Section 4.2).

In the first NEPP a stabilisation of CO_2 emissions by 2000 was earmarked. In the following years the plan was tightened, striving for a 3% reduction (Pettenger 2007: 58; Kok *et al.* 2002: 6; Tellegen 1989: 344).

This target could not be met, however, and in the 1990s, CO_2 emissions even rose (Pettenger 2007: 61; OECD 2003: 21). A possible reason for this could be that the NEPPs themselves were formally still not legally binding (Busch and Jörgens 2014: 245), although several energy-efficiency agreements (Long-Term Agreements, or LTAs) between government and the economy had shown measurable reduction effects (Rietbergen *et al.* 2002). Several laws were even passed[10] that could have further targeted greenhouse gas reductions, but the regulations were obviously not sufficient in themselves for attaining the goals set. Authors such as Pettenger are critical of the early efforts: 'It should be noted that the rhetoric has not always been matched to concrete policies and implementation' (Pettenger 2007: 57).

At first glance, this observation appears to hold into the 2000s. At the end of the first Kyoto period in 2012, the share of fossil fuels in gross final energy consumption was still over 90%. This raises the question of how the first burden-sharing target of a 6% reduction in CO_2 emissions might be achieved. In addition, there is the question of which measures should be discussed for the even more ambitious reduction targets for 2020.

Soon after the formulation of the Kyoto Protocol in 1998, the government passed the so-called *National Climate Policy Implementation Plan* (VROM 2009: 56). An important aspect of Dutch climate policy relates to the pursuit of measures not only domestically, but also by balancing a large part of the country's emissions through the use of market-oriented mechanisms abroad (clean development mechanisms, joint implementation, emissions trading, etc.) (ibid.).

Looking more closely at the level of renewable energies in the energy mix, the high share of fossil fuels in the Dutch energy mix, as already mentioned, is conspicuous. Renewable energies stand at a comparatively low value of 4% (see Section 4.3.1). Dutch energy consumption has mainly depended on oil and gas into the 2010s. In this context, it appears significant that even in 2015 the government still envisaged maintaining a high share of gas in the future, relying on its own resources for as long as possible, since the Netherlands has extensive gas fields both underground and offshore in the North Sea (Government of the Netherlands 2015).

With the 2009 Renewable Energy Directive, however, the Netherlands is now bound for the first time by EU law, and is therefore now obligated to increase its share of renewable energies from 2.4% in 2005 to 14% of its gross energy consumption by 2020 (EC 2009b).

The slow developmental history of the feeding-in of renewable energies to the electricity networks began to take shape in the Netherlands from the late 1990s. With the Electricity Act of 1998 producers of renewable energies were for the first time given feeding-in priority over conventional energies (van Leeuwen 2013).

Alongside an emphasis on flexible Kyoto mechanisms, the Dutch reduction target was largely achieved through efficiency. Efficiency-measure agreements (especially long-term agreements) between governments and economic representatives had been reached as early as 1992. In 2007 the government published an energy-efficiency action plan, which was created within the framework of an EU energy-efficiency directive from 2006 (IEA 2013). The government also set a new target in the 'Clean and Efficient' programme from 2007: 'The Netherlands is aiming to have the most efficient and cleanest energy systems in Europe by the year 2020' (VROM *et al.* 2007d: 3). The Netherlands' internal target was now to achieve an annual energy-efficiency increase of 2% between 2011 and

2020 (VROM *et al.* 2007d: 17). It should be noted though, that even by the end of 2011, the energy intensity of the Dutch economy was only just below the EU average (Donat *et al.* 2014: 15).

As part of the implementation of the 2009 EU directive, efficiency targets have become an international requirement. In 2011, the government coalition initiated a broad-based social negotiation process to achieve the greatest possible social consensus over new, long-term goals and measures (SER 2013: 1). The result was the 'Energy Agreement for Sustainable Growth ('Energieakkoord voor duurzame groei') published in September 2013 (ibid.: 3), in which the participating partners communicated previous and new common goals. In the energy accord, some of the targets set in the 2000s were revised downwards.

In September 2014, the national government adopted a so-called 'Climate Agenda' with the subtitle 'Resilient, Prosperous and Green' (I&M 2014) in order to be able to implement the targets of the 'energy accord' at the national level. Here, in addition to climate-mitigation targets, it is noticeable that adaptation measures are also of high importance. Authors such as Stead, analysing the Dutch climate discourse, have concluded that: 'In general, there has recently been more emphasis on adaptation than mitigation' (Stead 2014: 25).

Climate-adaptation policies in the Netherlands

Climate-adaptation discourse also gained momentum in the Netherlands during the first decade of the new millennium. Two central paths of development can be identified: first, the construction and continuation of a *national adaptation strategy* that initially focused on spatial development issues but which is now increasingly extending to other sectors; and second, the implementation of a new *Delta programme*, that once again deals in more detail with long-term resilience safeguards against water-related climate impacts (EC/EEA 2015, Stead 2014: 22ff.).

In 2007 the *Dutch national adaptation strategy* (*Nationale Adaptatiestrategie*, *NAS*) was published under the title 'Maak ruimte voor klimaat!' ('Make room for the climate!', VROM *et al.* 2007c). It was issued jointly by all national ministries concerned with spatial development, representatives of the water boards, provinces, and municipalities (VROM *et al.* 2007c, 2007b: 2). The strategy was intended to serve as a starting point for the discussion, development, and implementation of comprehensive climate-adaptation packages for the Netherlands (VROM 2007b: 2). Protection from flooding resulting from rising sea and river levels was named as a central challenge, along with the design of residential areas within the context of the increasing incidence of extreme temperatures and rainfall (ibid.: 5). The adaptation strategy was developed on the basis of the *National Programme for Spatial Adaptation to Climate Change* (*ARK*), adopted in 2006 (VROM *et al.* 2007a). The central aim of the programme was 'to climate proof the spatial planning in the Netherlands' (ibid.: ii). As the reason for its implementation, the ministries stated that climate change was no longer thought of as something that would be perceivable in the distant future, but that changes were already showing themselves: 'climate change is already happening now' (VROM *et al.* 2007a: 4).

The national adaptation strategy was developed together with relevant representatives from the kingdom, province, and municipality levels, as well as from the water boards. Additionally, in 2009 the provinces together with the national government signed an agreement to further stabilise spatial planning in the region, the so-called Inter-Provincial Accord (IPO) (Fleischauer 2009: 4). In this, the provinces and national government once again confirmed the targets of the national adaptation strategy. Building upon this, the provinces each committed to developing their own adaptation concept by 2015 (ibid.). By this point, in fact, almost all of the provinces were able to present their concepts (EC/EEA 2015).

The 2010s are noticeable for their thematic breadth, from specifically water-related problems and opportunities in spatial development to other themes and sectors. In 2012 the report of a national audit concluded that the adaptation policies in place in several areas which were considered to be vulnerable were insufficient (Algemene Rekenkamer 2012).

The early focus of climate discourse in the Netherlands on water-related issues is hardly surprising when one considers that a life below sea level and its associated challenges have been a very long design task for the inhabitants of the country (see Section 3.3, Van Koningsveld *et al.* 2008). To understand the Dutch adaptation discourses better, we must therefore go back a little further into their history. Even by around the thirteenth century, numerous innovations in land reclamation and cultivation can be found in the national archives (ibid.). These include technical innovations such as windmills, canals, dykes, and flood barriers, as well as social innovations such as the water boards founded at that time, which had already developed solutions for the coordination of water-related issues (ibid.: 371f.). Until this time it had been usual to build artificial hills, which would be surrounded during floods. With the development of larger settlements and towns, this early form of 'life with the water' was largely abandoned. The paradigm of *integration* thus gave way to that of *protection* and from here on the dominant theme over the centuries was of the notion of a 'struggle against water' (ibid.).

This supposed struggle reached a climax in the mid-twentieth century when, in 1953, the Netherlands was afflicted by very severe storm flooding, resulting in many deaths (Deltacommissie 2008: 5). This occurred during a period in which, in many European countries, a strong belief in technologically and socially formative capabilities was dominant (see also Nuissl 2007: 44; Healey 1997: 7ff.). At this point, representatives of the Dutch government summoned together a group of experts, which was to develop plans for 'hydraulic engineering works' (ibid.). The central aim of this first *Delta commission* was to prevent large-scale flooding and fatalities in the future. In the following decades, on the basis of the recommendations of the commission, major technological flood barriers were built at the large estuaries, now referred to as *Delta works* (ibid.).

The construction of the Delta works was still strongly oriented towards the centuries-old tradition of protection and the strict separation between water and humans. DeJonge, for example, still considered the construction phase of the Delta works a 'Period of Big Closures' (DeJonge 2009: 26). In recent times authors have increasingly noted a specific shift in the handling of water in the Netherlands. Wilson and Piper observed at the turn of the millennium a 'paradigm shift in the Dutch approach to water planning' (Wilson and Piper 2010: 87). The centuries-old tradition of 'managing water by keeping it out' (ibid.) has thus been abandoned in favour of a policy that creates more space for water. Wiering and Immink describe the changes as discursive displacements, according to which the 'struggle against the water' has faced increasing competition from integration strategies:

> In water management and in spatial planning there is a debate on the fundamental underlying discourses of the policy domains. In Dutch water management an emergent discourse of 'accommodating water', which is competing with the traditional 'battle against water' discourse, can be seen.
>
> (Wiering and Immink 2006: 423)

This particular point of view appears to express itself in the semantic references of recent water-related work and cooperation programmes such as 'Room for the River' (V&W 2006), 'Climate Buffers' (EC/EEA 2015) or 'Building with Nature' (DeVriend and Van Koningsveld 2012). The central decisions of the Room for the River programme adopted in 2006 by the Dutch government are directed at increasing the flow potential of Dutch rivers. The programme includes more than 30 different sub-projects following various rivers under the subtitle 'From higher dykes to river widening' (Room for the River 2012: 1). With the implementation of the programme, various measures from the repertoire of retreat and integration come into effect, such as the construction of inland dykes and the creation of drainage channels or additional flooding areas (ibid.: 19).

To be able to react to water-related changes in the long term and on a regular basis, in September 2007 the Dutch government initiated a new *Delta commission*. One year later its findings were presented under the title 'Working together with water' (original: 'Samen werken met water') (Deltacommissie 2008). The report recommended adaptation measures that would provide long-term security against floods and protect drinking water supplies until the end of the century end (ibid.: 11). The Delta commission thus considers climate change to be a serious threat for the Netherlands, making it imperative that comprehensive measures be taken (Deltacommissie 2008: 5). The recommendations of the Delta commission were not, however, concerned only with water-related integration strategies, as the title of the report might suggest (see Section 4.3). Rather, it proposed a comprehensive programme taking into account the complete spectrum of coastal and floodwater protection, and also provided recommendations for its further stabilisation through governance arrangements.

With the 'Delta Act on flood safety and fresh water supply' (EC/EEA 2015, henceforth 'Delta Act') adopted in 2012, the legal foundation for the stabilisation of the new *Delta programme* was achieved.

An important outcome in 2014 is presented by the 'decisions to keep the Netherlands safe and liveable' (I&M/EZ 2014), also knowns as the 'five key decisions' (EC/EEA 2015). These broadly take up the suggestions of the Delta commission from 2008 and elaborate them still further into specific work packages for the five central areas, namely flood-risk management, fresh-water provision, general urban and spatial planning, the Rhine–Meuse–Scheldt delta, and the Ijsselmeer region (see in more detail: ibid.). For flood-risk management, clear and comprehensive measures derived from protection, integration, and retreat strategies are now provided (I&M/EZ 2014: 16ff.).

On the basis of the available literature it can be determined that both adaptation to climate change and spatial development in general appear to have high significance in the Netherlands. Fleischauer thus came to the conclusion in 2009 that: 'the whole climate change adaptation process in the Netherlands is seen from a spatial planning perspective' (Fleischauer 2009: 10). The most important topic is that of dealing with water: 'Throughout history spatial planning and development was always connected with water management and the other way round' (ibid.). Diversification into other areas of focus occurred relatively late in the 2010s.

6.3.3 *Denmark: sustainable development and green expansion*

Ecological issues have a longer perceptual as well as practical tradition in Denmark. In the following, the climate-change-related discourse will be presented.

Climate-mitigation policies in Denmark

In recent years Denmark's climate-mitigation policies have attracted repeated attention, even in foreign-language media. During the negotiations in December 2012 on the Kyoto follow-up agreement, the Danes emerged as the 'world champion' (Donnebaum 2012; *Hamburger Abendblatt* 2012; MOZ 2012) after taking the lead in a climate-mitigation ranking undertaken by the NGO Germanwatch. In 2007, the German weekly news magazine *Der Spiegel* had already described Denmark as a 'climate-mitigation prodigy' (Schmitt 2007), as it was possible to refer to measures and their empirical success in this country as far back as the early 1990s.

In the EU's first Kyoto burden-sharing agreement, Denmark, like Germany, set a reduction target of 21% by 2012 compared with 1990 (EC 2002).[11] The (percentage) value was the same as the German target and was among the highest among the signatories (see also Section 6.2). The UNFCCC reports an actual reduction of 24.1% by the end of 2012 (UNFCCC 2014c: 9), thus also attaining the first objective within the framework of the international agreements. A reduction of 20% compared to 2005 has been earmarked for the next

Kyoto period by 2020 (EC 2009a). Denmark is thus pursuing one of the highest-targeted relative-reduction values within the European burden-sharing agreement for 2020 (ibid.).

As early as April 1990, the then coalition presented a climate programme with the title *Energie 2000. Action Plan for Sustainable Development* (Mez 1997: 21). The action plan for the first time gave explicitly defined reduction targets for various greenhouse gases. By 2005 CO_2 savings of 30% had been targeted (ibid.). The targets would be achieved primarily by a 15% reduction in primary energy consumption and by changes to the energy mix. Representatives from economic and civil society spheres were involved early on in the course of the planning of measures. To achieve a reduction in electricity consumption, for example, a 'power conservation council' (ibid.) was set up, in which larger electricity producers and other business interests were represented, to create a broad social consensus.

Measures could also be generated for sectors other than the electricity sector. Starting in 1993, residential thermal-insulation standards and targets for the district heating supply of specific settlements were put into effect (Mez 1997: 22). As early as 1992 the coalition passed a bundle of legislation that was adopted by the Folketing (the Danish parliament) as a 'CO_2 package' ibid.: 23). Further measures followed through the 1990s.[12] The aggregated measures showed their effect early on: CO_2 emissions were already in abeyance from 1992 onwards (ibid.: 24).

To better understand the aspirations of the Danish energy policy, one must look back a little further into the past. An important reference point is the oil crisis of 1973, whose consequences resulted in early changes in Danish energy policy (Abildgaard 2000: 40f.). This crisis had an even greater impact in Denmark than its European neighbours, as at the beginning of the 1970s the country had the highest share of oil (93%), and among industrialised countries the largest import dependency (Krawinkel 1998; Mez 1997: 20). The result was economic stagnation, and many private households had difficulties obtaining fuel oil. Krawinkel even describes how, during the crisis, people had to suffer the cold even in their homes (Krawinkel 1998). The then newly publicised ideas of the 'limits to growth' (Meadows *et al.* 1972; see also Section 4.2) and its possible impacts were therefore physically tangible to many Danes.

In 1976, following the experience of the crisis, the first national energy plan was adopted, initially focusing on the development of Denmark's own oil and natural gas reserves (ECF *et al.* 2013; Abildgaard 2000: 40). Already during the 1980s though, Abildgaard observes an additional shift in priorities, from security of supply to environmental and sustainability targets (Abildgaard 2000: 40). This culminated first of all in the government's decision to abolish nuclear power from 1986, and not to use it in the long term (ECF *et al.* 2013: 11). The rapid increase in the significance of environmental themes in the 1980s also suggests that a framework of socially shared knowledge had emerged in Denmark, making it possible for the global discussions on climate change in the late 1980s to resonate more here than in other countries. I shall examine this framework later in the empirical part of this study.

In the 2000s, the climate endeavours of the 1990s continued, but now within the context of international agreements. To better achieve the Kyoto targets, in 2003 the Folketing adopted a new climate strategy (KEMIN 2009: 124). An important element was the linking of efficiency principles with climate-mitigation targets (ibid.: 126). Numerous further measures followed (e.g. Green Vision Transport Plan, Green Growth Plan) and by 2011 Denmark's economy had achieved the second-lowest energy intensity in Europe (Eberle *et al.* 2014: 10).

With regard to renewable energies, Denmark has also been at the forefront. In 2012 Denmark had the highest share of renewables in Europe at 23%. Contributing to this, besides the aforementioned tax incentives for specific energy sources, were additional funding models introduced since the early 1990s (Eberle *et al.* 2014: 11). Danish climate targets have also continued to develop in recent years. In 2011 the government passed the *Danish Energy Strategy 2050* with the subtitle 'from coal, oil and gas to green energy' (KEMIN 2011). This provides for Denmark to be almost independent of fossil fuels by 2050 in both electricity generation and in transport (ibid.). In the *Climate Policy Plan* (KEBIN 2013) intermediate steps towards these objectives are further defined. By as early as 2030, oil and coal are to be banned from Denmark (ibid.: 14). The supply of electricity and heat is to be produced 100% from renewable energies (ibid.). In order to continue along this path, the Danish government has passed further laws.[13] In 2013 the Danish government summarised the core principles of its national climate policy thus:

> It is about organizing climate policy so that climate change mitigation concerns are integrated in solutions to achieve other high priority goals such as growth, private job creation, resource efficiency, natural values, or security of energy supply.
>
> (KEMIN 2013: 35)

Ecology and economics are thus hardly viewed as antagonistic any more. Normative notions of sustainable development appear to have been strongly internalised, as the documents discussed regularly demonstrate that, in Danish climate mitigation, environmental, social, and economic objectives are being addressed in equal measure.

Climate-adaptation policies in Denmark

The first works on behalf of the Danish Ministry for the Environment to discuss conceivable climate impacts for Denmark can be traced back to the early days of Danish climate-mitigation efforts (see for example Kjaergaard and Fenger 1990). National political and administrative endeavours to adapt to climate change did not take systematic shape until the 2000s, though. In 2004 a study was commissioned by the Ministry for the Environment, which discussed projected climate impacts and possible adaptation measures ('Adapting to the climate of the future', Miljøministeriet 2004). In October 2005 the Danish government, under the direction of the Danish Ministry for Climate and Energy, began preparations for a national climate-adaptation strategy (KEMIN 2009: 245; The Danish Government 2008: 5). This was finally adopted in March 2008. Already in its introduction, the authors stated the following conclusion:

> Climate Change is a reality, and an important part of global warming is due to human activity. [...] Climate Change is coming to Denmark, but it is still uncertain to what extent and when.
>
> (The Danish Government 2008: 6f.)

The adaptation strategies of Germany and the Netherlands previously considered emphasised that climate change impacts are already clearly observable, and that there is thus an acute need for adaptation. In the Danish strategy, by contrast, anthropogenic climate change is regarded as a given, but its impacts are viewed only as an uncertain future. The Danish authors attach great importance to the human element in the cause of climate change and emphasise the high importance of climate-mitigation measures. In addition, from the authors' point of view, adaptation measures in Denmark should also serve as a basis for sustainable development goals: 'The goal is a society that will continue to be environmentally, economically and socially sustainable' (The Danish Government 2008: 5).

The Danish strategy is based on the premise that adjustment must be viewed as a very long-term process (ibid.: 8). Building upon this, it initially focuses on three main areas of work: first, the promotion of science and research for the further generation of information; second, the implementation of organisational and institutional frameworks for adaptation processes; and third, the improvement of communication throughout the country (The Danish Government 2008: 8). The adaptation strategy discussed the possible consequences and adaptation stages for 11 sectors in which possible steps over the next 10 years were considered as technologically and socio-economically feasible. However, the adaptation strategy itself did not formulate any formally binding rules (see also Wejs *et al.* 2014: 492). Additionally, the authors frequently concluded that more detailed analyses were needed in each sector before further action is taken. In particular, analyses with regard to the socio-economic impact of measures were required (The Danish Government 2008: 9).

A year after the national adaptation strategy was published, the authors of the national UNFCCC report saw Denmark as well positioned with regard to future climatic changes. Among other things, their views were justified with recourse to the country's history of environmental spatial development:

> In relation to future global climate change, Denmark is a robust country. This is primarily because of a long tradition of legislation which prevents building in river valleys, along the coast and forests. Agricultural land is well-drained and many farmers are able to irrigate in dry periods. Moreover,

> the Danish population is aware of, and uses systematic warning systems of extreme weather events and the consequence thereof.
>
> (KEMIN 2009: 242)

One year later, an in-depth 'socio-economic screening' (DEA 2010) came to the conclusion: 'For almost all of the sectors it is assessed that climate change adaptation can take place within the existing regulatory framework' (ibid.: 3).

In 2012 the government finally published its preliminary results, along with a discussion of planned measures, in a national action plan for climate adaptation. The plan gives an overview of national measures both previously introduced and planned for the future, drawing on the defined strategic fields of action of the 2008 adaptation strategy. In addition, measures that can contribute to a 'green transformation' (ibid.: 20) are discussed, by which are meant adaptation measures that not only increase resilience, but also meet sustainability goals (ibid.: 20ff.). The action plan also envisages that the municipalities should develop their own climate-adaptation plans within two years (ibid.: 6).

In summary, the Danish national adjustment process is remarkable for starting from a broad examination of possible impacts and problems for a large number of different sectors. On the basis of close examination, weighing possible climate and adaptation impacts against potential socio-economic effects, national efforts focused on what was then seen as the priority issue of heavy rain. As the Danish adaptation strategy is designed for the long term, further problem areas could follow if scientific analysis suggests they should be a priority. Alongside state efforts, the nationally produced documents further emphasise the possibility of autonomous adaptation processes, to be carried out by actors from different sectors without the direct influence of the state, and which can be backed up by national communication campaigns. A strongly cross-sectoral way of thinking and working is also apparent as a characteristic feature, and sustainable development issues play an important role. The aim is to harmonise spatial development and climate adaptation with ecological, social, and economic development. Climate change and spatial development should thus ideally be able to generate synergistic development processes for the future.

6.3.4 *Poland: energy security and economic development*

For a better contextualisation of the ways of dealing with climate change in Poland, one should first remember that regular struggles for independence and sovereignty exist throughout Poland's history,[14] something that appears of high relevance to its climate policy.

Climate-mitigation policies in Poland

By adopting the Kyoto Protocol in 1998,[15] Poland, a former Eastern-bloc country, committed itself to a 6% reduction in CO_2 emissions by 2012 compared to 1988 (UNFCCC 1998: 20, 1996: 16). The UNFCCC reported an actual CO_2-reduction for Poland of 29.9% by the end of 2012 (UNFCCC 2014c: 9).

The negotiated value had thus been surpassed by a large margin. For the subsequent Kyoto period up to 2020, Polish negotiators explicitly brokered a positive value. According to Burden Sharing within the EU, Poland was permitted to increase its emissions up to a maximum of 15% compared to 2005 (EC 2009a). Let us take a closer look at the background to these figures.

Compared with western Europe in the 1980s, countries of the Eastern bloc had not only lower economic living standards, but also lower environmental standards (Bingen and Kielczewski 2011: 37f.). Furthermore, numerous authors describe a low level of awareness of environmental and climate-politics problems in Poland up until the 2000s (Kundzewicz and Matczak 2012: 304f.; Bingen and Kielczewski 2011: 37). Bingen and Kielczewski claim that, unlike in socialist Bulgaria during the 1980s, for example, there has never been a strong ecological movement in Poland despite high environmental impacts (Bingen and Kielczewski 2011: 37). Considering this, it is all the more pressing to ask how the previously mentioned climate-mitigation targets for Poland came about.

Several authors argue that Poland's climate measures have rarely thus far been intrinsically motivated, but have been driven primarily by international alliance obligations (ibid.: 303; McLean and Stone 2012: 110). Thus during the early 1990s, with the dissolution of the economic and military alliances of the Eastern bloc countries, the idea of a 'retreat to Europe' (Pysz 2011: 26) as a new foreign-policy objective gained wide support. Poland formally applied for membership of the European Union in April 1994 (McLean and Stone 2012: 110), for which a regular requirement of admission is implementation of the EU's common judicial canon (the Community acquis, or in French *acquis communautaire*). Poland and other candidates were thus faced with much higher environmental policy requirements. Apart from this, the acquis requires that member states enter into legal commitments with external states in accordance with international law (ibid.). Poland participated in the preparations for and follow-up work to the Kyoto Protocol until 1998, and ratified it in 2002 (UNFCCC 2014b).

It is not the place of this study to further explore the theory of Poland's extrinsic motivation, and more comprehensive historical source studies already exist. A counter-argument to this thesis could be made, however, that Poland signed the United Nations Convention on Climate Change at the 1991 Rio Conference, two years ahead of the start of negotiations on EU accession (UNFCCC 1992: 25).[16] It is also noticeable that the principle of sustainable development was even understood as a normative element in the new Polish constitution (Art. 5; see Zelljadt *et al.* 2014: 4). In the case of climate mitigation, though, it will become apparent in the following that, besides the international commitments mentioned above, hardly any other measures can be identified during the 1990s. Even in 2012 Kundzewicz and Matczak come to the conclusion that: 'Poland is therefore not a frontrunner in arguing for climate-change mitigation' (Kundzewicz and Matczak 2012: 303).

Already during talks on the first Kyoto Protocol, Polish negotiators secured wide scope for the country's own fossil-fuel development trajectory. One concession agreed was a transfer of the base year from 1990 to 1988 (UNFCCC 1998: 20, 1996: 16) which, according to McLean and Stone, meant the 6% reduction target was already achieved at the time of signing (McLean and Stone 2012: 109f.). This is due to the great economic downturn following 1988 and the rapid changeover of the economic system after 1990. For the years from 1988 to 1994, a reduction in greenhouse gas emissions has been calculated at 25% (McLean and Stone 2012: 109f.). The Kyoto Protocol's reduction value of minus 6% thus created an internationally legitimised margin for a new increase in emissions.

The above-mentioned reduction of nearly 30% by 2012 does, however, show that further reductions have been achieved since signing, and this indeed alongside sustained economic growth (UNFCCC 2014c: 9; Kühne 2002: 89). Polish economist Piotr Pysz later described the country as a successful 'pioneer of system transformation' (Pysz 2011: 26), as economic conditions have consolidated since 1992. In 2009, the year of strong economic crisis, Poland was the only EU state to record positive economic growth (ibid.).

Reasons for the reduction in emissions can be discerned in the system transformation itself. Kühne (2002: 83ff.) details factors that may have been relevant, among them the implementation of EU environmental standards. A high share of the reduction in emissions can, however, initially be attributed to the introduction of the competition and market mechanisms themselves. The energy efficiency of Polish industrial production increased by 59% between 1995 and 2010 (Zelljadt *et al.* 2014: 12). Furthermore, the composition of the economy shifted drastically: between 1988 and 1998 alone the share of the lower-emission services sector rose from 35% to 57% of GDP (Kühne 2002: 83ff.). A clear transition from an industrialised to a service economy can thus be observed during the 1990s (ibid.: 84ff.).

There were, however, developments in the course of the transformation process and afterwards that contributed to increases in emissions, leading to a renewed increase in emissions in the annual accounts from the 2000s onwards (Kundzewicz and Matczak 2012: 304). In the period following the ratification of the Kyoto Protocol, efforts can be identified to implement targeted climate-measures in Poland (Kundzewicz and Matczak 2012: 302f., Institute of Environmental Protection 2010: 39ff.). In 2003, following ratification and one year before formal accession to the EU, Poland's Council of Ministers adopted the first and to date only national climate-mitigation strategy (Zelljadt *et al.* 2014: 4; Ministry of the Environment 2003). Possible measures were only very generally formulated, however, before being assigned to various ministries for their further development. Eleven years later, Zelljadt *et al.* came to the conclusion that 'the document has little relevance to Poland's current climate policy situation. The applicability [...] is particularly low' (Zelljadt *et al.* 2014: 4).

Some legal implementations concerned with climate-relevant issues can be found in the 2000s, however (Institute of Environmental Protection 2010: 40ff.).

For example, in 2003 the revised woodlands programme sought to increase the proportion of Polish forests from 28% of its area to 30% by 2020, leading to an increase in natural sinks (ibid.). Many climate-related laws can be derived from EU directives. For example, the European Emissions Trading System was introduced after EU accession from 2005 onwards (ibid.). Poland must also achieve energy-efficiency goals set within the framework of the 20–20–20 targets, and increase its share of renewable energies (EC 2012; EC 2009b).

In 2009 the Council of Ministers submitted a draft on Poland's energy policy up to 2030 (Ministry of Economy 2009), which emphasised the continued high importance of coal in the energy mix (ibid.: 14; Kundzewicz and Matczak 2012: 304). In 2012 the proportion of coal in the gross energy consumption lay at 52%. In relation to this, the introduction of CCS technology has been discussed, and two test sites installed (at Bełchatów and Kędzierzyn) (Kundzewicz and Matczak 2012: 304).

Additionally, it is intended that greater greenhouse gas reductions will be achieved by introducing nuclear energy (Ministerstwo Gospodarki 2014: 1; Ministry of Economy 2009: 15ff.). To date, Poland has not produced any nuclear energy. In his 2006 governmental statement, Prime Minister Jarosław Kaczyński mentioned the use of nuclear power as a measure to diversify the Polish energy mix (Lang 2006: 5). Since then the Polish government has developed a nuclear energy programme that was presented by the Ministry of Economy in 2014 (Ministerstwo Gospodarki 2014), according to which by 2025, three years following the total withdrawal of Germany from nuclear power, Poland's first nuclear reactor should be on the grid (ibid.: 21).

In accordance with the EU 20–20–20 targets, Poland must increase its proportion of renewable energies from 7.2% in 2005 to 15% by 2020 (EC 2009b: Appendix I). In 2012 the proportion stood at 9%, attributable in the main to the use of bioenergy, a large part of which currently involves 'co-combustion', in which biomass is combusted together with coal (Schnell 2013: 86f.). Until the early 2010s there were hardly any measures promoting renewable energies that might have specifically increased the shares of different types of energy generation (see also Zelljadt *et al.* 2014; Olszewski 2013; Schnell 2013). Indeed, a system of certification was introduced after EU accession, establishing a guaranteed return for producers of electrical energy from renewable sources (Schnell 2013: 85), and in 2011 the first draft of a bill to promote renewable energies was also presented (ibid.). Nevertheless, some authors indicate that actors were hardly motivated to invest (Zelljadt *et al.* 2014: 13; Schnell 2013: 85f.).

In light of the foregoing, one can conclude that in Poland, unlike in the other countries considered, there seems to be little consensus on the necessity or feasibility of climate-mitigation measures. Since joining the EU in 2004, the Polish government representatives have also regularly assumed oppositional positions when joint EU climate-mitigation targets were on the agenda (Riesbeck 2014; Uken 2014; Kundzewicz and Matczak 2012: 303; Vorholz 2012). Polish government representatives have lately taken an obstructionist attitude during recent discussions for setting EU energy and climate targets up to 2030 (ibid.).

Several possible reasons are given in the literature for the ostensibly low importance of climate mitigation in Poland. Some authors cite explanations from modernisation theory, understanding there to be a need for many Poles to catch up on infrastructure and consumption (Kundzewicz and Matczak 2012: 303; Bingen and Kielczewski 2011: 37). It is already clear from the climate-policy discourses of other countries considered here that economic development and climate mitigation need not necessarily be regarded as opposed to one another. In Poland, though, discursive patterns seem to dominate which assume that economic development and climate mitigation are not really compatible. Additionally, Kundzewicz and Matczak cite the barely noticeable effects of climate change in Poland as a second explanation for its lesser significance, with the special situation of domestic energy production given as a third reason.

In fact, certain discernible peculiarities in the Polish energy discourse should be noted to gain an understanding of Polish climate policy. At least two principal themes can be identified. First, it can be observed that the Polish energy policy is regularly discussed in close connection with foreign policy and security policy considerations (see for example Ministerstwo Gospodarki 2014: 2; Council of Ministers 2012: 84f.; Ministry of Economy 2009: 8ff.; Lang 2007). Second, the high significance of coal in the Polish energy mix for its present and future provides a central theme (Olszewski 2013: 2; Schnell 2013: 86; Ministry of Economy 2009: 8ff.). Lang states: 'Poland defines its energy policy challenges in a context highly motivated by its political security' (Lang 2006: 2) and is 'at pains to overcome its situation of Russian energy dependence, in order to reduce the risk of social vulnerability' (ibid.: 3). Grzeszak also comes to the conclusion:

> In political discussions, Russia is regarded as the greatest danger to Poland's energy security. There is a conviction that the Kremlin seeks to gain dominance with the help of economic weapons such as oil and gas.
>
> (Grzeszak 2012: 2f.)

These statements suggest that the historical experience of the last centuries is still embedded as internalised knowledge stocks in Polish society, with a particularly significant role being taken by the leitmotiv of independence from Russia. I shall return to motives of security later in the empirical part of the investigation (see Section 9.1.3ff.).

Finally, attention should be given to actor-specific differences in the ways of dealing with climate mitigation in Poland. Although the importance of climate mitigation appears on the whole to be lower than in other countries, observations point to major differences between actors. Thus Kundzewicz and Matczak (2012: 306) state that in Poland since the 1990s, NGOs have appeared which advocate political measures towards climate mitigation. In 2002 they founded a national climate coalition. However, it can be shown that NGOs in Poland are paid little attention in relation to information on climate change (see ibid. and Section 8.2: Use of media).

Climate-adaptation policies in Poland

The first publications from domestic scientific actors in Poland on possible climate-change and adaptation strategies for the country had already appeared in the 1990s (Zeidler 1997; Wróblewski 1992). At the same time, several international research projects were carried out, such as the Dutch-financed 'sea-level rise programme for Poland's coast' (Zeidler 1997: 151) and the 'US Climate Change Country Study Program' (ibid.). In 2010, based on further studies, Poland's fifth national UNFCCC Report reported possible problems for Poland as a result of climate change (Institute of Environmental Protection 2010: 81ff.). Under consideration were Polish agriculture, coastal zones, water resources, and forestry. To date, the report has not yet explicitly presented measures under the heading of 'climate adaptation'. It does, however, make recommendations for future implementation. It points in addition to the then regular operations of coastal and flood protection. Since March 2003, Poland has had a national coastal protection programme which includes a long-term, 100-year perspective, as well as measures for the years 2004–2023. The main objective of this programme is to prevent erosion, protecting and preserving the coastline as it stood in 2000 for the long term (ibid.: 83f.). The strategy follows the traditional coastal protection paradigm (see also Section 4.3).

A national climate-change discourse in politics and administration developed at the end of the 2000s. In March 2010, the Council of Ministers decided to configure a national adaptation strategy that would also include action plans for the implemention of measures (SPA 2020[17]) (Ministry of the Environment 2013: 6). As justification for the implementation of the strategy, both climate-related and economic reasons were given:

> The results of scientific research clearly indicate that phenomena caused by climate change are a threat to the social and economic development of many countries in the world, including Poland [...]. The [...] strategy [...] has been prepared with a view to ensure the conditions of stable socio-economic development in the face of risks posed by climate change but also with a view to use the positive impact which adaptation actions may have not only on the state of the Polish environment but also on the economic growth.
>
> (Ministry of the Environment 2013: 5)

The authors of the SPA 2020 thus see climate change as a danger to the economic and social development of Poland, and at the same time emphasise the chances presented by an adaptation strategy for the further development of the country. They thus also stress the high significance of international efforts towards the implementation of their own strategies:

> The need to develop and elaborate appropriate and integrated adaptation plans results from Article 4 of the above Convention [UNFCCC] and the

> decisions of [the] Conference of the Parties to the Convention such as [the] Cancun Adaptation Framework and [the] 'Nairobi work programme on impacts, vulnerability and adaptation to climate change' of 2006. The programme has been adopted in the Convention's forum and provides for, *inter alia*, the need for countries to assess the possible adverse impacts of climate change on various fields of life and to create a strategy to respond to those impacts.
>
> (Ministry of the Environment 2013: 6)

Not least, the formulation of the EU White Paper intended to establish a European framework for climate-change adaptation is drawn upon as a reason to decide upon a Polish adaptation strategy: 'the Polish government has taken its stance on the White Book issue, recognizing the need to develop a strategy of adaptation' (ibid.).

The government assigned responsibility for the monitoring of the work on the strategy development to the Ministry of the Environment. Forming the basis of the adaptation strategy (SPA 2020) is the research project KLIMADA, which, beyond the current strategy, is to be concerned more extensively with climate-change and adaptation strategies for Poland (Klimada 2015; Ministry of the Environment 2013: 6).

The SPA 2020 first diagnoses observable climatic changes in Poland since the 1970s and investigates the country's resulting costs and losses (Ministry of the Environment 2013: 9ff.). According to its evaluation, extreme weather events such as heat waves, heavy rain, and droughts have increased. The majority of the damage is connected to flood, but the consequences of drought and storms are also mentioned (ibid.: 12f.). The SPA 2020 contains prospective discussion of climate projections for Poland up to 2030 and identifies 11 'vulnerable sectors' (ibid.: 22ff.), including coasts, water management, and spatial planning and urban development. In order to meet the discussed outcomes through adaptation measures, the SPA 2020 concludes by formulating 'action lines' with targets that should be met by 2020. Their implementation should be overseen and evaluated by the Ministry of the Environment (ibid.: 33ff.; Klimada 2015).

Responsibility for the implementation will thus be assigned to various national ministries and authorities. In formulating the goals and bundles of tasks, conformity with various existing development objectives for the country need be ensured. Therefore the action plans also include the objectives and task bundles listed in connection with Poland's other national development strategies (ibid.: 35ff.). In this way, the SPA 2020 should serve the cross-sectoral development of Poland. The main target is thus defined as follows:

> The major objective of SPA is to ensure sustainable development and efficient functioning of the economy and society in the conditions of climate change.
>
> (Ministry of the Environment 2013: 33)

The Polish adaptation strategy likewise recalls the guiding principle of sustainable development (see also Section 6.4.3), emphasising the economic and social development of Poland separately. The third dimension of sustainable development is also addressed, as the central challenge is the task of reconciling the economic dimension with the ecological dimension: 'a key remaining challenge [...] will be to ensure the economic growth with the preservation and efficient use of environmental resources' (ibid.: 33). Thus the adaptation strategy also shows that shared knowledge structures in Poland seem to be of greater discursive relevance, and that economic and ecological developments tend to be positioned in opposition to each other.

The SPA 2020 distinguishes between cross-sector and sector-specific task bundles. Among the overarching measures are included legislative changes, such as the implementation of the EU flood directive from the designation of flood risk areas to the 'creation of a legal basis for the resettlement of the population from areas at permanent risk' (ibid.: 34). Additionally, improvements in organisation and communication in dealing with natural hazards is to be achieved. Financing of the measures is to come from national and international, as well as private sources. EU funding programmes would play an important role, along with national programmes such as the national fund for environmental protection and water management. An explicit adaptation fund for Poland is not included, with funds being made available mostly from existing sources (Klimada 2015; Ministerstwo Środowiska 2015; Ministry of the Environment 2013: 53).

In summary, issues of economic and social development are of great importance in the Polish adaptation strategy. Aspects of spatial development are also extensively discussed from the outset. It is further striking that Poland began its own development of an adaptation strategy only after the formulation of the 2009 EU White Paper, but at the same time has becomes highly active with action plans for the implementation of specific measures by different ministries and institutions.

6.4 Interim conclusion

In this chapter, I have considered the political–administrative resonance of the originally scientifically constructed climate knowledge at the global, European, and national levels, and in particular through the processes of institutionalisation of climate-mitigation and adaptation practices.

According to the theory of cultural homogenisation, the emergence of common institutions should lead to the creation of a common reality, values, and identities. Regional or national borders may, in addition, lose their significance within the separation of climate cultures (see Sections 2.2 and 5.3.2, Erhardt-Martinez *et al.* 2015: 218f.; Weiss and Burke 2011). It can be argued here that, at the global level of the United Nations during the 1990s, a common global understanding of the problem arose with regard to the need for the political and administrative treatment of anthropogenic climate change. This is indicated by the increase in institutionalisation processes, such as the setting up of the UNFCCC in Rio, the agreement on joint climate-mitigation targets in Kyoto, Doha, and Paris, and the creation of global institutions to promote adaptation processes such as the global adaptation fund. It is clear, in addition, that beyond the national level, the European Union can also be described as a driving force of global climate policy, being well advanced within the global community in the formulation of common targets.

Frictions between various negotiating parties continue to arise, however, and not only between political–administrative actors, but also between groups of actors from the economic and civil society spheres. It thus remains unclear about which groups of actors the term 'homogenisation' can be used, or where exactly the possible cultural boundaries in dealing with climate change occur.

The development of national climate-related institutionalisation processes has also been closely followed here. The continuing relevance of national borders as lines of climate-culture demarcation has been indicated, for example in relation to the priority of climate-change issues and perception of it as a problem. Homogenisation and differentiation processes both seem to be present here, as when we consider the course of implementation of climate-mitigation policies: the older EU member states Germany, the Netherlands, and Denmark began implementation almost as early as 1990, while Poland was only able to offer a climate-mitigation strategy in the 2000s, shortly before joining the EU. From that point of view, the Oder-Neisse line might still demarcate an existing climate-culture border, while the other states may already have converged.

If we consider adaptation discourses, however, climate-culture boundaries seem to be rather different, with a variety of substantive priorities. While in Poland the theme of development as a whole is the focus of attention, economic and technical innovations are repeatedly apparent in the Netherlands. In the Danish discourse, on the other hand, 'green themes' such as sustainability and 'gentle' spatial development recur. This is also true for Germany.

In summary, it seems reasonable to conclude that neither national differentiation nor total cultural homogenisation processes should be taken as starting points. These categories appear in the present case to be comparatively one-dimensional. Instead, as already discussed, it is necessary to consider specific *aspects* of knowledge which different groups in the research area share in common relating to problems and opportunities as well as necessary measures for climate mitigation and adaptation. On this basis, it can then be examined to what extent national origin or field affiliation can be attributed to the separation of knowledge stocks and climate cultures. In each case it is then necessary to investigate in detail on the basis of which criteria, and at what point, difference or homogeneity can be assumed. In the next chapter I shall consider the approach and empirical findings of the present study's own survey.

Notes

1 For clarity, the % symbol is used here. More precisely, the figures deal with percentage points.
2 The European Union appears as an individual actor alongside the 191 individual countries in the protocol (UNFCCC 2014b).
3 Emerging nations, whose emissions are continually on the increase, did not assume any globally binding emissions obligations until the Paris Conference, amongst them China, which overtook the USA as the world's greatest emitter in 2009 (BMUB 2014a).
4 In 2002 the following agreement was reached among the EU15 nations with respect to reduction targets to be reached by the end of 2012: Germany 21%, the Netherlands 6%, and Denmark 21% (Council of the European Union 2002).
5 The German ratification of the Kyoto Protocol was executed on 31 May 2002 (UNFCCC 2014b).
6 Own calculation on the basis of all Kyoto greenhouse gases, net emissions per capita, including LULUCF (land use, land-use change, and forestry), excluding air traffic. Data source: www.eea.europa.eu/data-and-maps/data/data-viewers/greenhouse-gases-viewer (5 May 2015).
7 In 2011, eight out of 17 nuclear power plants were taken out of service. The remaining plants will be decommissioned in 2022 (Lechtenböhmer and Samadi 2012: 234; BMJ 2011).
8 The action plan names the German Weather Service (Deutscher Wetterdienst, DWD), the Centre for Climate Impacts and Adaptation in Germany (Kompass), the Climate Service Centre (CSC), the Potsdam Institute for Advanced Sustainability Studies (IASS) and a strategic alliance of public officials consisting of the Federal Office for Citizen Protection and Disaster Support, the German Weather Service, the Federal Environment Agency, and the Federal Agency for Technical Relief (Bundesregierung 2011: 18).
9 The Netherlands' ratification of the first Kyoto Protocol took place on 31 May 2002 (UNFCCC 2014b).
10 For example, a law on the development of minimum energy requirements for construction in 1992, the determination of energy-efficiency norms for household appliances in 1993, and the introduction of an energy tax in 1996 (Pettenger 2007: 62).
11 Ratification of the Kyoto Protocol by the Danish parliament took place on 31 May 2002 (UNFCCC 2014b).
12 Alongside consumption standards, climate-related fiscal instruments such as taxes, duties, and tariffs have been introduced, reformed, and extended by successive coalitions (Eberle *et al.* 2014: 7; DEA 2000: 2; Krawinkel 1998; Mez 1997).
13 For example, the installation of oil and gas heaters in new buildings has been prohibited since 2013. Since 2016 this regulation has also applied to all existing buildings, for which comprehensive subsidies have been provided (Eberle *et al.* 2014: 9).
14 In 1795, Poland was fully divided through the intervention of Russia, Prussia, and the Habsburg Monarchy. Sovereignty was first restored after the First World War in 1918. After renewed division of the entire Polish territory following 1939, the country was finally remodelled in 1944/1945 as a socialist people's republic, resulting in an asymmetrical relationship with Soviet Russia. In 1989 the transition to the current republican system was introduced.
15 Ratification of the Kyoto Protocol by Poland took place on 13 December 2002 (UNFCCC 2014b).
16 Ratification took place at the end of July 1994 (McLean and Stone 2012: 109).
17 SPA 2020 stands for *Strategiczny plan adaptacji dla sektorów i obszarów wrażliwych na zmiany klimatu do roku 2020 z perspektywą do roku 2030* (Ministerstwo Środowiska 2013), translated as *Polish National Strategy for Adaptation to Climate Change (NAS 2020) with the perspective by 2030* (Ministry of the Environment 2013).

References

Abildgaard, Jorgen (2000) 'Energiepolitik und Energieplanung in Dänemark'. In: Jänicke, Martin and Jörgens, Helge (ed.): *Umweltplanung im internationalen Vergleich. Strategien der Nachhaltigkeit.* Berlin/Heidelberg: Springer, pp. 39–52.

Algemene Rekenkamer (2012) *Adaptation to Climate Change: National Strategy and Policy.* Den Haag: Sdu.

Beck, Silke, Bovet, Jana, Baasch, Stefanie, Reiß, Philipp and Görg, Christoph (2011) *Synergien und Konflikte von Anpassungsstrategien und -Maßnahmen.* Umweltbundesamt Final Report: Dessau-Roßlau. Available at: www.umweltbundesamt.de/publikationen/synergien-konflikte-von-strategien-massnahmen-zur (accessed 9 October 2013).

Bingen, Dieter and Kielczewski, Dariusz (2011) 'Umweltprobleme und Umweltstandards'. In: *Polen. Informationen zur politischen Bildung* 311/2011, pp. 36–38.

BMJ Bundesministerium der Justiz (ed.) (2002) 'Gesetz zur geordneten Beendigung der Kernenergienutzung zur gewerblichen Erzeugung von Elektrizität'. In: *Bundesgesetzblatt 2002.* Part I. No. 26 (26 April 2002), pp. 1351–1359.

BMJ Bundesministerium der Justiz (ed.) (2010) 'Elftes Gesetz zur Änderung des Atomgesetzes'. In: *Bundesgesetzblatt 2010.* Part I. No. 62 (13 December 2010), pp. 1814–1816.

BMJ Bundesministerium der Justiz (ed.) (2011) 'Dreizehntes Gesetz zur Änderung des Atomgesetzes'. In: *Bundesgesetzblatt 2011.* Part I. No. 43 (5 August 2011), pp. 1704–1705.

BMJV Bundesministeriums der Justiz und für Verbraucherschutz (ed.) (2014) 'Gesetz für den Ausbau erneuerbarer Energien (Erneuerbare-Energien-Gesetz – EEG 2014)'. In: *Bundesgesetzblatt 2014* Part I No. 33 (24 July 2014), pp. 1066–1132.

BMU Bundesministerium für Umwelt, Naturschutz und Reaktorsicherheit (ed.) (2005) *Nationales Klimaschutzprogramm 2005. Sechster Bericht der Interministeriellen Arbeitsgruppe 'CO$_2$-Reduktion'.* Berlin: BMU. Available at: www.bmub.bund.de/fileadmin/bmu-import/files/klimaschutz/downloads/application/pdf/klimaschutzprogramm_2005_lang.pdf (accessed 26 January 2015).

BMUB Bundesministerium für Umwelt, Naturschutz, Bau und Reaktorsicherheit (ed.) (2014a) *Weiterentwicklung des internationalen Klimaschutzes – auf dem Weg zu einem Klimaabkommen für das 21. Jahrhundert. Position: 16 May 2013.* Available at: www.bmub.bund.de/P202/ (14 November 2014).

BMUB Bundesministerium für Umwelt, Naturschutz, Bau und Reaktorsicherheit (ed.) (2014b) *Aktionsprogramm Klimaschutz 2020.* Berlin: BMUB.

BMVBS (ed.) (2010) *Nationale Strategien europäischer Staaten zur Anpassung an den Klimawandel aus Perspektive der Raumordnung/Raumentwicklung.* BMVBS-Online-Publikation 20/2010. Available at: www.bbsr.bund.de/cln_016/nn_629248/BBSR/DE/Veroeffentlichungen/BMVBS/Online/2010/ON202010.html (accessed 15 November 2010).

Bundesregierung (ed.) (2007) *Eckpunkte für ein integriertes Energie- und Klimaprogramm.* Berlin: Bundesregierung.

Bundesregierung (ed.) (2008) *Deutsche Anpassungsstrategie an den Klimawandel.* Berlin: no publisher's imprint.

Bundesregierung (ed.) (2011) *Aktionsplan Anpassung der Deutschen Anpassungsstrategie an den Klimawandel.* Berlin: Bundesregierung.

Busch, Per-Olof and Jörgens, Helge (2014) 'From the Outside in: Explaining Convergence in the Legal Recognition of the Sustainability Principle'. In: Jörgens, Helge, Lenschow, Andrea, and Lifferink, Duncan (ed.): *Understanding Environmental Policy Convergence. The Power of Words, Rules and Money.* Cambridge/New York: Cambridge University Press, pp. 237–264.

CDU/CSU/SPD (ed.) (2013) *Deutschlands Zukunft gestalten. Koalitionsvertrag zwischen CDU, CSU und SPD. 18. Legislaturperiode*. Rheinbach: Union Betriebs GmbH.

Council of Ministers (ed.) (2012) 'National Development Strategy 2020. Active Society, Competitive Economy, Efficient State'. Attachment to Resolution No. 157 of the Council of Ministers of 25 September 2012. Warsaw: no publisher's imprint.

Council of the European Union (ed.) (2002) Council Decision 2002/358/EC of 25 April 2002 concerning the approval, on behalf of the European Community, of the Kyoto Protocol to the United Nations Framework Convention on Climate Change and the joint fulfilment of commitments thereunder. 15 May 2002. Available at: https://eur-lex.europa.eu/legal-content/EN/TXT/?uri=celex:32002D0358.

DEA Danish Energy Authority (2000) *Green Taxes in Trade and Industry – Danish Experiences*. Copenhagen: DEA.

DEA Danish Energy Agency (2010) *Socio-economic Screening of Climate Change Adaptation. Summary*. Available at: www.klimatilpasning.dk/media/7851/samfunds%C3%B8konomisk%20screening%20final%20juni%202010_smfatning_eng_final.pdf (accessed 1 April 2015).

Dehmer, Dagmar (2014) 'Energieeffizienz. Nur warme Worte'. *Der Tagesspiegel Online* (10 August 2014). Available at: www.tagesspiegel.de/politik/energieeffizienz-nur-warme-worte/10314422.html (accessed 2 February 2015).

DeJonge, Victor N. (2009) 'From a Defensive to an Integrated Approach'. In: Reinhard, Stijn and Folmer, Henk (ed.): *Water Policy in the Netherlands. Integrated Management in a Densely Populated Delta*. Washington DC: RFF, pp. 17–46.

Deltacommissie (2008) *Working Together with Water. A Living Land Builds for its Future. Findings of the Deltacommissie 2008 [Samen werken met water. Een land dat leeft, bouwt aan zijn toekomst. Bevindingen van de Deltacommissie 2008]*. Alkmaar: Hollandia Printing. Available at: www.deltacommissie.com/doc/deltareport_full.pdf (accessed 15 May 2012).

DeVriend, H.J. and Van Koningsveld, M. (2012) *Building with Nature: Thinking, Acting and Interacting Differently*. Dordrecht: Alphen a/d Rijn.

Donat, Lena, Eberle, Andrew, Velten, Eike Karola, Prahl, Andreas, and Wevers, Melissa (2014) *Assessment of Climate Change Policies in the Context of the European Semester. Country Report: The Netherlands*. Berlin: Ecologic Institute /Eclareon.

Donnebaum, Denise (2012) 'Klimakonferenz. Dänemark ist Weltmeister beim Klimaschutz'. *Die Welt* (3 December 2012). Available at: www.welt.de/111774784 (accessed 16 March 2015).

Eberle, Andrew, Velten, Eike Karola, Donat, Lena, Duwe, Matthias, and Poblocka, Anna (2014) *Assessment of Climate Change Policies in the Context of the European Semester. Country Report: Denmark*. Berlin: Ecologic Institute/Eclareon.

EC European Commission (ed.) (2000) Communication from the Commission to the Council and the European Parliament on EU policies and measures to reduce greenhouse gas emissions. *Towards a European Climate Change Programme (ECCP)*. [COM/2000/0088 final – not published in the Official Journal]. Available at: http://eur-lex.europa.eu/legal-content/EN/TXT/?uri=CELEX:52000DC0088 (accessed 16 January 2015).

EC European Commission (ed.) (2002) Council Decision 2002/358/EC of 25 April 2002 concerning the approval, on behalf of the European Community, of the Kyoto Protocol to the United Nations Framework Convention on Climate Change and the joint fulfilment of commitments thereunder. Available at: http://eur-lex.europa.eu/legal-content/EN/TXT/?uri=LEGISSUM:l28060 (accessed 16 January 2015).

EC European Commission (ed.) (2005) Communication from the Commission to the Council, the European Parliament, the European Economic and Social Committee and the Committee of the Regions. *Winning the Battle Against Global Climate Change* {SEC(2005) 180} COM/2005/0035 final. Available at: http://eur-lex.europa.eu/legal-content/EN/TXT/?uri=CELEX:52005DC0035 (accessed 16 January 2015).

EC European Commission (ed.) (2007) Green Paper from the Commission to the Council, the European Parliament, the European Economic and Social Committee and the Committee of the Regions. *Adapting to Climate Change in Europe – Options for EU Action* {SEC(2007) 849}. COM/2007/0354 final. Available at: http://eur-lex.europa.eu/legal-content/EN/TXT/?uri=CELEX:52007DC0354 (accessed 16 January 2015).

EC European Commission (ed.) (2009a) Decision No. 406/2009/EC of the European Parliament and of the Council of 23 April 2009 on the effort of member states to reduce their greenhouse gas emissions to meet the Community's greenhouse gas emission reduction commitments up to 2020. In: *Official Journal of the European Union.* Availableat:http://eur-lex.europa.eu/legal-content/EN/TXT/?uri=CELEX:32009D0406 (accessed 15 January 2015).

EC European Commission (ed.) (2009b) Directive 2009/28/EC of the European Parliament and of the Council of 23 April 2009 on the promotion of the use of energy from renewable sources and amending and subsequently repealing Directives 2001/77/EC and 2003/30/EC. In: *Official Journal of the European Union.* Available at: http://eur-lex.europa.eu/legal-content/EN/TXT/?uri=CELEX:32009L0028 (accessed 15 January 2015).

EC European Commission (ed.) (2009c) White Paper. *Adapting to Climate Change: Towards a European Framework for Action* {SEC(2009)386}{SEC(2009)387}{SEC(2009)388} COM/2009/0147 final. Available at: http://eur-lex.europa.eu/legal-content/EN/ALL/?uri=CELEX:52009DC0147 (accessed 16 January 2015).

EC European Commission (ed.) (2012) Directive 2012/27/EU of the European Parliament and the Council of 25 October 2012 on energy efficiency, amending Directives 2009/125/EC and 2010/30/EU and repealing Directives 2004/8/EC and 2006/32/EC in the *Official Journal of the European Union.* Available at: http://eur-lex.europa.eu/legal-content/EN/TXT/?uri=CELEX:32012L0027 (accessed 11 March 2015).

EC European Commission (ed.) (2013a) Communication from the Commission to the European Parliament, the Council, the European Economic and Social Committee and the Committee of the Regions. *An EU Strategy on Adaptation to Climate Change.* COM/2013/0216 final. Available at: http://eur-lex.europa.eu/legal-content/EN/TXT/?uri=CELEX:52013DC0216 (accessed 15 January 2015).

EC European Commission (ed.) (2013b) Green Paper. *A 2030 Framework for Climate and Energy Policies.* COM/2013/0169 final. Available at: http://eur-lex.europa.eu/legal-content/EN/TXT/?uri=CELEX:52013DC0169 (accessed 20 January 2015).

EC European Commission (2014) *Climate Action. 2030 Framework for Climate and Energy Policies.* Available at: http://ec.europa.eu/clima/policies/2030/index_en.htm (accessed 20 January 2015).

EC European Commission/EEA European Environment Agency (2015) *Climate Adapt.* European Climate Adaptation Platform: Netherlands. Available at: http://climate-adapt.eea.europa.eu/countries/netherlands (accessed 23 January 2015).

ECF European Climate Foundation/PBL Netherlands Environmental Assessment Agency/CIEP Clingendael International Energy Programme (2013) *Germany, Denmark and the United Kingdom: Lessons to be Learnt for the Netherlands?* Report on a Seminar for the SER Negotiations on the Netherlands' National Energy Agreement. Available at: www.pbl.nl/sites/default/files/cms/publicaties/PBL-2013-germany-denmark-and-the-united-kingdom-lessons-to-be-learnt-for-the-netherlands-1150.pdf (accessed 26 March 2015).

Eichhorst, Urda and Madry, Thomas (2013) *Die Kehrseite der Medaille. Anpassung an den Klimawandel.* Dossier. Bundeszentrale für Politische Bildung Online. Available at: www.bpb.de/gesellschaft/umwelt/klimawandel/38484/anpassung-an-den-klimawandel (accessed 14 May 2015).

Erhardt-Martinez, Karen, Rudel, Thomas K., Norgaard, Kari Marie, and Broadbent, Jeffrey (2015) 'Mitigating Climate Change'. In: Dunlap, Riley E. and Brulle, Robert J. (ed.): *Climate Change and Society. Sociological Perspectives.* New York: Oxford University Press, pp. 199–234.

Fischer, Robert and Foißner, Peter (2002) 'Raumordnung, Stadtentwicklung und Städtebau in den Niederlanden'. *Standort – Zeitschrift für angewandte Geografie* 4: 153–157.

Fleischauer, M. (2009) *The Netherlands – Country Report.* (Report on the Project 'National Climate Change Adaptation Strategies of European States from a Spatial Planning/Regional Development Perspective'). Dortmund: TU Dortmund. Available at: www.bbsr.bund.de/BBSR/DE/FP/MORO/Studien/2009/NatStrategienKlimawandel/Laenderstudie Niederlande.pdf?__blob=publicationFile&v=2 (accessed 23 January 2015).

Fritsch, Michael, Schindele, Yvonne, and Wyrwich, Michael (2010) 'Ein langer Weg. Anpassungsprobleme in der ostdeutschen Unternehmenslandschaft'. In: Bundeszentrale für politische Bildung Online 29 October 2010. Available at: www.bpb.de/geschichte/deutsche-einheit/lange-wege-der-deutschen-einheit/47208/anpassungsprobleme (accessed 6 May 2015).

Government of the Netherlands (2015) *Energy Policy. Natural Gas.* Available at: www.government.nl/issues/energy-policy/natural-gas (accessed 12 March 2015).

Grzeszak, Adam (2012) 'Energie – Herausforderungen für Polen'. In: Deutsches Polen Institut/Forschungsstelle Osteuropa/Deutsche Gesellschaft für Osteuropakunde (ed.): *Polen-Analysen No. 109. Der Energiesektor* (15 May 2012). Available at: www.laender-analysen.de/polen/pdf/PolenAnalysen109.pdf (accessed 10 April 2015).

Hamburger Abendblatt (ed.) (2012) 'Dänemark ist Weltmeister beim Klimaschutz'. *Hamburger Abendblatt Online* (3 December 2012). Available at: www.abendblatt.de/politik/ausland/article111773270/Daenemark-ist-Weltmeister-beim-Klimaschutz.html (accessed 16 March 2015).

Hansestadt Lübeck (ed.) (2014) *Thematischer Landschaftsplan. Klimawandel in Lübeck. Vorsorge- und Anpassungsmaßnahmen für die Landnutzungen. Entwurf.* Lübeck: Bereich Umwelt-, Natur- und Verbraucherschutz der Hansestadt Lübeck.

Hansestadt Rostock (ed.) (2013) *Rahmenkonzept zur Anpassung an den Klimawandel. Bearbeitungsstand 2012/2013.* Rostock: Amt für Umweltschutz.

Hardin, Garrett (1968) 'The Tragedy of the Commons. The Population Problem has no Technical Solution – it Requires a Fundamental Extension in Morality'. *Science* 162 (3859): 1243–1248.

Healey, Patsy (1997) *Collaborative Planning. Shaping Places in Fragmented Societies.* Vancouver: UBC Press.

I&M Ministry of Infrastructure and the Environment (2014) *Climate Agenda: Resilient, Prosperous and Green.* Den Haag: I&M.

I&M Ministry of Infrastructure and the Environment/EZ Ministry of Economic Affairs (2014) *Delta Programme 2015. Working on the Delta. The Decisions to Keep the Netherlands Safe and Liveable.* The Hague: I&M/EZ.

IEA International Energy Agency (2013) *Policies and Measures. Netherlands: Energy Efficiency Action Plan.* Available at: www.iea.org/policiesandmeasures/pams/netherlands/name-23949-en.php (accessed 13 March 2015).

Institute of Environmental Protection (Republic of Poland) (ed.) (2010) *Republic of Poland. Fifth National Communication under the United Nations Framework Convention on Climate Change*. Warsaw: Ministry of the Environment and the National Fund for Environmental Protection. Available at: http://unfccc.int/resource/docs/natc/pol_nc5.pdf (accessed 10 April 2015).

IPCC Intergovernmental Panel on Climate Change (ed.) (2014) *Climate Change 2014: Impacts, Adaptation, and Vulnerability. Part A: Global and Sectoral Aspects*. Contribution of Working Group II to the Fifth Assessment Report of the Intergovernmental Panel on Climate Change. Cambridge/New York: Cambridge University Press.

KEMIN The Danish Ministry of Climate and Energy (ed.) (2009) *Denmark's Fifth National Communication on Climate Change. Fifth Report on the National Implementation of the United Nations Framework Convention on Climate Change and the Kyoto Protocol*. Available at: http://unfccc.int/resource/docs/natc/dnk_nc5rev.pdf (accessed 10 March 2015).

KEMIN The Danish Ministry of Climate and Energy (ed.) (2011) *Energy Strategy 2050 – From Coal, Oil and Gas to Green Energy*. Copenhagen: Litotryk Copenhagen A/S.

KEMIN The Danish Ministry of Climate Energy and Building (2013) *The Danish Climate Policy Plan. Towards a Low Carbon Society*. Copenhagen: The Danish Government.

Kjaergaard, Jens J. and Fenger, Jes (1990) *Fremtidens Klima? Miljo-Thema No. 2 1990*. Copenhagen: Miljøministeriet/Miljøstyrelsen. Available at: www2.mst.dk/Udgiv/publikationer/1990/87-503-8125-3/pdf/87-503-8125-3.pdf (accessed 31 March 2015).

Klimada (2015) *Klimada. Adaptation to Climate Change*. Available at: http://klimada.mos.gov.pl/en/ (accessed 30 April 2015).

Kok, Marcel, Vermeulen, Walter, Faaij, André, and DeJager, David (2002) 'Towards a Climate-neutral Society'. In: Kok, Marcel, Vermeulen, Walter, Faaij, André, and DeJager, David (ed.): *Global Warming and Social Innovation. The Challenge of a Climate Neutral Society*. London: Earthscan, pp. 1–16.

Krawinkel, Holger (1998) 'Die dänische CO_2-Steuer für Industrie und Gewerbe'. In: Landeszentrale für Politische Bildung Baden-Württemberg (ed.): *Globalisierung als Chance*. Document 7. LpB Forum 20–22 November 1997, Haus auf der Alb, pp. 71–74. Available at: www.lpb-bw.de/publikationen/global/krawin.htm (accessed 26 March 2015).

Kühne, Olaf (2002) 'Ökologie und Ökonomie in Ostmittel- und Osteuropa. Sozialistisches Erbe und Systemtransformation'. *Zeitschrift für Wirtschaftsgeografie* 46 (2): 73–91.

Kundzewicz, Zbigniew W. and Matczak, Piotr (2012) 'Climate Change Regional Review: Poland'. *WIREs Climate Change* 3: 297–311. DOI: 10.1002/wcc.175.

Lang, Kai-Olaf (2006) 'Mit Sicherheit – Polens Energiepolitik zwischen europäischer Solidarität und russischer Abhängigkeit'. *Polen-Analysen No. 2* (5 May 2006). Available at: www.laender-analysen.de/polen/pdf/PolenAnalysen02.pdf (accessed 21 April 2015).

Lang, Kai-Olaf (2007) *Polens Energiepolitik. Interessen und Konfliktpotentiale in der EU und im Verhältnis zu Deutschland*. SWP-Study S13 June 2007. Berlin: Stiftung Wissenschaft und Politik. Deutsches Institut für Internationale Politik und Sicherheit, pp. 1–34.

Lechtenböhmer, Stefan and Samadi, Sascha (2012) 'Blown by the Wind. Replacing Nuclear Power in German Electricity Generation'. *Environmental Science & Policy* 25 (2013), 234–241.

Luitwieler, Felix (2000) 'National Environmental Policy Plan 3: Dauerhafter Fortschritt durch Kontinuität und Modernisierung der Umweltpolitik'. In: Jänicke, Martin and Jörgens, Helge (ed.): *Umweltplanung im internationalen Vergleich. Strategien der Nachhaltigkeit*. Berlin/Heidelberg: Springer, pp. 15–26.

Maubach, Klaus-Dieter (2014) *Energiewende. Wege zu einer bezahlbaren Energieversorgung*. Wiesbaden: Springer VS.

McLean, Elena V. and Stone, Randall W. (2012) 'The Kyoto Protocol: Two-Level Bargaining and European Integration'. *International Studies Quarterly* 56: 99–113 DOI: 10.1111/j.1468-2478.2011.00706.x.

Meadows, Donella H., Meadows, Dennis L., Randers, Jorgen, and Behrens, William W. (1972) *The Limits to Growth. A Report for the Club of Rome's Projection on the Predicament of Mankind*. New York: Universe Books.

Mez, Lutz (1997) 'Klimaschutzpolitik als CO_2-Minderungspolitik. Dänemark und Deutschland im nationalen Alleingang'. In: Metz, Lutz and Jänicke, Martin (ed.): *Sektorale Umweltpolitik: Analysen im Industrieländervergleich*. Berlin: Edition Sigma, 15–31.

Miljøministeriet (Danish Ministry of the Environment – Environmental Protection Agency) (ed.) (2004) *Environmental Essay. Adapting to the Climate of the Future*. Environmental Essays No. 29, 2004. Copenhagen: Ministry of the Environment. Available at: www2.mst.dk/Udgiv/publications/2004/87-7614-456-9/pdf/87-7614-457-7.pdf (accessed 31 March 2015).

Ministerstwo Gospodarki (ed.) (2014) *Polish Nuclear Power Programme*. Warsaw: no publisher's imprint. Available at: www.bmub.bund.de/fileadmin/Daten_BMU/Download_PDF/Umweltpruefungen/polnische_kernenergie_programm_en_bf.pdf (accessed 24 April 2015).

Ministerstwo Środowiska (ed.) (2013) *Strategiczny plan adaptacji dla sektorów i obszarów wrażliwych na zmiany klimatu do roku 2020 z perspektywą do roku 2030*. Warsaw: no publisher's imprint. Available at: http://klimada.mos.gov.pl/wp-content/uploads/2013/10/SPA2020.pdf (accessed 30 April 2015).

Ministerstwo Środowiska (2015) *SPA 2020*. Available at: www.mos.gov.pl/artykul/5146_spa2020/20109_spa2020.html (accessed 30 April 2015).

Ministry of Economy (Republic of Poland) (ed.) (2009) *Energy Policy of Poland until 2030*. Warsaw: no publisher's imprint. Available at: www.mg.gov.pl/files/upload/8134/Polityka%20energetyczna%20ost_en.pdf (accessed 24 April 2015).

Ministry of the Environment (Republic of Poland) (ed.) (2003) *Poland's Climate Policy. The Strategies for Greenhouse Gas Emission Reductions in Poland until 2020*. Warsaw: no publisher's imprint. Available at: www.mos.gov.pl/g2/big/2009_04/cf234906b019de170218bf79f913990c.pdf (accessed 28 April 2015).

Ministry of the Environment (Republic of Poland) (ed.) (2013) *Polish National Strategy for Adaptation to Climate Change (NAS 2020)*. Warsaw: no publisher's imprint. Available at: https://klimada.mos.gov.pl/wp-content/uploads/2014/12/ENG_SPA2020_final.pdf (accessed 28 April 2015).

MKRO Ministerkonferenz für Raumordnung (ed.) (2013) *Umlaufbeschluss/Circular Resolution from 6 February 2013. Raumordnung und Klimawandel*. Berlin: MKRO.

MOZ Märkische Onlinezeitung (ed.) (2012) *Studie: Dänemark ist Weltmeister im Klimaschutz*. 3 December 2012. Available at: www.moz.de/artikel-ansicht/dg/0/1/1065639 (accessed 16 March 2015).

Nuissl, Henning (2007) 'Entwicklung der raumbezogenen Planung in Deutschland'. In: Weiland, Ulrike/Wohlleber-Feller: *Einführung in die Raum- und Umweltplanung*. Paderborn: Verlag Ferdinand Schöningh, pp. 39–47.

OECD (ed.) (2003) *OECD Environmental Performance Reviews. Netherlands 2003*. Paris: OECD.

Olszewski, Michał (2013) Die Polen und die Energiewende. In: Deutsches Polen Institut/Forschungsstelle Osteuropa/Deutsche Gesellschaft für Osteuropakunde (ed.)

Polen-Analysen No. 124. Die Polen und die Energiewende (16 April 2013), pp. 2–5. Available at: www.laender-analysen.de/polen/pdf/PolenAnalysen124.pdf (accessed 10 April 2015).

Pettenger, Mary E. (2007) 'The Netherlands' Climate Change Policy: Constructing Themselves/Constructing Climate Change'. In: Pettenger, Mary E. (ed.): *The Social Construction of Climate Change: Power, Knowledge, Norms, Discourses*. Hampshire/ Burlington: Ashgate, pp. 51–73.

Pysz, Piotr (2011) 'Wirtschaftssystem und ordnungspolitische Prozesse seit 1990'. In: *Polen. Informationen zur Politischen Bildung* 311: 26–34.

Riesbeck, Peter (2014) 'Energie-Politik. Polen bremst EU beim Klimaschutz'. *Frankfurter Rundschau* (22 October 2014). Available at: www.fr-online.de/energie/energie-politik-polen-bremst-eu-beim-klimaschutz,1473634,28818910.html (accessed 17 April 2015).

Rietbergen, Martijn, Farla, Jacco C.M., and Blok, Cornelis (2002) 'Do Agreements Enhance Energy Efficiency Improvement? Analyzing the Actual Outcome of Long-Term Agreements on Industrial Energy Efficiency Improvement in The Netherlands'. *Journal of Cleaner Production* 10: 153–163.

Room for the River (ed.) (2012) *Room for the River. From Higher Dykes to River Widening. Working Together Towards a Safe and Attractive River Region*. Utrecht: Ruimte voor de Rivier. Available at: www.ruimtevoorderivier.nl/english/publications/ (accessed 2 March 2015).

Roose, Jochen (2012) 'Wollen die Deutschen das Klima retten? Mobilisierung, Einstellungen und Handlungen zum Klimaschutz'. *Forschungsjournal Soziale Bewegungen* 25 (2): 89–100.

Schmitt, Stefan (2007) 'Dänemarks Klima. Wackelkurs des Windkraft-Weltmeisters'. *Spiegel Online* (8 March 2007). Available at: www.spiegel.de/wissenschaft/mensch/daenemarks-klima-wackelkurs-des-windkraft-weltmeisters-a-470459.html (accessed 14 January 2015).

Schnell, Christian (2013) 'Die Zukunft der erneuerbaren Energien in Polen'. *Energiewirtschaftliche Tagesfragen* 63 (4): 85–88.

Schroer, Markus (2006) *Räume, Orte, Grenzen. Auf dem Weg zu einer Soziologie des Raums*. Frankfurt am Main: Suhrkamp.

Schultz, Stefan (2012) 'EU-Richtlinie für Energieeffizienz. Deutschland torpediert europäischen Stromsparplan'. *Spiegel Online* (2 June 2012). Available at: www.spiegel.de/wirtschaft/soziales/energieeffizienz-bundesregierung-torpediert-richtlinie-der-eu-a-836489.html (accessed 2 February 2015).

SER Social and Economic Council of the Netherlands (2013) *Summary of Energy Agreement for Sustainable Growth. (Energieakkoord voor duurzame grei)* (6 September 2013). Available at: www.ser.nl/~/media/files/internet/talen/engels/2013/energy-agreement-sustainable-growth-summary.ashx (accessed 13 March 2015).

Stead, Dominic (2014) 'Urban Planning, Water Management and Climate Change Strategies: Adaptation, Mitigation and Resilience Narratives in the Netherlands'. *International Journal of Sustainable Development and World Ecology* 21 (1): 15–27. DOI: 10.1080/13504509.2013.824928.

Tellegen, Egbert (1989) 'The Dutch National Environmental Policy Plan'. *The Netherlands Journal of Housing and Environmental Research*. 4 (4): 337–345.

The Danish Government (ed.) (2008) *Danish Strategy for Adaptation to a Changing Climate*. Copenhagen: Schultz.

UBA Umweltbundesamt (ed.) (2005) *Klimawandel in Deutschland. Vulnerabilität und Anpassungsstrategien klimasensitiver Systeme. Umweltforschungsplan des Bundesministeriums für*

Umwelt, Naturschutz und Reaktorsicherheit. Research Paper 201-41-253. Dessau: Umweltbundesamt. Available at: www.umweltbundesamt.de/publikationen/klimawandel-in-deutschland-vulnerabilitaet (accessed 20 January 2015).

Uken, Marlies (2014) 'Kein Klimagipfel – ein Kohlegipfel'. *Zeit Online* (24 October 2014). Available at: www.zeit.de/politik/ausland/2014-10/eu-gipfel-kohle-polen (accessed 17 April 2015).

UNFCCC United Nations Framework Convention on Climate Change (1992) *United Nations Convention on Climate Change*. Available at: https://unfccc.int/resource/docs/convkp/conveng.pdf (accessed 31 October 2014).

UNFCCC United Nations Framework Convention on Climate Change (ed.) (1996) Decision 9/CP.2. Communications from Parties included in Annex I to the Convention: Guidelines, Schedule and Process for Consideration. In: *Report of the Conference of the Parties on its Second Session, Held at Geneva from 8 to 19 July 1996. Addendum. Part Two. Action Taken by the Conference of the Parties at its Second Session* (UNFCCC/CP/1996/15/Add.1), pp. 15–17.

UNFCCC United Nations Framework Convention on Climate Change (ed.) (1998) *Kyoto Protocol to the United Nations Framework Convention on Climate Change*. Available at: http://unfccc.int/resource/docs/convkp/kpeng.pdf (accessed 31 October 2014).

UNFCCC United Nations Framework Convention on Climate Change (2012) *Doha Amendment to the Kyoto Protocol*. Available at: http://unfccc.int/files/kyoto_protocol/application/pdf/kp_doha_amendment_english.pdf (accessed 14 November 2014).

UNFCCC United Nations Framework Convention on Climate Change (2014a) *Background on the UNFCCC: The International Response to Climate Change*. Available at: http://unfccc.int/essential_background/items/6031.php (accessed 31 October 2014).

UNFCCC United Nations Framework Convention on Climate Change (2014b) *Status of Ratification of the Kyoto Protocol*. Available at: http://unfccc.int/kyoto_protocol/status_of_ratification/items/2613.php (accessed 31 October 2014).

UNFCCC United Nations Framework Convention on Climate Change (ed.) (2014c) *National Greenhouse Gas Inventory Data for the Period 1990–2012*. UNFCCC/SBI/2014/20 (17 November 2014). Available at: http://unfccc.int/resource/docs/2014/sbi/eng/20.pdf (accessed 3 March 2015).

UNFCCC United Nations Framework Convention on Climate Change (2015a) *Focus: Adaptation*. Available at: http://unfccc.int/focus/adaptation/items/6999.php (accessed 14 May 2015).

UNFCCC United Nations Framework Convention on Climate Change (2015b) *Cancun Adaptation Framework*. Available at: http://unfccc.int/adaptation/items/5852.php (accessed 14 May 2015).

V&W Ministry of Transport, Public Works and Water Management (2006) *Spatial Planning Key Decision 'Room for the River'. Investing in the Safety and Vitality of the Dutch River Basin Region*. Utrecht: Kris Kras Design.

Van Koningsveld, D.M., Mulder, J.P.M., Stive, M.J.F., Vandervalk, L., and Vanderweck, A.W. (2008) Living with Sea-Level Rise and Climate Change: A Case Study of the Netherlands. *Journal of Coastal Research* 24 (2): 367–379.

Van Leeuwen, Matthijs (2013) *European Renewable Energy Incentive Guide – The Netherlands*. Available at: www.nortonrosefulbright.com/knowledge/publications/66148/european-renewable-energy-incentive-guide-the-netherlands (accessed 11 March 2015).

Vorholz, Fritz (2012): 'Die Ostblocker. Polen bremst den europäischen Klimaschutz'. *Zeit Online* (15 March 2012). Available at: www.zeit.de/2012/12/Klimaschutz-Europa (accessed 17 April 2015).

VROM Ministry of Housing, Spatial Planning and the Environment/V&W Ministry of Transport, Public Works and Water Management/LNV Ministry of Agriculture, Nature and Food Quality (ed.) (2007a) *National Programme for Spatial Adaptation to Climate Change*. The Hague: VROM. Available at: www.climatechangesspatialplanning.nl/gfx_content/documents/documentation/National_Adaptation_Strategy_The_Netherlands.pdf (accessed 20 January 2015).

VROM Ministry of Housing, Spatial Planning and the Environment/V&W Ministry of Transport, Public Works and Water Management/LNV Ministry of Agriculture, Nature and Food Quality/EZ Ministry of Economic Affairs/IPO Association of Provincial Authorities/VNG Association of Netherlands Municipalities/UvW Association of Water Boards (ed.) (2007b) *Make Room for the Climate! Initial Steps Towards a National Adaptation Strategy*. Memorandum for policy discussion. Den Haag: VROM. Available at: www.climatechangesspatialplanning.nl/gfx_content/documents/documentation/ARK_make_room_for_climate.pdf (accessed 23 January 2015).

VROM Ministry of Housing, Spatial Planning and the Environment/V&W Ministry of Transport, Public Works and Water Management/LNV Ministry of Agriculture, Nature and Food Quality/EZ Ministry of Economic Affairs/IPO Association of Provincial Authorities/VNG Association of Netherlands Municipalities/UvW Association of Water Boards (ed.) (2007c) *Maak ruimte voor klimaat! Nationale adaptiestrategie – de beleidsnotitie*. Den Haag: VROM. Available at: www.rijksoverheid.nl/bestanden/documenten-en-publicaties/kamerstukken/2007/11/02/nationale-adaptatiestrategie-maak-ruimte-voor-klimaat-beleidsnotitie-pdf/arkbeleidsnotitiemr2nov07.pdf (accessed 23 January 2015).

VROM Ministry of Housing, Spatial Planning and the Environment/EZ Ministry of Economic Affairs/V&W Ministry of Transport and Public Works/LNV Ministry of Agriculture, Nature and Food Quality/FIN Ministry of Finance/BuZa Ministry of Foreign Affairs (2007d) *New Energy for Climate Policy. The 'Clean and Efficient' Programme*. Available at: www.rijksoverheid.nl/bestanden/documenten-en-publicaties/brochures/2010/11/25/new-energy-for-climate-policy/11br2007g486-2008122-154154.pdf (accessed 12 March 2015).

VROM Ministry of Housing, Spatial Planning and the Environment (ed.) (2009) *Fifth Netherlands National Communication under the United Nations Framework Convention on Climate Change*. Available at: http://unfccc.int/resource/docs/natc/nld_nc5.pdf (accessed 2 February 2015).

Weiss, Thomas G. and Burke, Martin J. (2011) 'Legitimacy, Identity and Climate Change: Moving from International to World Society?' *Third World Quarterly* 32 (6): 1057–1072.

Wejs, Anja, Harvold, Kjell, Vammen Larsen, Sanne and Saglie, Inger-Lise (2014) 'Legitimacy Building in Weak Institutional Settings: Climate Change Adaptation at the Local Level in Denmark and Norway'. *Environmental Politics* 23 (3): 490–508.

Wiering, Mark and Immink, Irene (2006) 'When Water Management Meets Spatial Planning: A Policy-Arrangements Perspective'. *Environment and Planning C: Government and Policy* 24: 423–438.

Wilson, Elizabeth and Piper, Jake (2010) *Spatial Planning and Climate Change*. New York: Routledge.

Wróblewski, Andrzej (1992) 'Analysis and Forecast of Long-Term Sea Level Changes along the Polish Baltic Sea Coast Part I: Annual Sea Level Maxima'. *Oceanologica*, 33: 65–85.

Zeidler, Ryszard B. (1997) 'Climate Change Vulnerability and Response Strategies for the Coastal Zone of Poland'. *Climatic Change* 36: 151–173.
Zelljadt, Elisabeth, Velten, Eike Karola, Prahl, Andreas, Duwe, Matthias, and Poblocka, Anna (2014), *Assessment of Climate Change Policies in the Context of the European Semester. Country Report: Poland.* Berlin: Ecologic Institute/Eclareon.

7 Empirical approach

The aim of this study is to investigate vulnerability and resilience constructions of actors in spatial planning in Europe, and to find explanations for their differences. Actors from spatial planning were thus surveyed via an online questionnaire (Schnell *et al.* 2008: 214; de Vaus 2001: 50). Europe-wide studies are associated with particular challenges. Besides the control of classic challenges such as coverage, sampling, non-response, or measurement errors (Groves *et al.* 2009: 49ff.; Schwartz *et al.* 1998: 144), transnational comparability (equivalence) must be established (Rippl and Seipel 2008: 67ff.; Van de Vijver 2003; van Deth 1998). However, this then presupposes a high relevance of national difference, which should not be examined until later with regard to its culture-spatial relevance. Nonetheless, in selecting actors from the field of spatial development it is important to take account of possible institutional differences at the point of selection. Actors from the realm of politics and administration are obviously active within different political systems. It should become clear in the course of this chapter what steps have been taken in order to justify each of the methodical equivalence postulates. In Section 7.1 the study's survey design will be explained, first outlining the appropriate target population in terms of the basic population for the study, and then explaining the move from the gross to the net sample within the selected population. Special features of the data used will likewise be made clear. In section 7.2, the constructs and indicators used in the questionnaire will be presented.

7.1 Survey design

7.1.1 Basic population: determination of the target population

The first issue in the design of the survey was the determination of a suitable target population, which here is represented by actors in the field of spatial development. Their precise determination is discussed in the following steps: (1) the selected area of study, (2) the functional area of spatial development and the relevant societal fields (administration, politics, economy, civil society), and (3) the thematically equivalent positions of actors in these fields.

(1) The area of study. Research design quality depends on the extent to which possible interference factors can be minimised (Schnell *et al.* 2008: 217). It should be considered that even varying landscape configurations might play a role in the genesis of vulnerability and resilience constructions (see also Section 3.3). Outside of laboratory experiments, this central problem of culture-comparative social-science climate research cannot be categorically eliminated and must therefore be taken into account particularly when analysing the data.[1]

Nevertheless, keeping attributes of all research areas constant could be a useful strategy to reduce the problem (Schnell *et al.* 2008: 221). The present study aims to achieve this by concentrating on the most similar landscapes possible. Coastal regions thus present an interesting research field for comparative investigation into the treatment of climate change, as several central problem constructions are regularly discussed here, such as sea-level rise and flooding caused by tides and storms (see Chapter 3). In this study, the coastal municipalities of Germany, the Netherlands, Denmark, and Poland will therefore be focused upon.

Schernewsky (2004) points out that an authoritative definition of coasts does not exist.[2] Coastal municipalities will thus be precisely defined as those within areas up to a maximum of 10km away from the coastline. To introduce a settlement-specific constant, only those towns and municipalities with a population of 5000 or more will be referenced. On the basis of these criteria, a first selection stage could be made with the assistance of a geographical information system.

(2) The functional area of spatial development. This study concentrates on the *area of spatial development* (see Section 4.1). The issue of focusing on concrete actors can thus be elaborated. Gustedt (2010: 20) points out that differences in planning traditions in European countries has led to the varying participation of actors in spatial development processes. To be able to justify the postulate of equivalence (Van de Vijver 2003; Van Deth 1998), actors must nevertheless be selected whose positions in spatial development are similar to one another. So that the social relevance of the research results can be assumed, the actors surveyed should be involved as far as possible in spatial development decision-making processes in European coastal regions, or be actively influencing them. My preliminary considerations thus follow the social-constructivist concept of expertise used by Hitzler and his colleagues: it must be possible for respondents to be ascribed 'institutionalised competence to construct reality' (Hitzler *et al.* 1994) in the spatial development of European coastal regions. To meet the limited scope of this study, I have proceeded on the basis of theoretical considerations.

Section 4.1 has already discussed how new forms of communication and coordination between the state, economy, and civil society have gained in significance (see also Section 7.1; Benz *et al.* 2007: 13). In social-science climate research too, multiple issues connected to the term climate-change 'governance' have been discussed (Heimann and Mahlkow 2012; Bauer *et al.* 2011; Termeer *et al.* 2011; Frommer 2010). For the selection of appropriate target individuals, these observations mean that currently both state and non-state actors are relevant to spatial interpretation and action (Frommer 2010: 66; Zürn 2008: 559), thus *actors from different societal fields* are selected.[3] It is assumed that social relevance for spatial action is present if specific organisational positions are used as a criterion for selection.

(3) Field-specific equivalent positions. This criterion allows for the discussion of *functionally equivalent positions* in the fields of administration, politics, economy, and civil society. Within *administration* one can differentiate for the European coastal regions under consideration between general spatial planning and spatially relevant specialist planning (Birkmann *et al.* 2013: 8; Runkel 2005: 281ff.; Turowski 2005: 898). Actors are selected from both levels of planning. *General spatial planning* includes the overall planning of structures and their quality, up to the specific usage of land (Turowski 2005: 895). *Specialist planning* has sectoral spatial planning relevance including, for example, planning of formal environmental law, watercourses, means of transport, or flood protection (Runkel 2005: 281ff.). Handling climate change can involve diverse departments, from flood protection, agricultural and environmental protection to transport (Frommer 2010: 38). In order not to complicate matters unnecessarily, I have focused only on specialist planning relevant to coastal regions, namely coast and flood protection. In addition, those environmental departments responsible for climate and the protection of natural landscape are included.

Following *administration* within the scheme of classification, under *politics* the focus will be on actors from spatial planning and environmental committees. Representing *civil society*, the target group is formed by those responsible for each of the ten largest environmental protection NGOs in each country with responsibility for spatial planning, protection of the natural landscape, or climate change. Under *economy*, contacts from chamber of commerce departments responsible for regional economic development and the environment (as specialists responsible for spatial planning) are consulted.

The basic population is defined as follows: all of the publicly available[4] actors from spatial development relating to the coastal municipalities of the Netherlands, Germany, Denmark, and Poland having the following functions:

Administration: General spatial planning/spatially relevant specialist planning

Management and assistants from spatially relevant general planning:

- Scheduling and preparation of plans[5]

Management and assistants from the following spatially relevant specialist planning:

- Coastal protection, flood protection
- Environment: Protection of natural landscape, climate mitigation, and climate adaptation

Table 7.1 NUTS (Nomenclature of Territorial Units for Statistics) levels

EU classification	*Germany*	*Netherlands*	*Denmark*	*Poland*
LAU1/LAU2	Gemeindeebene (*Municipality level*), Kreisfreie Städte (*Urban districts*)	Gemeenten (*Municipality level*)	Kommuner, Sogne (*Municipality level*)	Gminy (*Municipality level*)
NUTS3	Landkreise ((*Rural*) *districts*)			Poviaty ((*Rural*) *districts*)
NUTS2	Regionen, Planungsräume (SH) (*Regions, planning areas*) StAUN (MV) (*State agencies for environment and nature*)	Watershappen, Plusregio (*State agencies for environment and nature*)		Podregiony (*Subregions*), Coastal protection
NUTS1	Länderebene (*State level*)	Provincies (*Regions, Provinces*) (NUTS2), Rijkswaterstat (*Regional agency for environment/water*)	Regioner (*Regions, Provinces*)	Voivodeships, Regiony (*Regions, Provinces*)
NUTS0	Bundesebene (*Federal/National level*)	*National level*	(*National level*)	(*National level*)

Politics: Spatial planning, construction, and environment committees
Members of political committees with responsibility for:

- Spatial planning/construction
- Environment

Civil society: Environmental protection NGOs
Persons responsible for:

- Spatial planning
- Environment: Protection of natural landscape, climate mitigation, and climate adaptation

Economy: Chambers of industry and commerce
Contacts for:

- Spatial planning: Local economic policy
- Environment

Challenge: vertical planning levels. A particular challenge is presented by the fact that spatially relevant (specialist) planning is closely bound to national conditions, particularly in politics and administration. The central challenge becomes apparent when one considers that specific task fields and functions can be located at different vertical levels in each country. Using the traditional terms in Anglo-American political science, one must precisely differentiate between *polity equivalence*, *policy equivalence*, and *politics equivalence* when considering actors from different national backgrounds (Schubert and Bandelow 2009: 5). These considerations show that a total functional equivalence is not achievable.

To be able to find approximate functional equivalents, all relevant planning levels relating to coastal municipalities are thus taken into account. Table 7.1 lists these according to the NUTS (Nomenclature of Territorial Units for Statistics) classification for each of the national coasts considered, as well as an English translation of the levels. This shows that in selecting the basic population, I have sought to avoid a container-space selection concept for municipalities. In fact, actors are considered who are of central functional relevance for the coastal towns and communities and not simply located within the municipalities, but rather at decision-making levels above them.

7.1.2 Selected population: gross and net samples

The survey was conducted by means of a written online questionnaire. The intended basic population (see section 7.1.1) is a *rare population*.[6] To be able to capture rare populations as precisely as possible, selection procedures must be tailored to the specific characteristics of the target population (Schnell *et al.* 2008: 294ff.; Diekmann 1999: 345ff.). Although clearly demarcated by institution, there were no complete lists of the members of the target population when the study began, and so it was necessary to compile a separate register for the population to be selected. To accomplish this, extensive internet and telephone research was combined to collect postal addresses for the relevant actors at the respective institutions on the register.

Using a geographical information system, all 439 coastal municipalities with more than 5,000 inhabitants were then defined for the Netherlands, Germany, Denmark, and Poland. Subsequently, actors from the horizontal levels (politics, administration, NGOs, and chambers of commerce) and vertical levels (politics and administration only, see Section 7.1.1) were captured with individual e-mail addresses, following the departmental criteria listed in Section 7.1.1. For politics and administration, the postal addresses of members from spatial planning and environmental committees, as well as those of managers and assistants of each (specialist) planning area, had to be ascertained for each of the 439 municipalities and the above-mentioned planning levels. Chambers of industry and commerce are partly superordinate to several municipalities, and so the institutions and their representatives responsible for the coastal municipalities were researched, following the specifications of Section 7.1.1. NGOs responsible for environmental protection do not conform to the national organisational structure either. The ten largest organisations in each country were thus researched and specific positions within them selected. Table 7.2 shows the composition of the gross sample by country and field.

As a relatively small population was involved, the survey was carried out as a full population survey. The collection of data was effected by means of a written online questionnaire. Personalised weblinks to programmed questionnaires were sent to the target individuals, and the questionnaires remained accessible online from January to April 2012. To achieve the optimal response rate, the procedure was oriented to the tailored design approach of Dillman (Dillman 2007; Porst 2001). The initial contact and the questionnaire were each authored in German (Germany), English (the Netherlands, Denmark), or Polish (Poland), and the questionnaire was optionally available in each of these languages.

The survey was targeted at experts, predominantly in positions of responsibility which were directly attributable to their work environment. The total time required to complete the questionnaire was about 40 minutes. The response rate came to 13%, although the figure varies by country. For Germany, the return of 21% can be assessed as satisfactory under the circumstances. The response rate was 12% for Poland, 9% for the Netherlands, and 7% for Denmark. There is hardly any consensus in the literature about what response rate should be expected for online surveys.[7] With these response rates, however, distortions through unit non-response can be assumed to occur.[8]

The selection population is based on online lists and information obtained by telephone. It has already been noted that the current validity could not always be taken for granted. Changes in position, as well as errors in the information given, can lead to individuals who do not belong to the basic population being targeted who then end up being included in the selection population, and ultimately in the data record. In order to minimise distortions caused

Table 7.2 Gross sample, response rates, and final data set after adjustment

	Germany		*Netherlands*		*Denmark*		*Poland*		TOTAL	
	#	%	#	%	#	%	#	%	#	%
Gross sample										
Administration	1315	55	332	25	778	42	986	80	3411	50
Politics	814	34	738	55	581	32	85	7	2218	33
NGOs	188	8	213	16	387	21	104	8	892	13
Economy	64	3	66	5	95	5	52	4	277	4
Total/Gross [Overall Proportion]	2381	[35]	1349	[20]	1841	[27]	1227	[18]	6798	[100]
Net sample										
Response	501	(21)	116	(9)	104	(7)	145	(12)	866	(13)
Net case numbers (Data set after adjustment)	473	(20)	115	(9)	103	(7)	139	(11)	830	(12)

Source: own calculations.

Table 7.3 Socio-demographic composition of the final data set

	Germany	*Netherlands*	*Denmark*	*Poland*	TOTAL
Age					
Under 25	1	0	1	1	3
	(0.2%)	(0.0%)	(1.0%)	(0.7%)	(0.4%)
25 to 34	20	17	9	35	81
	(4%)	(15%)	(9%)	(25%)	(10%)
35 to 44	55	20	26	20	121
	(12%)	(17%)	(25%)	(14%)	(15%)
45 to 54	161	28	19	29	237
	(34%)	(24%)	(18%)	(21%)	29%)
55 to 64	167	33	30	32	262
	(35%)	(29%)	(29%)	(23%)	32%)
65 and older	44	10	14	9	77
	(9%)	(9%)	(14%)	(6%)	(9%)
No information given	25	7	4	13	49
	(5%)	(6%)	(4%)	(9%)	(6%)
Gender (0=female, 1=male)					
Female	124	24	26	56	230
	(26%)	(21%)	(25%)	(40%)	(28%)
Male	337	84	74	74	569
	(71%)	(73%)	(72%)	(53%)	(69%)
No information given	12	7	3	9	31
	(3%)	(6%)	(3%)	(6%)	(4%)

continued

Table 7.3 Continued

	Germany	*Netherlands*	*Denmark*	*Poland*	TOTAL
Education level					
Without qualifications	1 (0.2%)	0 (0.0%)	0 (0.0%)	0 (0.0%)	1 (0.1%)
Level 1: Secondary-modern school certificate	14 (3.0%)	0 (0.0%)	0 (0.0%)	0 (0.0%)	14 (1.7%)
Level 2: High-school diploma/Apprenticeship or equivalent	61 (13%)	2 (2%)	3 (3%)	0 (0.0%	66 (8%)
Level 4: Advanced technical-college certificate/Higher-education entrance qualification or equivalent	78 (16%)	9 (8%)	6 (6%)	7 (5%)	100 (12%)
Level 5: University degree or equivalent	307 (65%)	101 (88%)	92 (89%)	123 (88%)	623 (75%)
No information given	12 (3%)	3 (3%)	2 (2%)	9 (6%)	26 (3%)
Field affiliation					
Administration	264 (56%)	46 (40%)	61 (59%)	117 (84%)	488 (59%)
Politics	159 (34%)	40 (35%)	16 (16%)	10 (7%)	225 (27%)
NGOs	38 (8%)	22 (19%)	22 (21%)	8 (6%)	90 (11%)
Economy	12 (3%)	7 (6%)	4 (4%)	4 (3%)	27 (3%)
Total	473 (100%)	115 (100%)	103 (100%)	139 (100%)	830 (100%)

by overcovering, actors were therefore removed from the data set if their area of responsibility did not correspond to the predetermined functions of the basic population. Additionally, actors who did not belong to the intended coastal municipalities were removed from the sample. The final data set consists of n = 830 actors. Table 7.3 shows its socio-demographic composition.

7.2 Questionnaire: constructs and indicators

In the following, the most important indicators used in the online questionnaire will be detailed. For the capturing of vulnerability and resilience constructions, a number of indicators were developed and (pre-)tested. As previously noted, the interviewees could choose between German, English, and Polish versions. Reverse translation was undertaken in each case to ensure linguistic equivalence (Harkness *et al.* 2010; Smith 2003). Here the most important constructs and their indicators are presented in English. When recording explanatory factors, common indicators could also be drawn upon from the literature.[9]

Vulnerability constructions

Table 7.4 shows the constructs and indicators for the capture of actors' vulnerability constructions (on the scientific discourse, see Chapter 3). Captured first of all were constructions concerning *climate change development and cause*, indeed the idea itself that climate change is occurring and its cause can be attributed to humans. To this end, a five-point Likert scale was used, ranging from 1 'Strongly disagree' to 5 'Strongly agree'.

Additionally, various problem grade attributions were captured, namely with regard to *climate change as a problem in general*, and related to different *places and regions*. A five-point Likert scale was used here, ranging from 1 'No problem at all' to 5 'Very serious problem'. In this way, problem grades ascribed to concrete impacts of climate change for each reference area were also captured. Also surveyed was the extent to which advantages and disadvantages were expected to result from climate change.

Resilience constructions

Table 7.5 shows the captured constructs and indicators for resilience constructions. To capture preferences for the discussed climate-mitigation and adaptation measures, five-point Likert scales were used (on the discourse, see Chapter 4). The survey first assessed the *necessity* of various practices until 2025 from 1 'Unnecessary' to 5 'Very necessary'; second, the ascribed *feasibility* was captured on a five-point Likert scale from 1 'Not at all feasible' to 5 'Very easily feasible'. The year 2025 was chosen to allow reasonable time for current spatial development decision processes to be pictured (see also Heimann 2013: 23). In each case, *familiarity* with climate-mitigation and adaptation measures against sea flooding, inland flooding, heat, and storms was also captured on a scale of 1 'Not at all familiar' to 5 'Very familiar'.

Table 7.4 Vulnerability constructions: constructs and indicators

Construct	*Indicator question/item*	*Scale*
	Do you agree or disagree with the following statements?	Five-point scale:
The course of climate change	… Climate change is already taking place.	1 'Strongly disagree' to
Anthropogenic cause	… Climate change is caused by humans.	5 'Strongly agree'
Ascribed problem grade	Do you think climate change is a serious problem	Five-point scale:
… of climate change in general	… in general.	1 'No problem at all' to
… of climate change for the reference area	… for [programme output: reference area].	5 'Very serious problem'
… globally	… for the world.	
Advantages or disadvantages ascribed to climate change	By 2025/2100: Will there be more advantages or disadvantages for the Coastal Municipalities of Central Europe as a result of climate change?	1 More advantages 2 More disadvantages 3 Neither more advantages nor disadvantages
Ascribed problem grade of possible climate-change impacts (each by the year 2025 and 2100)	By 2025/2100: Please estimate how some aspects of the climate may change in the Costal Municipalities of Central Europe/[reference area]. Do you agree with the statements? How serious do you think problems will be? How confident are you in your estimation? Please select three options in each line.	Five-point scale: 1 'No problem at all' to 5 'Very serious problem'
	… Rise in sea level	
	… More storm floods	
	… More torrential rain/Heavy rainfalls	
	… Highest levels of continental waters (lakes, rivers, creeks) rising	
	… More floodplains	
	… Loss of land to water	
	… More heatwaves	
	… More periods of drought	
	… Increased intensity of storms	
	… High fluctuations of groundwater levels	

Table 7.5 Resilience constructions: constructs and indicators

Construct	*Indicator question/item*	*Scale*
	Please assess the necessity and feasibility of the following measures for [program output: reference area] by 2025	Five-point scale: 1 'Unnecessary' to 5 'Very necessary' 1 'Not at all feasible' to 5 'Very feasible'
Preference for climate-mitigation measures	Sufficiency … Individual motorised traffic reduction … Better infrastructure for bicycle traffic … Spatial combination concepts for living and working	
	Consistency: Renewable energies … Securing areas for renewable energies (e.g. wind and solar power) … Renewable energies in private households and private companies (solar power)	
	Efficiency … General energy efficiency measures … Energy optimised buildings (e.g. insulation) … Increase of building density (e.g. decrease the distances between buildings) in order to develop energy efficient settlements	
	Ecological sinks … Creation of public areas with plants and/or grass … Creation of private areas with plants and/or grass	
	Technological sinks … Technical solution: Carbon Capture Storage technology (CCS) (technological capturing of carbon dioxide (CO_2) from large point sources, such as fossil-fuel power plants, and storing it underground)	

continued

Table 7.5 Continued

Construct	*Indicator question/item*	*Scale*
Preference for measures against sea flooding	Protection	
	… Improvement of dykes and protective constructions (e.g. new installation, dyke enlargement)	
	Retreat	
	… Displacement of dykes inland	
	… Deconstruction of dykes in order to reintroduce nature at the coast	
	… Withdrawal/deconstruction of settlement areas located near the sea	
	Integration	
	… Water-adapted settlement structures (e.g. floating houses, installation of anti-flood doors)	
	… Local population safety rooms	
Preference for measures against inland flooding	Protection	
	… Improvement of dykes and protective inland constructions	
	… Installation of watersheds to preserve water	
	Retreat	
	… Reintroduction of natural watercourses (e.g. rivers and creeks)	
	… Deconstruction of settlement areas located near inland waters (e.g. rivers and creeks)	
	… Removal of sealed surfaces to slow down water flow	
	Integration	
	… Roofs and facades with plants and/or grass to slow down water flow	
	… Water-adapted settlement structures, (e.g. floating houses, anti-flood doors, etc.)	

Preference for measures against heat	Building optimisation ... Energy optimised buildings
	Renaturation ... New structural corridors for fresh- and cold-air supply ... Reduction of building density (e.g. increased distances between buildings)
	Surface design ... Removal of sealed surfaces ... Roofs and facades with plants and/or grass
Preference for measures against storms	Technological structure reinforcement ... Technical adaption of buildings ... Changes to regulations concerning the wind loads of buildings
	Shelters ... Provision of local population safety rooms

Value orientations, ecological beliefs, and place identification (VBI)

Table 7.6 shows the captured constructs and indicators for values, ecological beliefs, and place identification:

Values: Shalom Schwartz has suggested various measuring instruments for the evaluation of his *value orientations*, such as the 57-item Schwartz Value Survey (SVS), or the Portraits Value Questionnaire (PVQ) with 40 or 21 items (Schmidt *et al.* 2007; Schwartz 2003, 1992). Additionally, the World Values Survey uses a 10-item variant (WVS 2006). It was necessary to weigh the questionnaire length against the potential gain in validity. The 10-item variant of the World Values Survey was therefore chosen in order to minimise the length of the questionnaire and the dropout rate of respondents. Each value type was thus represented by an item as shown in the table.

It is important to emphasise here that Schwartz also suggests centring the values for the analysis on individuals.[10] However, as this study focuses on the socially shared values of groups rather than individuals, no individual centring was carried out.

Beliefs: In the 1970s, Dunlap and his colleagues suggested a scale for measuring NEP beliefs, revising it in 2000 (Dunlap *et al.* 2000; Dunlap and Van Liere 2008 [1978]), which has been used for this study.

As already discussed in section 5.2.2, the 2000 scale captures individual characteristics along the pole 'anthropocentrism (HEP) vs. ecocentrism (NEP)'. Here the interviewees assess their degree of agreement with various statements on a scale from 1 'Strongly disagree' to 5 'Strongly agree'. The scale from 2000 is based on 15 items which should capture the previously mentioned four dimensions (ecological limits, balance of nature, human dominance, and ecological catastrophe) (see also Section 5.2.2). In order not to stretch out the questionnaire, a maximum of three items per dimension were captured. The items are shown in the table. Dunlap *et al.* (2000: 430) observe that distinct independent dimensions are also frequently assumed in the literature. Thus is it is important to test how reliably the items taken together capture 'NEP orientation'. The scale used indicates a good total reliability (Cronbach's alpha 0.746, Heimann 2017).

Place identification: To capture interviewees' degree of place identification, this investigation makes use of items from the previously discussed Lalli scale (1992). To reduce the length of the questionnaire, it was decided that only one item should be used for each of Lalli's postulated dimensions. As previously discussed, Lalli's instrument was initially developed to measure city identification, although he sees the possibility of its use in research on other scales. I thus combine the scale with an item from the International Social Survey Programme (ISSP), within which framework a survey on national identity was carried out in 2003. With the ISSP item, place identification was captured on various scales, ranging from the municipality to the European level (ISSP 2003: 3).

The scale used here demonstrates very good reliability (Cronbach's alpha: 0.84; see Heimann 2017), with one exception: the item on 'External Evaluation'

Table 7.6 Items for the detection of values, beliefs, and identities

Schwartz-values

Standard type	*Value type*	*Item*
Self-transcendence	Benevolence	It is important to this person to help the people nearby; to care for their well-being
	Universalism	Looking after the environment is important to this person; to care for nature
Self-enhancement	Power	It is important to this person to be rich; to have a lot of money and expensive things
	Achievement	Being very successful is important to this person; to have people recognise one's achievements
	Hedonism	It is important to this person to have a good time; to 'spoil' oneself
Openness to change	Stimulation	Adventure and taking risks are important to this person; to have an exciting life
	Self-direction	It is important to this person to think up new ideas and be creative; to do things one's own way
Conservation	Conformity	It is important to this person to always behave properly; to avoid doing anything people would say is wrong
	Tradition	Tradition is important to this person; to follow the customs handed down by one's religion or family
	Security	Living in secure surroundings is important to this person; to avoid anything that might be dangerous

Source: WVS 2006.

Notes

Question: We would like to know more about basic value orientations. Please read the following description of people and please indicate for each description is that person like you or not?' Six-level scale from 1 'Not at all like me' to 6 'Very much like me'.

continued

Table 7.6 Continued

Beliefs: new ecological paradigm	
Dimension	*Item*
Ecological limits	We are approaching the limit of the number of people the earth can support (Agreement) The use of the Earth's natural resources is already reaching its limits (Agreement) (Menzel 2007) Human ingenuity will insure that we do NOT make the Earth unlivable (Rejection)
Balance of nature	When humans interfere with nature, it often produces disastrous consequences (Agreement) The balance of nature is very delicate and easily upset (Agreement) The balance of nature is strong enough to cope with the impacts of modern industrial nations (Rejection)
Human dominance	Humans have the right to modify the natural environment to suit their needs (Rejection)
Ecological catastrophe	If things continue on their present course, we will soon experience a major ecological catastrophe (Agreement)

Source: Menzel 2007; Dunlap *et al.* 2000.

Notes
Question: To what extent do you agree with the following statements? We expressly want to know your personal opinion. Please choose one option in each line. Five-level scale from 1 'Strongly disagree' to 5 'Strongly agree'.

Place identification

Dimension	*Item*
General place identification (ISSP 2003)	
Identification with reference area	'How close do you feel to [reference area]?'
Urban identity scale (Lalli 1992)	
External evaluation	'Externally [reference area] is seen as possessing prestige.'
Familiarity	'When I amble through [reference area] I feel very strongly that I belong here.'
General attachment	'I feel really at home in [reference area]'
Continuity with personal past	'Lots of things in [reference area] remind me of my own past.'
Commitment/Future	'My personal future is closely tied up with [reference area].'

Source: ISSP 2003; Lalli 1992.

Notes

Question, ISSP item: 'How close do you feel to [reference area]?' Originally, this scale was used with places/regions on various scales, such as 'Your place of residence', '… the [federal] state in which you live', '… country', '… Europe'. (Scale from 1 'Not close at all' to 5 'Very close').

Question, Lalli Items: 'To what extent do you agree with the following statements?' Five-level scale from 1 'Strongly disagree' to 5 'Strongly agree'.

shows a very low inter-item correlation (0.36). Cronbach's alpha improves if this dimension is not taken into account (alpha = 0.86). The corresponding item was therefore excluded in the construction of the Likert scale.

Climate-change-related communication

Actors' climate-communication behaviour was also surveyed, for which original indicators were developed. Individual *media use* was captured, based on questions about what specific media the interviewees had already used to inform themselves about climate change. Of interest here was the field affiliation and spatial reach of the media in question. In addition, the *direct communicative exchange on climate change with other actors* was investigated. Here actors were asked with which other actors they had already exchanged information about climate change, focusing specifically on their *field affiliation* and *spatial reach* (see Chapter 8).

Socio-demographic variables

Various indicators were additionally surveyed for the capture of socio-demographic variables, including age, gender, highest achieved level of education (according to country-specific qualifications), and field affiliation. As these can be regarded as common indicators of empirical social research, they will not be discussed any further here.

7.3 Interim conclusion

This chapter has illustrated the empirical approach of the study. The basic population and the formulation of the gross and net samples have been considered. The constructs and indicators captured by the questionnaire have also been discussed.

The basic population consists of actors from spatial development with relation to the coastal municipalities of Germany, the Netherlands, Denmark, and Poland. According to the 'narrow' governance concept, actors from administration, politics, the economy, and NGOs were selected. The specific positions of the actors in their fields have been comprehensively elucidated. It was also discussed that, in order to be able to justify the equivalence postulate, not only actors on location, but also actors with decision-making responsibilities at higher governance levels were taken into account. In this way, the basic population adheres to a relational understanding of space, beyond the container space notion of coastal municipalities.

The basic population is a rare population for which there was no pre-existing register. It was therefore necessary to produce this. The selection of the net sample proceeded in several stages. First the coastal municipalities were selected with the aid of a geographical information system, to be able to research relevant, field-specific actors. The gross sample comprised $n = 6798$ actors. For the fieldwork, the tailored design approach of Dillman was employed. The response amounted to 13%, and the corrected net sample finally comprised $n = 830$ actors.

The most important constructs and indicators of the questionnaire have been discussed. For the questionnaire itself, the actors could choose between versions in German, English, and Polish. To establish linguistic equivalence, reverse translation was performed.

Notes

1 For example, in the context of determining climate-adaptation cultures, the analysis of differing preferences for concrete problem constructions, such as flooding or storms, will be the focus. Only those actors who generally support measures such as floods or storms will be taken into account and compared (see Chapter 10).
2 A comprehensive discussion is available in Kannen (2000).
3 These are sometimes referred to as 'spheres' (Frommer 2010: 66). This concept of governance requires functionally differentiated societies and thus raises new questions of scaling in intercultural comparison. In particular, the question arises as to the degree to which these fields themselves structure the knowledge and form the culture-spatial relations of actors (see also the empirical section of this study).
4 See also the discussion on the intended basic and target populations in section 7.1.2.
5 Following Turowski, the scheduling of plans means in the case of Germany the installation of planning instruments through communal urban land use planning, regional and state planning, as well as their preparation through the regional planning activities of the German Länder and the Bund (federation) (Turowski 2005: 896). Spatially relevant planning is thus closely connected to national conditions. This study originates within the German system and I have subsequently researched functional equivalents for the remaining nations in the study. For comprehensive comparisons, see: for the Netherlands, ARL (2003: 1ff., 22ff.), Fischer and Foißner (2002); for Poland, ARL (2001: 29ff., 40ff.), Bielawska-Roepke (2010: 51). In the case of Denmark, no direct comparisons were available and it was necessary to continue without this form of assistance here (see Danish Ministry of the Environment, 2007).
6 Schnell *et al.* (2008: 294) speak of a rare population when, as a proportion of the total population, it contributes less than 5%; see also Glowsky (2011: 97).
7 Thus Maurer and Jandura name response rates of between 5% and 80% in a meta-analysis (Maurer and Jandura 2009: 66).
8 Thus, for example, the perception of climate change as a problem was highest in Denmark, which seems entirely plausible given the local discourse (see Section 6.3; keyword: 'climate-mitigation prodigy' (Schmitt, 2007). Nevertheless, the high value could also be connected to the low level of response. Individuals with a higher awareness of the problems and relevance of climate change may have been more prepared to participate in the survey.
9 The German and Polish questionnaire equivalents for my own and the modified indicators are given in the German version of this book (see Heimann 2017).
10 The guidelines and SPSS syntax of Shalom Schwartz can be found on the website of the European Social Survey: www.europeansocialsurvey.org/docs/methodology/ESS1_human_values_scale.pdf (11 December 2015).

References

ARL Akademie für Raumforschung und Landesplanung (ed.) (2001) *Planungsbegriffe in Europa. Deutsch-Polnisches Handbuch der Planungsbegriffe*. Hannover: Akademie für Raumforschung und Landesplanung.

ARL Akademie für Raumforschung und Landesplanung (ed.) (2003) *Planungsbegriffe in Europa. Deutsch-Niederländisches Handbuch der Planungsbegriffe*. Hannover: Akademie für Raumforschung und Landesplanung.

Bauer, Anja, Feichtinger, Judith, and Steurer, Reinhard (2011) *The Governance of Climate Change Adaptation in 10 OECD Countries: Challenges and Approaches*. Wien: InFER.

Benz, Arthur, Lütz, Susanne, Schimank, Uwe, and Simonis, Georg (2007) *Handbuch Governance. Theoretische Grundlagen und empirische Anwendungsfelder*. Wiesbaden: VS Verlag für Sozialwissenschaften.

Bielawska-Roepke, Patrycja (2010) 'Polnisches Raumplanungsgesetz. Aktuelle Vorschläge zur Änderung'. *Planerin* 3: 51–53.

Birkmann, Jörn, Böhm, Hans Reiner, Buchholz, Frank, Büscher, Dirk, Daschkeit, Achim, Ebert, Sebastian, Fleischhauer, Mark, Frommer, Birte, Köhler, Stefan, Kufeld, Walter, Lenz, Susanne, Overbeck, Gerhard, Schanze, Jochen, Schlipf, Sonja, Sommerfeldt, Petra, Stock, Manfred, Vollmer, Maike, and Walkenhorst, Oliver (2013) 'Glossar Klimawandel und Raumentwicklung'. ARL E-Paper No. 10. Hannover. Available at: http://nbn-resolving.de/urn:nbn:de:0156-73571 (accessed 15 February 2013).

Danish Ministry of the Environment (2007) *Spatial Planning in Denmark*, Copenhagen: The Ministry of the Environment.

de Vaus, David A. (2001) *Research Design in Social Research*. London/Thousand Oaks/New Delhi: Sage.

Diekmann, Andreas (1999) *Empirische Sozialforschung. Grundlagen, Methoden, Anwendungen*. Reinbek bei Hamburg: Rowohlt.

Dillman, Don A. (2007) *Mail and Internet Surveys. The Tailored Design Method*. Hoboken: John Wiley & Sons.

Dunlap, Riley E. and Van Liere, Kent D. (2008 [1978]) 'The New Environmental Paradigm'. *Journal of Environmental Education* 40 (1): 19–28.

Dunlap Riley E., Van Liere, Kent D., Mertig, Angela G., and Jones, Robert Emmet (2000) 'Measuring Endorsement of the New Ecological Paradigm: A Revised NEP Scale'. *Journal of Social Issues* 56 (3): 425–442.

Fischer, Robert and Foißner, Peter (2002) 'Raumordnung, Stadtentwicklung und Städtebau in den Niederlanden'. *Standort – Zeitschrift für angewandte Geografie* 4: 153–157.

Frommer, Birte (2010) *Regionale Anpassungsstrategien an den Klimawandel – Akteure und Prozess*. Darmstadt: IWAR.

Glowsky, David (2011) *Globale Partnerwahl. Soziale Ungleichheit als Motor transnationaler Heiratsentscheidungen*. Wiesbaden: VS/Springer.

Groves, Robert M., Fowler, Floyd J., Couper, Mick P., Lepkowsky, James M., Singer, Eleanor, and Tourangeau, Roger (2009) *Survey Methodology*. Hoboken: John Wiley & Sons.

Gustedt, Evelyn (2010) 'Planungssysteme und Planungskulturen'. *Nachrichten der ARL* 3: 20–21.

Harkness, Janet A., Edwards, Brad, Hansen, Sue Ellen, Miller, Debra R., and Villar, Ana (2010) 'Designing Questionnaires for Multipopulation Research'. In: Harkness, Janet A., Braun, Michael, Edwards, Brad, Johnson, Timothy P., Lyberg, Lars, Mohler, Peter Ph., Pennel, Beth-Ellen, and Smith, Tom W. (ed.): *Survey Methods in Multinational, Multiregional, and Multicultural Contexts*. Hoboken: John Wiley & Sons, pp. 33–57.

Heimann, Thorsten (2013) 'Konsens oder Dissens. Herausforderungen des Klimawandels für die deutschen Küstenstädte und –gemeinden aus Expertensicht'. In: Heimann, Thorsten and Christmann, Gabriela B. (2013): *Klimawandel in den deutschen*

Küstenstädten und –gemeinden. Befunde und Handlungsempfehlungen für Praktiker, pp. 22–72. Available at: www.irs-net.de/publikationen/sonderpublikationen/index.php (accessed 11 April 2013).

Heimann, Thorsten (2017) *Klimakulturen und Raum. Umgangsweisen mit Klimawandel an europäischen Küsten*. Wiesbaden: Springer VS.

Heimann, Thorsten and Mahlkow, Nicole (2012) 'The Social Construction of Climate Adaptation Governance. Cultural Differences in European Coastal Areas'. Conference Paper. Amsterdam: The Governance of Adaptation.

Hitzler, Roland, Honer, Anne, and Maeder, Christoph (ed.) (1994) *Expertenwissen: die institutionalisierte Kompetenz zur Konstruktion von Wirklichkeit*. Opladen: Westdeutscher Verlag.

ISSP International Social Survey Programme (2003) *ISSP 2003 – National Identity II: Questionnaire*. Available at: www.issp.org/page.php?pageId=4 (accessed 26 November 2015).

Kannen, Andreas (2000) *Analyse ausgewählter Ansätze und Instrumente zu Integriertem Küstenzonenmanagement und deren Bewertung*. Kiel: Christian-Albrechts-Universität.

Lalli, Marco (1992) 'Urban-Related Identity: Theory, Measurement, and Empirical Findings'. *Journal of Environmental Psychology* 12: 285–303.

Maurer, Marcus and Jandura, Olaf (2009) 'Masse statt Klasse? Einige kritische Anmerkungen zu Repräsentativität und Validität von Online-Befragungen'. In: Jackob, Nicolaus, Schoen, Harald, and Zernack, Thomas: *Sozialforschung im Internet. Methodologie und Praxis der Online-Befragung*. Wiesbaden: VS Verlag für Sozialwissenschaften, pp. 61–74.

Menzel, Susanne (2007) *Learning Prerequisites for Biodiversity Education. Chilean and German Pupils' Cognitive Frameworks and Their Commitment to Protect Biodiversity*. Göttingen: Universität Göttingen. Available at: https://ediss.uni-goettingen.de/handle/11858/00-1735-0000-0006-ACF0-7 (accessed 11 December 2015).

Porst, Rolf (2001) *Wie man die Rücklaufquote bei postalischen Befragungen erhöht*. ZUMA How-to Series, No. 09. Available at: http://nbn-resolving.de/urn:nbn:de:0168-ssoar-201415 (accessed 10 April 2012).

Rippl, Susanne and Seipel, Christian (2008) *Methoden kulturvergleichender Forschung*. Wiesbaden: VS Verlag für Sozialwissenschaften.

Runkel, Peter (2005) 'Fachplanungen, raumwirksame'. In: ARL – Akademie für Raumforschung und Landesplanung (ed.) *Handwörterbuch der Raumordnung*. Hannover: ARL, pp. 893–898.

Schernewsky, Gerald (2004) *Definitionen der Küstenzone*. Website. Available at: www.ikzm-d.de/inhalt.php?page=1,2 (accessed 2 December 2010).

Schmidt, Peter, Bamberg, Sebastian, Davidov, Eldad, Herrmann, Johannes, and Schwartz, Shalom H. (2007): 'Die Messung von Werten mit dem "Portrait Value Questionnaire"'. *Zeitschrift für Sozialpsychologie* 38 (4): 261–275.

Schmitt, Stefan (2007) 'Dänemarks Klima. Wackelkurs des Windkraft-Weltmeisters'. *Spiegel Online* (8 March 2007). Available at: www.spiegel.de/wissenschaft/mensch/daenemarks-klima-wackelkurs-des-windkraft-weltmeisters-a-470459.html (accessed 14 January 2015).

Schnell, Rainer, Hill, Paul B., and Esser, Elke (2008) *Methoden der empirischen Sozialforschung*. München: Oldenbourg Verlag.

Schubert, Klaus and Bandelow, Nils C. (2009) 'Politikfeldanalyse. Dimensionen und Fragestellungen'. In: Schubert, Klaus and Bandelow, Nils C. (ed.) *Lehrbuch der Politikfeldanalyse 2.0*. München: Oldenbourg Verlag, pp. 1–24.

Schwartz, Shalom H. (1992) 'Universals in the Content and Structure of Values: Theoretical Advances and Empirical Tests in 20 Countries'. *Advances in Experimental Social Psychology* 25: 1–65.

Schwartz, Shalom H. (2003) 'A Proposal for Measuring Value Orientations across Nations. Chapter 7: Questionnaire Development Package of the European Social Survey'. Available at: www.europeansocialsurvey.org/docs/methodology/core_ess_questionnaire/ESS_core_questionnaire_human_values.pdf (accessed 6 October 2015).

Schwartz, Norbert, Groves, Robert M., and Schuman, Howard (1998) 'Survey Methods'. In: Gilbert, Daniel, Fiske, Susan and Gardner, Lindzey (ed.): *The Handbook of Social Psychology*. New York: Oxford University Press, pp. 143–179.

Smith, Tom W. (2003) 'Developing Comparable Questions in Cross-National Surveys'. In: Harkness, Janet A., van de Vijver, Fons J.R., and Mohler, Peter P.H. (ed.): *Cross-Cultural Survey Methods*. Hoboken: John Wiley & Sons, pp. 69–92.

Termeer, Catrien, Dewulf, Art, Rijswik, Helena van, Buuren, Arwin van, Huitema, Dave, Meijerink, Sander, Rayner, Tim, and Wiering, Mark (2011) 'The Regional Governance of Climate Adaptation: A Framework for Developing Legitimate, Effective, and Resilient Governance Arrangements'. *Climate Law* 2: 159–179.

Turowski, Gerd (2005) 'Raumplanung (Gesamtplanung)'. In: ARL – Akademie für Raumforschung und Landesplanung (ed.): *Handwörterbuch der Raumordnung*. Hannover: ARL, pp. 893–898.

Van de Vijver, Fons J.R. (2003) 'Bias and Equivalence: Cross-Cultural Perspectives'. In: Harkness, Janet A., van de Vijver, Fons J.R., and Mohler, Peter P.H. (ed.): *Cross-Cultural Survey Methods*. Hoboken: John Wiley & Sons, pp. 143–155.

Van Deth, Jan W. (1998) 'Equivalence in Comparative Political Research'. In: Comparative Politics: The Problem of Equivalence'. London/New York: Routledge, pp. 1–19.

WVS World Values Survey (2006) *Questionnaire Wave 5 (2005–2009). Germany 2006. Poland 2005. United Kingdom 2005*. Available at: www.worldvaluessurvey.org/WVS DocumentationWV5.jsp (accessed 11 December 2015).

Zürn, Michael (2008) 'Governance in einer sich wandelnden Welt: eine Zwischenbilanz'. In: Schuppert, Gunnar Folke and Zürn, Michael (ed.) *Governance in einer sich wandelnden Welt. PVS – Politische Vierteljahresschrift, Special Issue*, 41/2008. Wiesbaden: VS Verlag für Sozialwissenschaften, pp. 553–580.

8 Communication about climate change in European coastal areas

Before exploring actors' knowledge constructions in the following chapters, here the ways in which they communicate with one another on climate change (Section 8.1) and which forms of media they use (Section 8.2) will be considered.

Section 2.2 discussed in depth how representatives of social globalisation theories, such as Robertson or Albrow, assume that communication is increasingly global and that traditional territorial boundaries such as the nation state have become less significant. It was argued from the perspective of communicative constructivism that processes of externalisation, internalisation, and objectification of knowledge might now transcend classical territorial boundaries. This would call into question the boundedness of cultural spatialisations to historically defined territories (see Section 2.2).

Is this also true for communication about climate change? What is the extent of this within the research area? What is the relevance of traditional territorial boundaries? Can one speak of a 'deterritorialisation' of communication? Are there any specific groups of actors for whom global communication on climate change appears particularly pronounced? These questions will be answered in the following.

It should be emphasised that only existing communication on climate change will be described in this work, in order to examine the premise of the existence of global communication more closely. The exact relationships between *communicative processes* and the *content-related genesis of specific stocks of knowledge* would be the subjects of study in their own right, which would in addition require reflection on a comprehensive repertoire of sociological and extra-sociological communications and media-impact research, from network theory to media usage to reception theories (see for example Jäckel 2012; Bonfadelli 2004). Such a claim should certainly not be made for the present work.

8.1 Personal exchange about climate change

To what extent can one speak of a 'compression of the world' (Robertson 1992: 8; see also Section 2.2) when one considers the communicative extent of personal interaction between actors on climate change? Here respondents were required to indicate with which groups of people they have already communicated about climate change (e.g. in person, via telephone, e-mail, etc.), and in particular where these groups are located.

Considerably more than two-thirds of all respondents had already exchanged views on climate change locally with actors in the immediate vicinity of their city or community (78%) or regionally (73%). A little more than half of interviewees state having communicated nationally with other actors on climate change (54%). More than a quarter (27%) have also had transnational contact with other actors, although this is often restricted to within the rest of Europe (within Europe: 26%, outside: 14%).

The results thus demonstrate high relevance for local and regional interpersonal communication between actors. Communication with more geographically distant actors is to a lesser extent present, involving fewer actors the greater the distance. It should be borne in mind that no substantive conclusions can be drawn about the genesis of common climate-change knowledge. If more than two-thirds of all actors are involved in climate-change communication at the local and regional level, and a further one-quarter partake in transnational exchange, it is quite conceivable that homogenous stocks of knowledge are shared among larger groups of actors through direct and indirect communication. But other formations remain possible. Chapters 9 and 10 will return to specific, content-related stocks of knowledge and the factors that explain their separation, and consider the relevance of traditional (national) borders as possible separating factors.

What differences in communicative exchange about climate change can be observed between the groups of actors? Comparing field affiliation in Figure 8.1, a step-like pattern is shown at almost all levels, omitting the local one. Climate-change exchange in person is highest among the NGOs, followed by actors in the economy, politics, and administration, respectively. It is also apparent that differences between fields are smallest at the local level, where information exchange on climate change occurs to the same degree across most fields (politics: 88%, economy: 82%, NGOs: 86%), except for administration, where it appears to take place at a lower rate (72%).

It has already been observed that with increasing distance between actors, the frequency of exchange decreases. This is also the case here, although differences between fields are apparent. In three of the groups, communication decreases significantly with increasing distance, namely administration, politics, and the economy, where at most only one-third of actors communicate with others about climate change at the international level (economy 30%, politics 29%, administration 22%). By contrast, among the group of NGOs, every second actor interacts with others at the international level (54%). The 'compression of the world' postulated by Robertson thus appears more evident with regard to personal communication about climate change among actors from NGOs.

Are there differences according to national affiliation? Figure 8.1 also provides answers to this. As before, a high relevance of local and regional personal

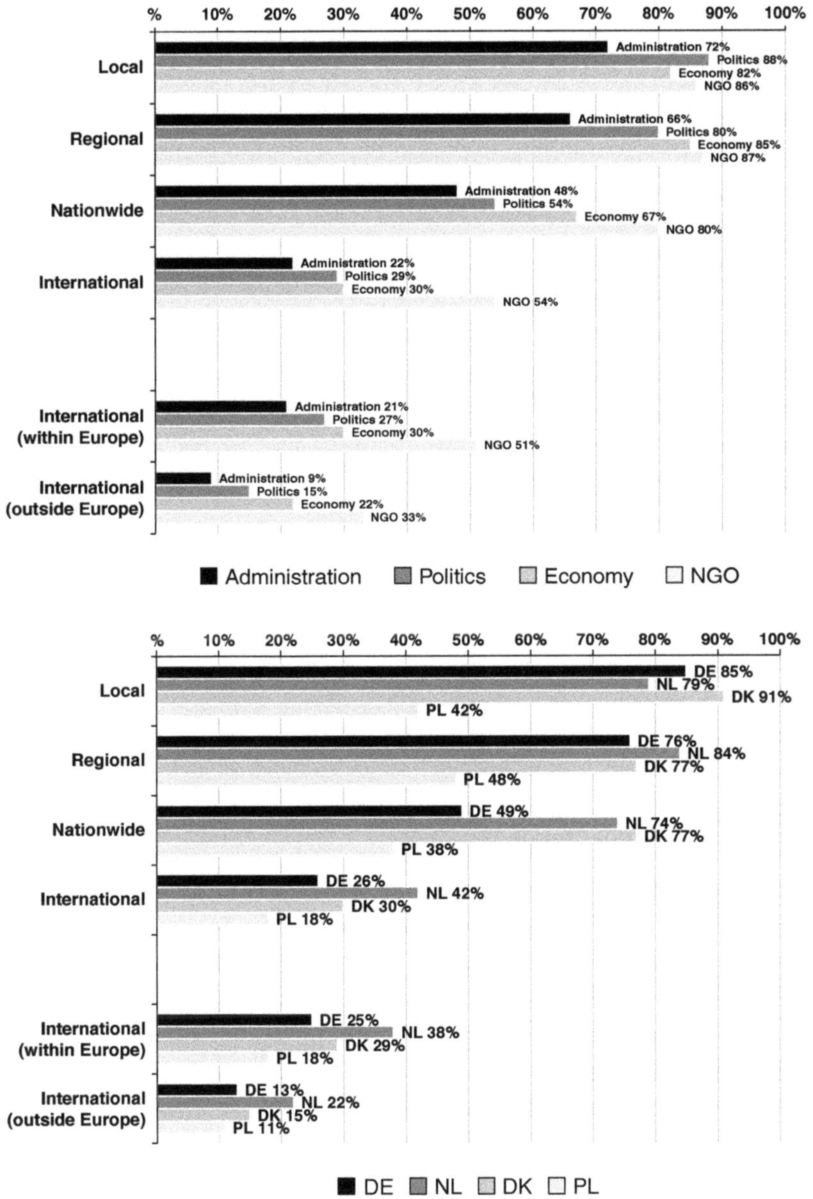

Figure 8.1 Territorial range of communication about climate change: field affiliation and national background.

Source: own calculations.

Notes

Question G1a/G1b: With whom have you communicated professionally about climate-related issues? From which fields have your communication partners come from? Please note all fields you have used (e.g. face-to-face, via telephone, e-mail, etc.).

Basis: All respondents; $n = 830$.

exchange is indicated, as well as a decreasing frequency of exchanges with increasing distance. Comparing among the groups, it is apparent that Polish interviewees were the most set apart from the other nationalities. Significantly lower levels of exchange about climate change occur across all levels than in the other countries. Considering exchange at the individual territorial level, half of the interviewees had already exchanged information about climate change regionally, i.e. across the Voivodship: 48%. In other countries, the maximum values are significantly higher: in Denmark, for example, almost all respondents had exchanged views on climate change regionally: 91%. High figures are also found in Germany (state-wide: 85%) and the Netherlands (province-wide: 84%).

Additionally, national exchanges take place significantly more frequently in Denmark (77%) and the Netherlands (74%) than in Germany (49%) and Poland (38%). Considering that a German federal state such as Lower Saxony is larger in terms of area than the Netherlands, country size could obviously also be of relevance here. Whether this observation is due to these countries' smaller size or to other factors could not be fully determined on the basis of the available data.

Transnational communication on climate change also demonstrates a consistent pattern, with the Dutch continuing to communicate somewhat more internationally (42%). Close behind are Danish (30%) and German actors (26%). Polish respondents provided the lowest figures again, with a maximum of only one-fifth engaging in international climate-change exchange (18%).

8.2 Use of media in relation to climate change

Beyond personal communication among actors, other media forms also provide sources of climate-change information. Their use will be investigated more closely in the following.

First of all, there are actors who state that they are not informed about climate change through media (Table 8.1). This group accounts for a total of 7% of respondents. Smaller differences are observable by field affiliation, and somewhat greater discrepancies according to national background. Among NGO actors, no one claims not to have used any media. In the other fields too, the values are low, with the largest being in administration (administration: 10%, politics: 2%, economy: 4%). By contrast, examining national background showed a clear difference in Polish respondents, where one in four respondents (25%) claim not to have used any media with information on climate change. In the other countries, the figures are very low (DE: 4%, NL: 0%, DK: 2%).

The regular reports of the IPCC, which are freely available on their website (see also Chapter 3), are the most consistently written and probably the most important media on climate change worldwide. Altogether, a third of actors (30%) have already read IPCC reports. Clear differences are demonstrated between fields and nationalities, however. Comparing fields, IPCC reports are most frequently read by NGOs, indeed by a majority: almost two-thirds of actors

Table 8.1 Information media according to societal field and national background (%)

	Administration	*Politics*	*Economy*	*NGO*
	n=*488*	n=*220*	n=*27 (!)*1*	n=*90*
No media used	10	2	4	0
Media companies, regional	53	68	74	54
Media companies, supra-regional	54	70	67	64
Political and administrative sources, regional	41	67	48	59
Political and administrative sources, supra-regional	53	60	63	71
NGO sources	33	48	44	82
Economic sources	18	25	48	19
Professional academic publications	53	46	56	68
IPCC reports	26	24	48	61

	DE	*NL*	*DK*	*PL*	*TOTAL*
	n=*473*	n=*115*	n=*103*	n=*139*	n=*830*
No media used	4	0	2	25	7
Media companies, regional	68	51	57	31	58
Media companies, supra-regional	64	65	61	42	60
Political and administrative sources, regional	55	53	63	22	50
Political and administrative sources, supra-regional	58	75	61	34	57
NGO sources	44	49	54	25	43
Economic sources	22	27	14	17	21
Professional academic publications	56	57	60	32	53
IPCC reports	28	44	51	12	30

Source: own calculations.

Notes
Question K1: Which information sources have you used to inform yourself about climate change?
*1 The group of economic actors, at a sample size of *n* = 27, was relatively small. More pronounced random variation for relative frequencies is therefore possible for this group.
Percentages add up to more than 100% because respondents could indicate more than one source of information.

(61%). This is also true of a large proportion – nearly half – of actors in the economy.[1] In administration and politics, on the other hand, only one in four had previously taken a look at IPCC reports (administration: 26%, politics: 24%).

Within Europe, the frequency of contact with IPCC reports also differs according to national affiliation. In Denmark, every other respondent has previously informed themselves through IPCC reports on climate change (51%). The proportion is also comparatively high in the Netherlands (44%). It is significantly lower in Germany (28%), and lower still in Poland, where only about one in eight (12%) has previously read IPCC reports. The same pattern repeats itself for other scientific publications on climate research, coastal protection, and spatial development: 32% of Polish respondents claim to have informed themselves about climate change through such academic media (compared to DE: 56%, NL: 57%, DK: 60%).

Both regional and supra-regional traditional media outlets (daily newspapers, radio, TV) are highly relevant sources of climate-change information. In all fields, more than half of the actors (>50%) inform themselves through these communication channels. This is also true across all nationalities, again except for one exception: only 31% of Polish interviewees inform themselves about climate change through regional media outlets, while 42% of Polish respondents use supra-regional media outlets as sources of information. This indicates the lower significance of climate-change issues in Poland than in other countries. It should further be noted that newspapers, radio, and TV are often bound to national language and space, so that sources of difference in climate knowledge are also conceivable here.

Data on the use of sources by different fields was also collected. These provide additional indications of the extent to which actors from administration, politics, the economy, and NGOs use their own media as well as that from other fields. For example, it is apparent that a majority of actors from NGOs inform themselves with media from almost all fields, with the exception of economic media, which only 19% of those surveyed have used. On the other hand, a moderate proportion of actors from the economy (44%) have read information produced by NGOs. Interestingly, it also appears that there are varying levels of media usage from the actors' own fields to keep informed about climate change. Among NGOs, a majority of those surveyed (82%) inform themselves on climate change through media from their own field. By contrast, the other fields show lower frequencies for use of their own media (economy 48%, politics and administration between 41% and 60%). If substantive, field-specific differences become observable in media discourses, it would be conceivable that field-specific differences in the actors' knowledge constructions would frequently emerge, for example with regard to NGOs in contrast to the economy. Such further investigation was outside the scope of this study, and future studies could therefore provide further clarification, in particular on further triangulations using media discourse analyses (see for example Christmann *et al.* 2012).

8.3 Interim conclusion

This chapter has investigated whether, and to what extent, one can speak in relation to communication about climate change of a 'compression of the world', or even of communicative deterritorialisation. At the same time, it examined the extent to which baseline conditions make it possible to talk of transterritorial climate cultures rather than local ones (see Chapter 2).

The results consistently show that the conditions are met for information on climate change to be shared globally. Thus IPCC reports or other academic publications on climate change are read throughout Europe. Within the framework of the study there are also actors who exchange information about climate change worldwide. Thus the condition is met, that knowledge constructions on climate change can be globally externalised, internalised, and objectified.

Despite the global reach of the problem as scientifically defined, and despite the global availability of media providing information about climate change, communication within areas at the local level continue to be highly significant. This shows itself particularly in the case of personal exchanges on climate change. The majority of actors communicate about climate-specific issues mainly with other actors from their own city or community, region, and (albeit within the limitations identified) the national arena. This occurs despite the availability of so-called global media such as e-mail, telephone, or the internet. Global communication clearly has a certain importance, but local and regional communication continues to be of greater significance. One may speak of the *transnationalisation* of communication, but *deterritorialisation* is not the right term to describe the forms of communication here, as the local and regional territory appears to be of continuing high relevance.

In addition, differences in frequency of communication on climate change are evident in terms of both field affiliation and national affiliation. In regard to fields, differences are particularly observable among actors from non-governmental organisations, whose networks are significantly more international in scope than those of other groups. Considering national affiliation, Polish respondents differ the most, showing a lower frequency of personal communication on climate change at all levels. Both in the reception of media and in personal communication among individual actors, the issue of climate change appears to be of lesser importance in Poland than in the other countries.

One must again recall that no conclusions can be drawn at this point about the separating force that traditional territorial borders might have on socially shared climate knowledge. Nor, therefore, on the constellation of climate cultures. Some territorial borders might continue to be of importance, in addition to which alternative delimitations, especially between non-governmental organisations and the other actors, might also be possible. In the following chapters, the internalised knowledge constructions of the actors will be examined in depth.

Note

1 Due to the comparatively small sample of members of chambers of industry and commerce, greater random variance is conceivable. Furthermore, the response rate is not lower here than in other fields (see also Chapter 7), meaning that systematic distortion resulting from unit non-response is possible. The probability of systematic differences between fields caused by unit non-response is, however, rather small.

References

Bonfadelli, Heinz (2004) *Medienwirkungsforschung 1: Grundlagen und theoretische Perspektiven*. Stuttgart: UTB.

Christmann, Gabriela B., Heimann, T., Mahlkow, N., and Balgar, K. (2012) 'Klimawandel als soziale Konstruktion?' *Zeitschrift für Zukunftsforschung 1*. (urn:nbn:de:0009–32–34126) Available at: www.zeitschrift-zukunftsforschung.de/ausgaben/2012/1/3412/Christmann_et_al (accessed 28 October 2013).

Jäckel, Michael (2012) *Medienwirkungen. Ein Studienbuch zur Einführung*. Wiesbaden: VS Verlag für Sozialwissenschaften.

Robertson, Roland (1992) *Social Theory and Global Culture*. London/Thousand Oaks/New Delhi: Sage.

9 Vulnerability constructions in European coastal areas

In Chapter 3, scientific constructions of vulnerability to climate change were given detailed consideration. The following chapter now turns to the question of which knowledge stocks spatial development actors from the areas under investigation have internalised, and in particular the extent to which problems or sometimes benefits of climate change are perceived.

The first question to arise is to what degree the actors regard climate change as a problem at all in general (Section 9.1.1). Next to be examined is to what extent they are convinced of the existence of climate change, that it is human caused, and that it is associated with local and global problems. This also raises the question of which climate constructions are systematically shared by actors, and what group differences are to be discerned (Section 9.1.2). Following this, the factors that contribute systematically to differentiating actor groups are considered (Section 9.1.3).

The internalised knowledge constructions of specific impacts of climate change are subsequently investigated. First of all, there is the issue of the extent to which advantages or disadvantages of climate change are seen (Section 9.2.1). Next, I investigate the degree to which physical impacts of climate change such as land loss, flood, heatwaves, or storms are seen as short- or long-term problems for coastal cities and communities in the research area (Section 9.2.2).

9.1 Knowledge constructions of climate-change development, causes, and problems

9.1.1 *Climate change as a general problem*

The actors were surveyed about how they perceived climate change as a problem in general. Figure 9.1 shows the proportion of actors who consider climate change as a great or very great problem.[1] Differences are apparent among both field affiliations and national origins.

In the case of field affiliation, it is initially noticeable that environmental-conservation NGOs have the largest share of actors who view climate change as a major problem (87%). This is hardly surprising given their role as *environmental protection* organisations. In the economy, slightly more than half of actors

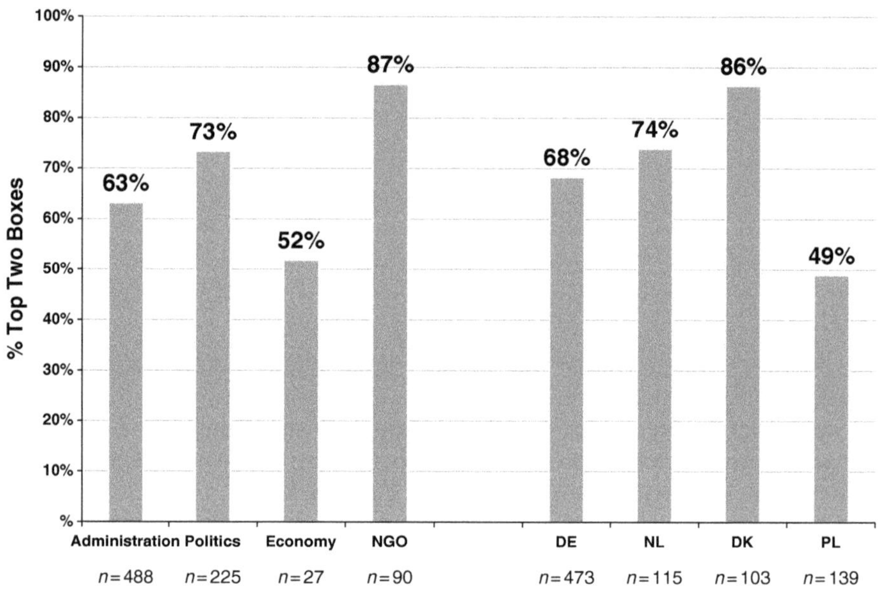

Figure 9.1 Climate change as a major problem in general.

Source: own calculations.

Notes

Question: Do you think climate change is a serious problem? Five-level scale from 1 = No problem at all to 5 = Very serious problem.

Basis: All respondents; *n* = 830. Top two boxes: Positions 4 and 5.

conceive of it as a major problem (52%),[2] that is, significantly less than is the case among NGOs. In between lie the actors from administration (63%) and politics (73%). Overall, however, the majority of actors in all fields view climate change as a major problem.

Comparing by national origin, Danish respondents have a level of awareness of the problem similarly high to that observed among NGOs. Here there is high consensus that climate change is a major problem (86%). In the Netherlands (74%) and Germany (68%) there are also majorities that share this view, albeit at somewhat lower levels. An even greater difference is to be seen among Polish respondents, where only half consider climate change a major problem (49%).

From the answers to this general question one can already conclude that, in dealing with climate change, systematic similarities and differences in the knowledge of the actors are possible that simultaneously follow the postulates of both global convergence and differentiation (see Section 2.2). Thus one can observe clear differences between actors from different countries with regard to the general problem constructions, particularly between Denmark and Poland. But as problem constructions are found to be highest in Germany, the Netherlands, and Denmark, homogeneity among these older EU countries can also simultaneously be postulated. Additionally, the field-specific differences indicate that systematic climate-culture borders can also be drawn between fields. Thus clear differences are to be seen in the general problem construction between actors from the economy and NGOs.

9.1.2 *Shared constructions of climate-change development, causes, and problems: cluster analysis*

Here, first of all, the extent to which actors can be grouped or differentiated from one another according to *socially shared knowledge constructions* can be investigated, according to the relational understanding of cultural space (see also Chapter 2). To this end I, as a social-scientific observer, construct a four-dimensional social space. Each dimension corresponds to a variable that records the possible core positions on climate-change topics presented in Chapter 3. The first cultural space dimension maintains the *belief that climate change is occurring* (see also Section 7.2). Actors are positioned on the second dimension according to their belief that *climate change is human caused.* Finally, the third and fourth dimensions refer to place-related problem attributions. Along the third dimension, actors are arranged according to their perception that *climate change is a global problem.* Finally, the fourth dimension records the belief that *climate change is a local problem* for the actors' respective reference area.

Four-dimensional space is not usually considered conceivable. However, when understood as a matter of positioning actors relationally, according to their knowledge-based similarity, by means of the four variables in four-dimensional space, such positioning can be performed without great difficulty. By means of a hierarchical, agglomerative cluster analysis, groups with similar knowledge constructions can be determined on this basis. The levels of similarity and difference along each dimension are first summarised as a distance matrix. This contains the knowledge-based proximity and distance ratios of all actors in the form of squared Euclidean distances.[3] Cluster amalgamation is subsequently carried out using Ward's procedure.[4] This results in groups with shared similar knowledge constructions being differentiated from other groups. Consequently, properties of the groups can be described on the basis of the four variables/cultural space dimensions. Figure 9.2 shows the arithmetic mean of the four variables per group.

Four groups can be identified in the present example,[5] each sharing similar problem constructions about climate change. These are those who are glocally convinced, globally convinced, climate sceptics, and anthropogenic sceptics.

- *Glocally convinced* indicate high-level conviction and problem constructions along all dimensions (all values Ø>4). They believe firmly that climate change is occurring and is human caused. They also believe that climate change is a major problem both at a global scale and for their own reference area.

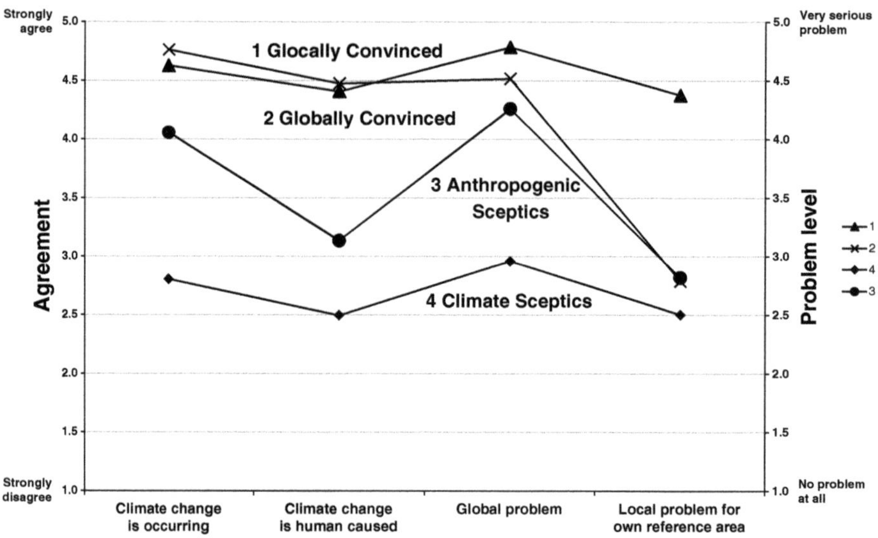

Figure 9.2 Climate-change problem constructions: cluster description.

Source: own calculations.

Notes
Question: Do you agree or disagree with the following statements? 'Climate change is already taking place'; 'Climate change is caused by humans'. Do you think climate change is a problem in general? Do you see climate change as a serious problem: 'for the world', 'for [reference area]'? Scale from 1 = No problem at all to 5 = Very serious problem.
Basis: All respondents; $n = 830$. Hierarchical cluster analysis, proximity measure: squared Euclidean distance; grouping method: Ward.

- *Globally convinced* also believe that climate change is occurring, that its cause is anthropogenic, and that it is a worldwide problem. They are to be differentiated from the 'glocals' because they see major problems at the global level but regard the problem as relatively minor in relation to their own reference area (local problem $\varnothing < 3$).

By contrast, two groups can be distinguished that each share sceptical positions.

- *Climate sceptics* show low conviction values ($\varnothing < 3$) along all four assessment dimensions. They are neither certain that climate change is occurring, nor that it is caused by humans. Additionally, they see little problem posed globally or to their own reference area.
- Finally, there is a group of *anthropogenic sceptics*. These are convinced of the occurrence of climate change and associate it with global problems, but doubt that its cause is human. Additionally, they hardly view it as a problem for their own reference area.

At this point, a 'global analysis' (Rippl and Seipel 2008: 151) will be presented that considers the distribution of the groups within the sample as a whole (Figure 9.3).[6] It will then be possible to examine the factors explaining the different groupings in more detail.[7]

Those convinced of climate change form the majority in the global sample with 63%. By far the largest group are the glocally convinced (45%). Climate change is thus often viewed as a major problem as well as one that is human caused. Close to one in every five is also among the group of globally convinced (18%), who believe in the occurrence of climate change but do not see any major problems for their own environment. Nearly one in three of the respondents can be assigned to one of the two sceptical groups, both of which are equally represented (at 15% each).

9.1.3 *Explanations of differences*

What factors can explain the observed differences in climate knowledge between groups? The first question to ask is whether actors' existing background knowledge can give rise to individual group affiliation. As described in Chapter 5, the issue arises as to which other factors contribute to the separation of the individual groups. Here the focus will be on the separating force of national origin and field affiliation.

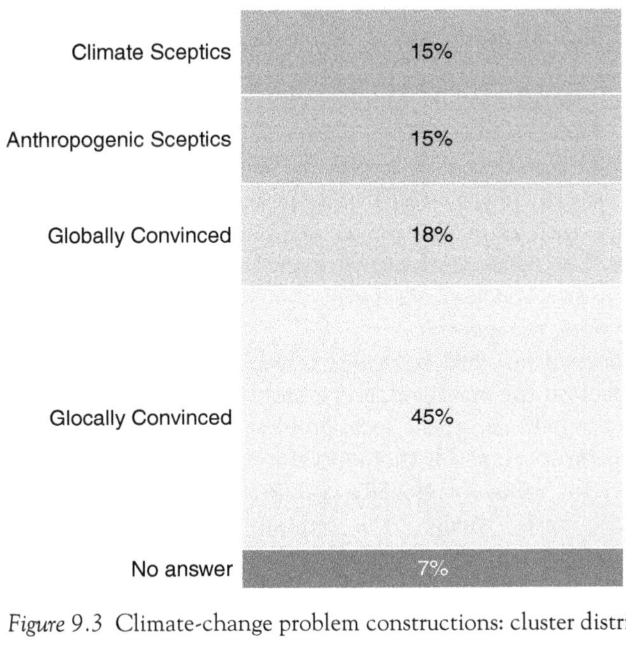

Figure 9.3 Climate-change problem constructions: cluster distribution – total.

Source: own calculations.

Note

Basis: All respondents; $n = 830$.

Background knowledge

On the basis of multinomial logistical regression, the extent to which actors' background knowledge can explain individual group memberships will first be examined, adducing values, ecological beliefs, and local identification as clarifying factors. The central thesis is therefore that 'common values, worldviews and place identification lead to common fears' (Section 5.2.4). Thus the detailed hypotheses derived in Chapter 5 can now be applied to the groups found here.

- With regard to value orientations, the first thesis was that the higher the *universal-altruism orientation*, the higher the probability of sharing a high problem perception of climate change. With increasing universal-altruism orientation, the probability of belonging to the convinced groups would be expected to rise.
- Higher *openness to change* should also increase the probability of belonging to these groups.
- It was additionally formulated that conservatism in the form of *traditionalist/conformity orientation* would lead to low levels of openness to the relatively new topic of climate change, and thus would necessarily increase the probability of belonging among the sceptics.

No further value hypotheses were formulated based on the review of the literature (see Section 5.2.1). There were, however, hypotheses regarding the New Ecological Paradigm and local identification:

- It is expected that actors who assume nature to be unstable and who see a stronger connection between humans and nature must also have internalised stronger climate-change threat constructions. With the increased influence of the *New Ecological Paradigm (NEP)*, the probability of belonging to the convinced groups, with an overall greater awareness of climate change, should therefore rise. The opposite should be observable at the other end of the spectrum: *human-exceptionalism orientation* should shift the probability towards belonging among the sceptics.
- For *local identification* two possible directions of effects were conceivable, namely, the halo effect on the one hand, and a problem-sensitisation effect on the other. With the halo effect, the local problem should decrease with increasing local identification, and it should in this case become less likely that one belongs to the group of glocally convinced. According to the problem-sensitisation thesis, though, the opposite would occur: with increasing local identification, the probability should increase that one belongs to the group of the glocally convinced, with a particular sensitivity regarding the reference area.

Table 9.1 shows the results of the multiple logistic regression. In each case, the beta coefficients indicate the direction of displacement for the probability of

Table 9.1 Multinomial regression models (1–3): problem constructions

	Model 1: Values			*Model 2: Values + NEP + LI*			*Model 3: Values + NEP + LI + CV*		
	$\beta^{Sig.}$ *[odds ratio]*			$\beta^{Sig.}$ *[odds ratio]*			$\beta^{Sig.}$ *[odds ratio]*		
Reference group: Glocally convinced	*Globally convinced* n = *141*	*Anthropogenic sceptics* n = *119*	*Climate sceptics* n = *110*	*Globally convinced* n = *141*	*Anthropogenic sceptics* n = *119*	*Climate sceptics* n = *110*	*Globally convinced* n = *133*	*Anthropogenic sceptics* n = *116*	*Climate sceptics* n = *105*
	Glocally convinced n = *350*			*Glocally convinced* n = *349*			*Glocally convinced* n = *341*		
Background knowledge									
Values									
Self-transcendence I: Universalism	–0.35** [0.70]	–0.63*** [0.53]	–0.81*** [0.44]	–0.25* [0.78]	–0.44*** [0.64]	–0.51*** [0.60]	–0.27* [0.77]	–0.44*** [0.65]	–0.54*** [0.58]
Self-transcendence II: Human altruism	–0.12	–0.10	–0.24* [0.79]	–0.07	–0.05	–0.14	–0.11	–0.06	–0.19
Self-enhancement	0.1	0.32* [1.37]	–0.34* [0.71]	0.11	0.33* [1.39]	–0.34* [0.71]	0.1	0.32* [1.38]	–0.3* [0.74]
Openness to change	0.03	0.06	0.02	0	–0.01	–0.12	0.03	–0.01	–0.07
Preservation I: Security	–0.05	–0.07	0.00	–0.07	–0.10	–0.02	–0.05	–0.11	–0.01
Preservation II: Tradition/Conformity	0.12	0.22* [1.243]	0.41*** [1.50]	0.12	0.18	0.36** [1.43]	0.1	0.19	0.37** [1.45]
Beliefs									
New Ecological Paradigm (NEP)				–0.62*** [0.54]	–1.16*** [0.31]	–2.09*** [0.12]	–0.61** [0.54]	–1.19*** [0.31]	–2.11*** [0.12]
Identity construction									
Local identification				–0.12	–0.01	–0.02	–0.13	0.03	0.03
Control variables (CV)									
Age							0.01	0.00	0.02
Gender (0 = female, 1 = male)							–0.12	–0.15	–0.45
Constants	0.74	0.73	3.41***	2.89**	4.23***	9.62***	2.89*	4.58***	8.95***
Model accuracy									
Pseudo R^2 (Nagelkerke)	0.153			0.273			0.281		
LR test	108.80***			207.54***			207.22***		
N	720			719			695		

Source: own calculations.

Notes
β ***$p<0.001$; **$p<0.01$; *$p<0.1$. Values in brackets [odds ratio].

transitioning between group memberships for each variable. As a reference group, the group of glocally convinced was chosen, as they demonstrate high values along all four dimensions, thus ensuring good comparability with the other groups.

To enable sequential observation, several models were calculated. This also allows for the detection of mediator and moderator effects between the variables (Urban and Mayerl 2011: 312). First, two exclusively knowledge-oriented models are used. The first contains only the discussed Schwartz-value orientations as explanations (Model 1). The second combines these with the New Ecological Paradigm (NEP) and local identification (Model 2). In the third model, the influence of age and gender is also controlled for (Model 3).[8]

Considering the overall model quality, the Likelihood Ratio test (LR test) can be used for all three models to reject the hypothesis that they do not provide a better explanation for the differentiation of the groups than the null model; indeed this is so to a high level of significance ($p < 0.001$).[9] The stated variance shares are still low for the first model (Model 1: Pseudo $R^2_{\text{Nagelkerke}} = 15.3\%$), but have more reasonable values in Models 2 and 3 (Model 2: 27.3%; Model 3: 28.1%), meaning that one can speak overall of a satisfactory model quality.

The question next arises as to which individual variables provide significant contributions to the differentiation of the groups:

- For values, *universalist* and *traditionalist orientations* make significant explanatory contributions to the differentiation of the groups.
- In addition, *self-enhancement* and *human-altruism orientations* can also explain group affiliation.
- For ecological beliefs, *NEP orientation* is able to contribute significantly to explaining group affiliation.
- By contrast, *openness to change* barely helps to explain group affiliation.
- *Local identification* also hardly explains individual group affiliation. Neither a halo effect nor a problem-sensitisation effect is observed in connection with local identification.

To go into more depth, the probability of affiliation with the glocally convinced also increases, as expected, with increased universalist orientation. All beta values appear negative here. An interesting detail is revealed by the effect coefficients (odds ratio) shown in brackets. As value orientations are all measured using the same scale, these give an impression of the relative effect strengths (see for example Behnke 2015: 73). It becomes clear that universalist orientation can explain group differences with respect to the glocally convinced. Increased universalist orientation is accompanied by an increased belief that climate change is occurring, is human caused, and presents a problem (Model 1: globally convinced: $\beta = -0.35^{**}$ [0.70]; anthropogenic sceptics: $\beta = -0.63^{***}$ [0.53]; climate sceptics: $\beta = -0.81^{***}$ [0.44], Models 2 and 3 analogous).

Additionally, an increased conservative orientation leads to greater antipathy with regard to climate change, and thus to a reduction in the perceived problems. As expected, traditionalist and conformity orientations contribute significant explanatory force to differentiation of the sceptical groups. Thus, in contrast to the glocally convinced, the likelihood of affiliation to the group of anthropogenic sceptics (Model 1: $\beta = 0.22^{*}$) or to the climate sceptics (Model 1: $\beta = 0.41^{***}$) shifts accordingly.

As previously postulated, NEP orientation is also able to make significant distinctions in group affiliation. The same pattern of effect strength appears here as was previously observed for universalism orientation: with decreasing climate-change conviction and problem perception comes a decrease in ability to differentiate (Model 2: $\beta = -0.62^{***}$ [0.54]; $\beta = -1.16^{***}$ [0.31]; $\beta = -2.09^{***}$ [0.12], Model 3 analogous).

The influence of the self-enhancement orientation has not been considered in formulating the hypothesis. Here, however, it can be seen that these models contribute to the differentiation of the groups. Thus with an increasing self-enhancement orientation, the probability of belonging to the group of anthropogenic sceptics rather than the glocally convinced increases (Model 1: $\beta = 0.32^{*}$, Models 2 and 3 analogous). Actors who value power and hedonism positively are accordingly less convinced that human activity contributes to climate change. The suspicion suggests itself that such actors have internalised a comparatively positive human image – somewhat in the tradition of Adam Smith's *Wealth of Nations* (Smith 1789) – that makes unintended negative side-effects of human activity, even those which come about in relation to self-interest, appear less likely. Further research would be required to be able to reach any unequivocal conclusions on this.

In summary, the groups can be differentiated by background knowledge as follows:

- Compared with the glocally convinced, the globally convinced are distinguished by somewhat lower universalism and NEP orientations.
- The two sceptical groups are also characterised by a higher degree of conservatism in the form of traditionalist and conformity orientations. In addition, climate sceptics demonstrate a lower self-enhancement orientation on average, while anthropogenic sceptics show a higher self-enhancement orientation.

Field affiliation and national background

Alongside a consideration of background knowledge, there is the question of the degree to which field affiliation and national origin contribute to the separation of the groups. Before focusing on the individual group-affiliation probabilities from the logistic regression, the distributions of the groups as relative frequencies at the macro level, as shown in Figure 9.4, should be considered. The overall impression is one that suggests systematic differences in distribution corresponding to both field affiliation and national origin.

With respect to field affiliation, major differences in the incidence of the glocally convinced are the most striking. These have internalised the highest levels of belief and problem perception regarding climate change. It is therefore not surprising that the glocally convinced are markedly over-represented in environmental-conservation NGOs compared to other groups (67%). Following with only a slight difference between them are the economy (52%) and politics (47%). The lowest proportion of glocally convinced is in administration (40%). The proportion of the two sceptical groups is accordingly very low among NGOs (climate sceptics: 7%, anthropogenic sceptics: 10%). The remaining three groups are similar in their distribution, although the proportion of climate sceptics is still greatest in the economy (19%).[10]

Comparing the frequency distributions by country, it becomes clear that much of the distribution is similar. The slightly larger proportion of glocally convinced is again noticeable in Denmark (56%), with slightly lower values in Germany (48%) and the Netherlands (45%), and slightly lower or higher proportions of sceptics, respectively. These three countries thus appear to differ only slightly from each other and are comparatively similar. There are clear differences to the Polish group composition, however, where the proportion of glocally convinced lies at only 28%. Only about one in four in Poland is convinced of significant global or local problems and anthropogenic climate change, whilst the two sceptical groups are over-represented compared to the other three countries, accounting for 45% of the population, almost half of Polish respondents.

With regard to vulnerability constructions, the results thus indicate regional differences in the distribution of climate cultures, particularly between Poland and the older EU countries under consideration. Additionally, major differences can be observed between actors from NGOs, the economy, and other fields.

Scheuch (1989: 152) reminds us of the risk of reaching false ecological conclusions when looking merely at frequency in national comparisons. For instance, the differences observed might also be caused by third-party variables which have not been controlled for. In this study in particular, the different compositions of the country samples could contribute to the differences observed in Figure 9.4 with regard to field affiliation, gender, and age (see also Section 7.1). Thus fields, nationality, and background knowledge need to be controlled for (in relation to false ecological conclusions, see also Rippl and Seipel 2008: 153ff.; Welzel 2003; Przeworski and Teune 1970).

Three more logistic regression models serve this purpose. In order to eliminate the 'blind spot' of missing third-party variables in the frequency tables, the ability of field affiliation and national background to differentiate climate cultures is controlled for in a common model (Model 4: field + country + social demography: Table 9.2). Here it must be understood that for the multiple logistic regression, each of the reference groups could serve as a category for comparison. Germany (national background) and administration (field affiliation) were selected here, as they have the largest group sizes. The discussion below will be restricted throughout to the discriminatory power of these two groups.[11]

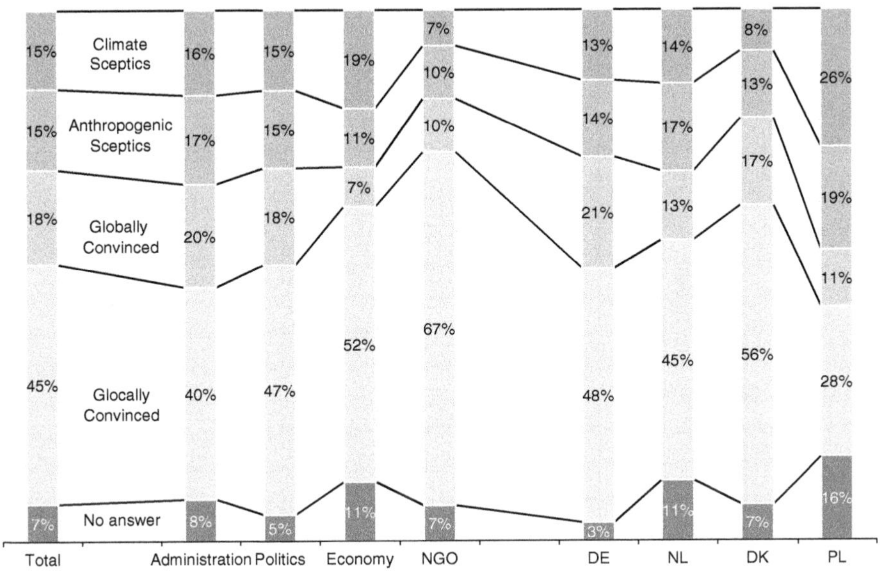

Figure 9.4 Climate-change problem constructions: field and national affiliation.

Source: own calculations.

Note
Basis: All respondents; $n = 830$.

Furthermore, I am interested in questioning whether the patterns described for the explanation of climate-culture differences according to the background knowledge of Models 1–3 remain constant if field affiliation and national origin are also included. Their mutual influence is therefore controlled for in the multi-variate analysis (Meulemann 2002: 16). If no major changes in the sequential analysis are observable, this suggests that the effects in question are independent of field and nationality. Otherwise, it will be necessary to take country- or field-specific contextual effects into account as well when assessing the direction and strength of the effects of the independent variables. To this end, a further model is calculated which controls for background knowledge together with field affiliation (Model 5). Finally, a common model is used that additionally controls for national background (Model 6).

Let us first consider the extent to which the results observed in the frequency table are consistent when field affiliation and national origin are controlled for at the individual level.[12]

If the greater differences observed between NGOs and the other groups in Figure 9.4 are not based on the age, gender, or national origin of respondents, then significant differences would still be observed with respect to the NGOs when administration is chosen as a reference group. A similar picture is yielded

Table 9.2 Multinomial regression models (4–6): problem constructions

	Model 4: Field + Country + CV			Model 5: Field + Knowledge + CV			Model 6: Field + Country + Knowledge + CV		
	β^{Sig} [odds ratio]			β^{Sig} [odds ratio]			β^{Sig} [odds ratio]		
Reference group: Glocally convinced	*Globally convinced* n = *137*	*Anthropogenic sceptics* n = *123*	*Climate sceptics* n = *115*	*Globally convinced* n = *133*	*Anthropogenic sceptics* n = *116*	*Climate sceptics* n = *105*	*Globally convinced* n = *133*	*Anthropogenic sceptics* n = *116*	*Climate sceptics* n = *105*
	Glocally convinced n = *362*			*Glocally convinced* n = *341*			*Glocally convinced* n = *341*		
Background knowledge									
Self-transcendence I: Universalism				–0.24* [0.79]	–0.42*** [0.66]	–0.51*** [0.60]	–0.24* [0.79]	–0.42*** [0.66]	–0.5*** [0.60]
Self-transcendence II: Human altruism				–0.11	–0.06	–0.21	–0.12	–0.06	–0.21
Self-enhancement				0.07	0.31* [1.36]	–0.33* [0.72]	0	0.38* [1.46]	–0.27
Openness to change				0.04	0	–0.07	0.11	–0.09	–0.12
Preservation I: Security				–0.06	–0.12	0	–0.03	–0.15	–0.01
Preservation II: Tradition/ Conformity				0.07	0.18	0.34* [1.41]	0.07	0.15	0.27* [1.31]
New Ecological Paradigm (NEP)				–0.61** [0.55]	–1.19*** [0.30]	–2.11*** [0.12]	–0.67*** [0.51]	–1.14*** [0.32]	–2.05*** [0.13]
Local identification				–0.13	0.03	0.01	–0.14	0.05	0.04
Field affiliation									
Politics	–0.24	–0.11	–0.1	–0.1	–0.02	0.16	–0.07	0.04	0.29
Economy	–1.2	–0.66	0	–1.26	–0.94	–0.27	–1.22	–0.94	–0.16
NGO	–1.18** [0.31]	–0.92* [0.40]	–1.28** [0.28]	–1.00* [0.37]	–0.45	–0.78	–0.89* [0.41]	–0.47	–0.65
(*Reference group*: Administration)									

National background									
NL	–0.3	0.37	0.31				–0.63** [0.53]	0.24	–0.18
DK	–0.31	–0.28	–0.49				–0.23	0.11	–0.13
PL	–0.13	0.78* [2.16]	1.26*** [3.51]				–0.35	0.73* [2.07]	0.81* [2.25]
(*Reference group*: DE)									
Control variables (CV)									
Age	0	–0.01	0.02	0.01	0	0.02	0	0	0.02
Gender (0 = female, 1 = male)	0.21	0.3	0.33	–0.12	–0.15	–0.46	–0.11	–0.1	–0.38
Constants	–0.68	–0.94*	–2.28***	2.91*	4.68***	9.16***	3.45**	4.28***	8.78***
Model accuracy									
Pseudo R^2 (Nagelkerke)	88			295			312		
LR Test	61.72***			219.71***			234.09***		
N	737			695			695		

Source: own calculations.

Notes
β ***$p < 0.001$; **$p < 0.01$; *$p < 0.1$. Values in brackets [odds ratio].

when considering national origin: if there are similarities between Germany, the Netherlands, and Denmark, as well as greater differences especially in regard to Poland, then significant differences would need to occur between Germany and Poland, but not between Germany and the Netherlands and Denmark.

The results concur exactly with these ideas. Thus the probability of belonging to the glocally convinced when compared to administrative actors shifts significantly in the direction of the NGOs (Model 4: globally convinced: $\beta=-1.18^{**}$ [0.31]; anthropogenic sceptics: $\beta=-0.92^{*}$ [0.40]; climate sceptics: $\beta=-1.28^{**}$ [0.28]). As expected, among the sceptical groups there is also a significant shift towards those with a Polish background when compared to Germany (Model 4: anthropogenic sceptics: $\beta=0.78^{*}$ [2.16]; sceptics: $\beta=1.26^{***}$ [3.51]). Thus, the observed results also hold true under mutual control of field affiliation and national origin.

Additionally, the influence of background knowledge when the actors' field affiliation and national origin are controlled for should be considered. In this way, the reasons for field- and nationality-specific differences in the distribution of climate cultures can be identified.

Regarding field affiliation first, the explanatory patterns found in Model 3 also hold for background knowledge with field affiliation controlled for (compare Model 3 with Model 5). At the same time, the ability of NGO background to explain the separation of the two sceptical groups, which could still be seen in Model 4, now falls to insignificant levels in Model 5 (Model 5: anthropogenic sceptics: $\beta=-0.45$; climate sceptics $\beta=-0.78$). In addition, the values of the beta coefficients increase even further between Model 3 and Model 5 for NEP orientation, while other coefficients remain mostly stable at this point. This indicates a mediator effect between NEP orientation and NGO affiliation (Urban and Mayerl 2011: 304). The observation that NGOs have a greater proportion of glocally convinced actors can therefore be largely explained by a higher internalisation of the NEP among their members. Moreover, the aggregate level also clearly indicates that NGOs display higher proportions of actors demonstrating high levels of internalisation of the NEP compared to other fields (NEP: administration: 74%, politics: 81%, economy: 67%, NGO: 84%). The distributions at the aggregate level are given in Table 9.3.

If national background is controlled for, as in Model 6, the effect of nationality when comparing Poland and Germany is still observable. The beta coefficients are slightly lower for Poland, indicating differences in knowledge distribution between these countries. This can be seen in more detail in Table 9.3. Here Poland demonstrates the lowest levels of NEP dissemination, which, considering both the individual and aggregate levels, can be identified as a reason for the lower proportions of the convinced groups in Poland (NEP: DE: 83%, NL: 69% DK: 81%, PL: 60%). The same pattern also emerges for universalism orientation, which is lowest in Poland (77%).

In addition, Polish respondents appear to be significantly more conservative. The differences in traditionalist and conformity orientations are very clear, being highly pronounced in Poland (49%) and significantly lower in the other

Table 9.3 Background knowledge: distribution by country and field

	Administration	*Politics*	*Economy*	*NGO*
Values				
Self-transcendence I: Universalism	80%	86%	81%	90%
Self-transcendence II: Human altruism	75%	87%	81%	84%
Self-enhancement	33%	36%	37%	19%
Openness to change	45%	54%	48%	57%
Conservation I: Security	44%	33%	37%	33%
Conservation II: Tradition/Conformity	34%	40%	41%	20%
View of nature and society				
New Ecological Paradigm (NEP)	74%	81%	67%	84%
Identity construction				
Local identification	64%	84%	78%	76%
n	488	225	27	90
	Germany	*Netherlands*	*Denmark*	*Poland*
Beliefs				
Self-transcendence I: Universalism	82%	85%	92%	77%
Self-transcendence II: Human altruism	81%	76%	84%	72%
Self-enhancement	36%	22%	26%	31%
Openness to change	40%	69%	63%	55%
Conservation I: Security	36%	37%	46%	48%
Conservation II: Tradition/Conformity	34%	24%	27%	49%
View of nature and society				
New Ecological Paradigm (NEP)	83%	69%	81%	60%
Identity construction				
Local identification	78%	72%	58%	58%
n	473	115	103	139

Source: own calculations.

Notes
High expressions (proportion of top three boxes). Basis: All respondents; $n = 830$.

countries (DE: 34%, NL: 24%, DK: 27%). Given that in all models higher traditionalist and conformity orientations lead to higher probability of being climate sceptic, it can be concluded that the higher level of conservativism among Polish respondents provides a further explanation for the higher level of climate scepticism in Poland.

Since the beta coefficients in Model 6 fall for Poland, while remaining largely significant, it must also be noted that there are further, unobserved factors responsible for the more divergent distributions in Poland compared to the other groups. There is opportunity for further research here to investigate these unobserved factors for a more comprehensive elucidation of the variances.

To summarise, the results for the vulnerability constructions considered indicate greater differences between NGOs and the other fields, which can be explained by their differences in background knowledge. Regarding the ability of national origin to differentiate groups, there is greater homogeneity in comparison to Poland between Germany, Denmark, and the Netherlands. The most significant differences exist mainly with Poland, whose origins can likewise be traced back to the previously discussed differences in background knowledge. At the same time, it is clear that other factors continue to be of relevance, and these require further future investigation.

9.2 Vulnerability constructions concerning climate-change impacts

What specific physical impacts of climate change do actors view as a problem in the short and long term? Are some topics more urgent to these actors than others?

Figure 9.5 shows the extent to which specific climate change impacts are already seen as posing a problem in the short term up until 2025. Once again, the proportions of actors who suppose a great or very great problem are displayed. Greater short-term problem constructions are only viewed by, at most, half the actors (48%). On average, the problems associated with various impacts of climate change appear to be considered as moderate up to 2025.

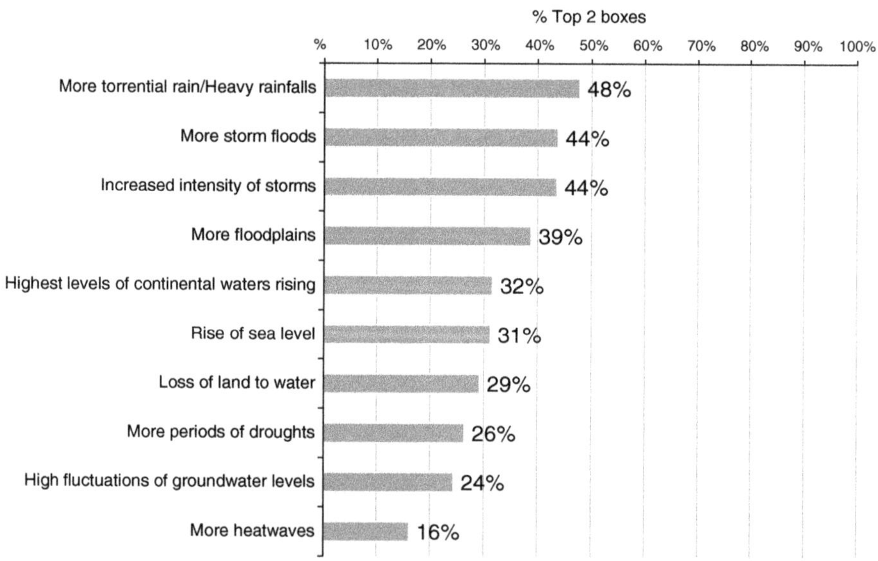

Figure 9.5 By 2025: climate-change impacts as a problem within the entire reference area.

Source: own calculations.

Notes

Question: By 2025: Please estimate how some aspects of the climate may change in the Coastal Municipalities of Central Europe. How serious do you think problems will be? Five-level scale from 1 = No problem at all to 5 = Very serious problem.

Basis: All respondents. Top two boxes: Positions 4 and 5.

Additionally, the proportions are arranged in descending order, so that at this point it is possible to see which climate change impacts seem most problematic for the interviewees up to 2025. According to this, the most significant problem is considered to be heavy precipitation (48%). This result appears to conform to current scientific observations showing an increase of severe precipitation in the research area over the past decades (see Section 3.3). On the other hand, scientists have yet to show any clear trends in regard to the development of storms in Europe, although some studies predict a future increase in wind speeds (ibid.). The associated problems up to 2025 appear relatively significant to actors, and 44% of respondents view an increase in storm flooding and wind speeds as posing a great to very great problem by 2025. Flooding also appears comparatively high on the list of problems (39%).

Increasing sea levels or even land loss, on the other hand, are viewed by less than a third of respondents as a major problem for the near future (sea-level rise: 31%; land loss: 29%). In Section 3.3 it was reported that the average length of summer heatwaves has doubled since 1880 in western Europe. On the whole, however, problems associated with heatwaves appear to be of little relevance to actors for the short-term future (16%).

In a separate survey up to the year 2100 (not shown here; see Heimann 2017: 251), the actors predominantly view the old challenges to the coast as persisting alongside new ones. There is again only a minority who regard heatwaves as a serious problem (40%). On the other hand, the majority of the respondents assume there will be serious problems caused by flooding (66%), storms (63%), and land loss (54%) by the end of the century.

9.3 Interim conclusion

This chapter has investigated the similarities and differences in actors' climate-change vulnerability constructions, and the factors that contribute to systematic differences.

In evaluating climate change as a general problem, commonalities and differences were discernible according to field affiliation and national background. Clear differences in the general problem construction emerged between actors from the economy and from NGOs. Viewed by national background, a majority of actors from the old EU nations – Germany, the Netherlands, and Denmark – regarded climate change as a serious problem. This view was most strongly expressed in Denmark. On the other hand, climate change was markedly less seen as a serious problem in Poland.

In addition, four groups could be distinguished according to the concept of culture as relational space, each with shared vulnerability constructions, and which could be found across the research area. These are: climate sceptics, anthropogenic sceptics, the globally convinced, and the glocally convinced. NGOs displayed significant differences in their composition, with the glocally convinced being represented much more frequently than in the other fields. Additionally, there were again major differences between Poland and the remaining countries. Here, the sceptical groups were represented much more frequently. Individual group memberships could be explained by background knowledge, in particular by internalisation of the New Ecological Paradigm as well as traditionalist and self-enhancement value orientations.

Two things can now be confirmed. First, that the deterritorialisation theory cannot be fully refuted, and second, that territorially oriented effects of national background are evident. There are groups across the entire research area who hold similar vulnerability constructions, indicating the existence of transnationally homogenous groups, as has been postulated by proponents of cultural deterritorialisation in connection with climate change (see Section 5.3.2). This is particularly true for NGOs, where vulnerability constructions are systematically shared. As this effect was also observable in the logistic regression with national background controlled for, it can be assumed that the observed climate-culture boundaries with regard to socially shared vulnerability constructions exist transnationally among NGOs and the other fields. In addition, the Oder-Neisse appears as a climate-culture boundary, as the proportion of actors on the Polish side who view climate change with scepticism, or who see it as a problem of little consequence, is markedly higher than in the west European countries.

Notes

1 These results were published in a similar form in relation to national background, but considering a group of academics, in Christmann *et al.* (2012: 26).
2 Owing to its small group size, the economy is subject to greater random fluctuation.
3 Squared Euclidean distances are used as they separate bigger differences more clearly than smaller ones. In this way, knowledge-related similarities (but not equivalences!) appear as common groups, while greater knowledge-related differences are represented by greater distances. Furthermore, it is the absolute distances between actors and not the concurrence of knowledge profiles that is of interest (Backhaus *et al.* 2008: 394).
4 Ward's procedure is used since it forms groups on the basis of minimising the sum of squares of differences as a heterogeneity measure, and in comparison with other algorithms in empirical tests provided the most sensible partitionings (Backhaus *et al.* 2008: 424). Before using the procedure, however, an outlier diagnosis was performed on the basis of the single-linkage method. No cases were removed in the present analysis.
5 A plethora of possible tests are available for determining the optimal number of clusters (see for example Milligan and Cooper 1985). A selection was made during this study largely on the basis of fusion values in the form of error variance by reference to the elbow criterion (see explanation below). Additionally, the clusters must make sense and exhibit a plausible relationship to the discourses presented in Chapters 3 and 4. Detailed discussion of the determination procedures for all clusters can be found in the German version of this book (see Heimann 2017). An English version of the Appendix is available from the author on request.
6 Global analyses in transnational studies are sometimes carried out with weighted data to compensate for possible distortions caused by different group sizes, such as those for different national origins (see for example Glowsky 2011: 125). In the present study, however, this approach would presuppose that the selected weighting criteria,

such as national origin, are the decisive criteria for data structuring. This study argues that alternative criteria for the structuring of cultures, such as secondary socialisation, are also conceivable. In the case of weightings, therefore, there is the danger of applying a priori categorical assumptions that do not exist in the data structures. Applying weightings at this point therefore does not seem sensible.

7 It should be noted that in the sample Germans are more frequently represented than actors from the other countries. If national differences were observable, a weighting of the data would be conceivable to arrive at a more precise distribution ratio in regard to national differences. However, the extent to which national differences can be assumed is meant to be examined at a later stage. Alternative weightings would also be conceivable, according to field affiliation or socio-demographic criteria, for instance. Owing to this plurality, weightings appear to have no value here, as their application would always be selective and risk distorting the findings.

8 Detailed results of a (multi-)collinearity test can be found in the German version of this book (Heimann 2017). They show that there is no risk of multi-collinearity. An English version of the Appendix is available from the author on request.

9 At this point, it should be pointed out that the *applicability of statistical significance tests in full population surveys* is questioned by some authors (for more detailed discussion, see for example Behnke 2005; Broscheid and Gschwend 2005). Nevertheless, the question whether 'what we see is due to chance' (Kohler and Kreuter 2012: 236) is also raised in the case of a full population survey. Broscheid and Gschwend (2005: 0–17) remind one that social processes, explainable in the form of theories or models, always represent slices of social reality and are thus essentially of a stochastic nature. Additionally, unintended factors can appear more or less at random as disturbances in the case of full population surveys (ibid.: 0–18). Since causal explanations, such as those in the logistic regression model here, can only ever take account of sections of social reality, the extent to which the observations might be due to chance must therefore be examined (Kohler and Kreuter 2012: 207). Significance tests are accordingly used in this study.

10 It should, be noted once again, however, that the group size of the economic actors is significantly lower at $n = 27$ than in the other groups. The possibility of greater random fluctuations can thus not be excluded.

11 Although a German perspective is thus consciously taken here, it should be made explicit that this does not necessarily lead to false, ethnocentric conclusions (Rippl and Seipel 2008: 30), but instead turns the attention towards the capacity of the German borders to separate climate cultures.

12 Considering the overall model quality, Model 4 differs significantly from the null model (LR Test: 61.72***), but also displays very low total variance (Pseudo $R^2_{\text{Nagelkerke}}$: 8.8%). It can thus be concluded, with little surprise, that factors not yet taken into account beyond national origin, field affiliation, age, and gender might still contribute to the explanation of differences. Taking into account background knowledge, Models 5 and 6 ultimately also show comparatively high total variance explanation values (Model 5: 29.5%, Model 6: 31.2%).

References

Backhaus, Klaus, Erichson, Bernd, Plinke, Wulff and Weiber, Rolf (2008) *Multivariate Analysemethoden. Eine anwendungsorientierte Einführung*. Berlin/Heidelberg: Springer Verlag.

Behnke, Joachim (2005) 'Lassen sich Signifikanztests auf Vollerhebungen anwenden? Einige essayistische Anmerkungen'. *Politische Vierteljahreszeitschrift* 46 (1): O.1–O.15.

Behnke, Joachim (2015) *Logistische Regressionsanalyse. Eine Einführung*. Wiesbaden: Springer VS.

Broscheid, Andreas and Gschwend, Thomas (2005) 'Zur statistischen Analyse von Vollerhebungen'. *Politische Vierteljahreszeitschrift* 46 (1): O.16–O.26.

Christmann, Gabriela B., Heimann, T., Mahlkow, N., and Balgar, K. (2012) 'Klimawandel als soziale Konstruktion?' *Zeitschrift für Zukunftsforschung 1*. (urn:nbn:de:0009–32–34126) Available at: www.zeitschrift-zukunftsforschung.de/ausgaben/2012/1/3412/Christmann_et_al (accessed 28 October 2013).

Glowsky, David (2011) *Globale Partnerwahl. Soziale Ungleichheit als Motor transnationaler Heiratsentscheidungen*. Wiesbaden: VS für Sozialwissenschaften/Springer.

Heimann, Thorsten (2017) *Klimakulturen und Raum. Umgangsweisen mit Klimawandel an europäischen Küsten*. Wiesbaden: Springer VS.

Kohler, Ulrich and Kreuter, Frauke (2012) *Datenanalyse mit Stata. Allgemeine Konzepte der Datenanalyse und ihre praktische Anwendung*. München: Oldenbourg Verlag.

Meulemann, Heiner (2002) 'Perspektiven und Probleme der internationalen Umfrageforschung'. In: *Statistisches Bundesamt, Spektrum Bundesstatistik*. Volume 20, pp. 13–38.

Milligan, Glenn W. and Cooper, Martha C. (1985) 'An Examination of Procedures for Determining the Number of Clusters in a Data Set'. *Psychometrika* 50 (2): 159–179.

Przeworski, Adam and Teune, Henry (1970) *The Logic of Comparative Social Inquiry*. New York/London/Toronto/Sydney: John Wiley & Sons.

Rippl, Susanne and Seipel, Christian (2008) *Methoden kulturvergleichender Forschung*. Wiesbaden: VS Verlag für Sozialwissenschaften.

Scheuch, Erwin K. (1989) 'Theoretical Implications of Comparative Survey Research: Why the Wheel of Cross-Cultural-Methodology Keeps on Being Reinvented'. *International Sociology* (4) 2: 147–167.

Smith, Adam (1789) *An Inquiry into the Nature and Causes of the Wealth of Nations*. London: Ward, Lock & Co.

Urban, Dieter and Mayerl, Jochen (2011) *Regressionsanalyse: Theorie, Technik und Anwendung*. Wiesbaden: VS Verlag für Sozialwissenschaften/Springer.

Welzel, Christian (2003) 'Irrtümer bei der Interpretation des "ökologischen Fehlschlusses": Zu Aussagekraft aggregierter Umfragedaten'. In: Pickel, Susanne, Pickel, Gert, Lauth, Hans-Joachim, and Jahn, Detlef (ed.) *Vergleichende politikwissenschaftliche Methoden. Neue Entwicklungen und Diskussionen*. Wiesbaden: Springer Fachmedien.

10 Resilience constructions in European coastal areas

Having shed light on shared vulnerability constructions in the previous chapter, I shall now consider which practical measures (resilience constructions) are preferred by actors in the research area. On the one hand, the question is centred on the extent to which resilience constructions (see Section 2.1.4) are socially shared and thus emerge as climate cultures. Furthermore, the extent to which the factors discussed in Chapter 5 contribute to the explanation of systematic differences will be investigated.

As already mentioned, it must be assumed that climate cultures can only ever be investigated aspectually (see Chapter 2). This is also true for socially shared resilience constructions. Following the central challenges outlined in scientific and spatial research discourses (see Chapters 3 and 4), the following will explore on the one hand which climate-mitigation cultures are to be found in the research area (Section 10.2) and, alongside this, which adaptation cultures are present (Sections 10.3 to 10.6). Climate adaptation is always bound together with particular climate-specific challenges. As previously described in Chapter 4, the focus will be on practices against sea flooding, inland flooding, heat, and storms. Before exploring each of these in detail, I shall first compare the general preference patterns of the fields of application being considered here below.

10.1 A general comparison of climate mitigation and adaptation

10.1.1 Necessity and feasibility of climate mitigation and adaptation in general

The respondents evaluated the necessity and feasibility of specific climate-mitigation and adaptation measures for the near future until 2025 (see also Section 7.2). The first step therefore is to analyse how actors relate climate-mitigation and adaptation measures to each other in general, and whether there are any differences in their assigned priorities. Figure 10.1 shows the mean values of all respondents, displaying the measures according to their affiliation to climate mitigation (diamond) or adaptation (cross). Additionally, the matrix can be divided into four fields, each of which should make it clear whether

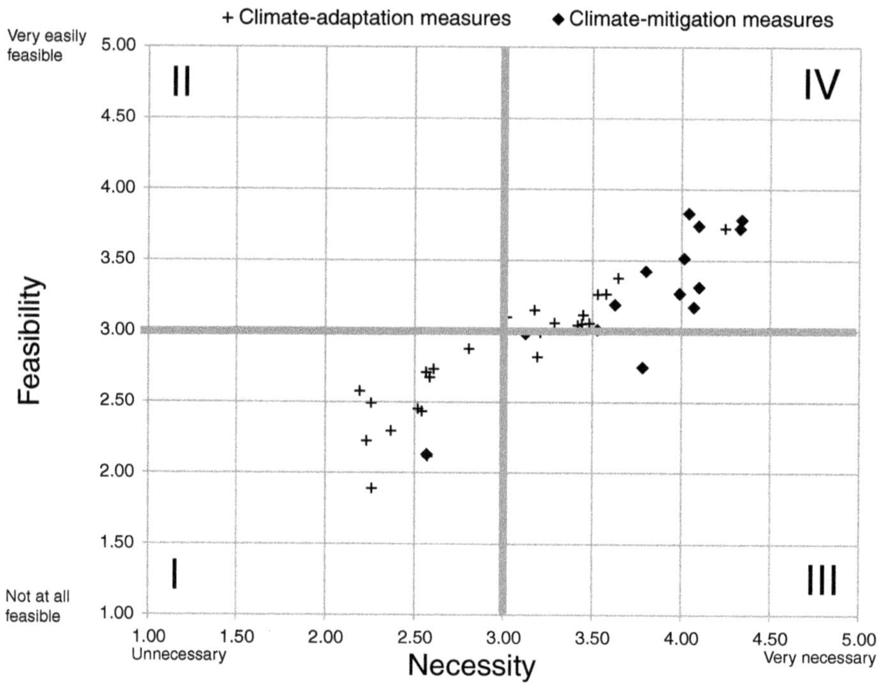

Figure 10.1 Necessity and feasibility of climate mitigation and adaptation measures: total.

Source: own calculations.

Notes

Question: Please assess the necessity and feasibility of the following measures for [reference area] by 2025.

Basis: Respondents familiar with the respective measures (as per questions MW1, MR1a, MR1b, and MS1; in each case positions 2–5).

measures are on average considered neither necessary nor feasible (Field I), feasible but unnecessary (Field II), necessary but unfeasible (Field III), or both necessary and feasible (Field IV) (Heimann 2013).

Assessing the respondents' evaluations generally, the prioritisation of climate-mitigation over adaptation measures can be observed. Most of the climate-mitigation measures are positioned in Field IV, and are thus regarded on average as predominantly necessary and feasible. On the other hand, most of the adaptation measures appear to be of secondary importance. Significantly fewer measures are included in Field IV, and many also appear in Field I, being therefore considered neither necessary nor feasible.

Additionally, an initial assessment needs to be made as to whether differences exist in the evaluations that can be attributed to either an actor's field affiliation or national background. Table 10.1 shows the averages for all

Table 10.1 Necessity and feasibility of climate-mitigation and adaptation measures: arithmetic mean according to field and country

	Necessity		*Feasibility*	
	Mitigation	*Adaptation*	*Mitigation*	*Adaptation*
Administration	3.8	2.9	3.1	2.7
Politics	3.9	2.9	3.3	2.8
Economy	3.8	3.0	3.3	2.9
NGO	4.1	3.2	3.5	3.1
ANOVA	**	**	***	***
DE	3.8	2.9	3.3	2.9
NL	3.9	3.0	3.2	2.9
DK	4.0	2.7	3.4	2.8
PL	3.7	3.1	2.8	2.5
ANOVA	*	***	***	***
TOTAL	3.8	2.9	3.2	2.8

Source: own calculations.

Notes
ANOVA $***p<0.001$; $**p<0.05$; $*p<0.1$.
Question: Please assess the necessity and feasibility of the following measures for [reference area] by 2025. Basis: Respondents familiar with the respective measures (as per questions MW1, MR1a, MR1b, and MS1, in each case positions 2–5).

climate-mitigation and adaptation measures per group, as well as the results of the *F*-test of single-factor variance analyses, in order to check whether the mean values differ significantly from each other (Level of significance: $***p<0.001$; $**p<0.05$; $*p<0.1$).

Looking at the results of the variance analysis, there are significant differences between groups both for climate-mitigation and adaptation measures. Differences in the assessment of the necessity for climate-mitigation measures appear somewhat smaller than in the assessment of feasibility, however.

Considering the values according to field, evaluations of the necessity and feasibility of both climate-mitigation and adaptation measures have the highest average among NGOs (NGO: Climate-mitigation$_{\text{necessity/feasibility}}$ $\varnothing=4.1/3.5$; Climate-adaptation$_{\text{necessity/feasibility}}$ $\varnothing=3.2/3.1$). On the whole, however, differences between the fields seem only slightly pronounced.

Considering national background also reveals both similarities and differences between countries. First of all, the graphs are the most similar for German, Danish, and Dutch respondents. This is mainly due to Polish respondents' more pessimistic assessments of the feasibility of measures (climate mitigation $\varnothing=2.8$; climate adaptation $\varnothing=2.5$). With regard to necessity, climate-mitigation measures are also on average the lowest in Poland ($\varnothing=3.7$). This trend does not continue for assessments of the necessity for climate-adaptation measures, however. Indeed, comparing countries, necessity is even assessed as highest in Poland on average ($\varnothing=3.1$). On average, adaptation measures are also given higher priority in the Netherlands (adaptation: necessity Ø = 3.0) than in Denmark (Ø = 2.7) and Germany (Ø = 2.9). Conversely, a somewhat greater priority for climate-mitigation measures is observable among Danish and Germans respondents, who on average judge significantly more measures to be necessary and feasible.

On the whole, major differences occur between Poland and the remaining countries, although this applies mostly to climate-mitigation measures and to the assessment of measure feasibility. On the other hand, the necessity of numerous adaptation measures is, as mentioned, barely discussed in Poland.

Considering the matrix, it is also noticeable that there are linear correlations between evaluations of the necessity of measures and their feasibility. To avoid redundancy, therefore, the following will be limited to evaluations of the necessity of measures.

10.1.2 *Necessity of general measure strategies*

Figure 10.2 shows the arithmetical means of the assessments of necessity for the general measures strategies. In addition, the results of an *F*-test of single-factor variance analyses are shown to verify whether the mean values differ significantly from one another (levels of significance: ***$p < 0.001$; **$p < 0.05$; *$p < 0.1$). Additionally, for climate mitigation I differentiate here between the two basic strategies of 'greenhouse gas reduction' and 'greenhouse gas sinks'. In each case, the basic population is formed of those respondents who are at least minimally familiar with the respective topics (positions 2–5). In the two profiles of mean values in Figure 10.2, values greater than 3 indicate a predominant group tendency to assess a measure as necessary, while values less than 3 indicate the opposite.

Considering field-specific differences in the upper mean-value profile, no major deviations appear. Measures against heat are considered hardly necessary by all groups (heat: Ø < 3), while all other measures are predominantly judged to be necessary (Ø > 3).

One group in particular regularly views measures to be of higher necessity than the others: NGOs are the most supportive in evaluating the necessity of measures. The difference is most evident in climate-mitigation measures to reduce greenhouse gas emissions (ANOVA: significance <0.001), with the exception only of politics, which in the case of storms sees the greatest necessity.

To what extent do differences exist according to national background? First, the central tendency remains that comparatively little necessity is ascribed for heat measures, while the other general strategies are supported. However, some differences are also to be seen, such as a somewhat higher evaluation of the need for measures against heat in general in the Netherlands and in Poland. Polish respondents also ascribe comparatively higher necessity to measures against inland flooding, closely followed by the Dutch. Due to the geography of the Netherlands, it is hardly surprising that the Dutch also evaluate measures against sea tides as being more necessary.

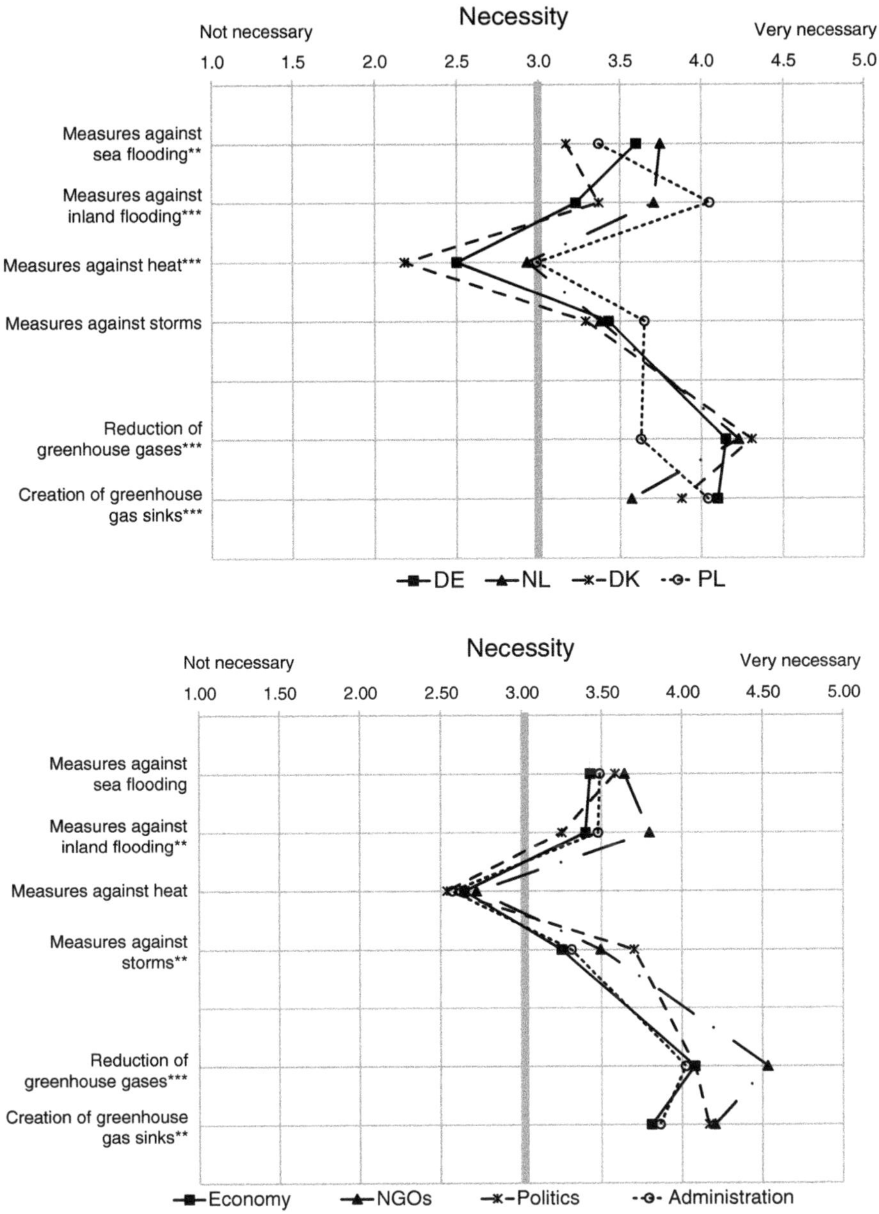

Figure 10.2 Necessity of climate-mitigation and adaptation measures in general by 2025: fields and countries.

Source: own calculations.

Notes
Question: Please assess the necessity of the following measures in [reference area] by 2025.
Basis: Respondents familiar with the respective measures (as per questions MW1, MR1a, MR1b, and MS1; in each case positions 2–5). ANOVA F Test: ***$p < 0.001$, **$p < 0.01$, *$p < 0.05$.

Taking the available results together, it can be concluded that, besides the fact that different forms of background knowledge have relevance for different evaluations, knowledge about physical conditions could be relevant to the assessment of the necessity of measures. As physical contexts can vary systematically within the research area, an analytical strategy is required to meet these variations. To this end, this chapter first examines the extent to which measure strategies for sea flooding, inland flooding, heat, and storms, as well as greenhouse gas reduction and sinks are preferred *in general*. In the following sections, the specific combinations of methods preferred by actors regarding respective climate impacts will be examined in more detail. Only those actors who presuppose a minimal necessity for measure strategies in general will be included in the analysis.

Actors are thus filtered in two stages. The first stage excludes actors who are minimally familiar with the measures. The second stage takes all actors into account who are minimally convinced of the necessity for measures in general.

10.2 Climate-mitigation strategies

10.2.1 Categorical framework: ecological modernisation, sufficiency, ecological sinks, and technological sinks

In order to determine the different climate-mitigation cultures, actors' measure preferences with regard to core strategies are summarised. To investigate the relational cultural spaces of *socially shared* knowledge, each category is investigated with regard to 'on which basis the actors symbolically organise their reality' (Reckwitz 2000: 143f.). Chapter 4 previously examined the central strategies discussed in the discourses of spatial development on climate change. The question poses itself whether the classification schemes used here have also been internalised by the respondents and are reflected in their answers. To what extent can a basic canon of socially shared cognitive schemata be assumed to exist among actors? After answering this, the socially shared central categories thus discovered are then used as a basis for determining climate-mitigation cultures in a cluster analysis.

Chapter 4 discussed the three core strategies that have to date been distinguished in climate-mitigation discourse for the reduction of greenhouse gases, namely *efficiency* and *consistency* strategies, developed from ecological modernisation discourses, and *sufficiency* strategies, which emerged from older ecology discourses closely connected to critiques of civilisation. Additionally, two strategies for absorbing greenhouse gases from the atmosphere have been discussed: *ecological* and *technological* sinks. The question arises whether the respondents share this categorical framework of *five core strategies*.

To answer this question, I shall now consider the evaluations by respondents regarding the necessity of climate-mitigation measures up to 2025 in a principal component analysis (Table 10.2). The schema can be represented in a five-factor solution with a number of restrictions.

Table 10.2 Factor analysis: strategies for climate mitigation

	Ecological modernisation	*Sufficiency I: Transport + Buildings*	*Sufficiency II: Increase in building density*	*Ecological sinks*	*Technological sinks*
	1	*2*	*3*	*4*	*5*
Reduction of individual motorised traffic	0.39	**0.72**	0.20	0.15	–0.04
(More local public transport (LPT))	**0.42**	**0.74**	0.10	0.19	–0.04
Improvement of infrastructure for cycling	0.38	**0.70**	0.00	0.21	–0.01
Combination of living and working spaces	0.24	**0.72**	0.29	0.13	0.17
Securing areas for renewable energies	**0.80**	0.20	0.16	0.17	0.12
Renewable energies in private households and businesses	**0.78**	0.35	0.13	0.20	0.05
(Energy efficiency measures in general)	**0.81**	**0.40**	0.11	0.06	–0.02
(Energy optimised buildings)	**0.80**	**0.41**	0.11	0.07	0.01
Increased building density	0.24	0.25	**0.91**	0.14	–0.02
New creation of public green spaces: necessity	0.12	0.22	0.06	**0.91**	0.09
New creation of private green spaces: necessity	0.17	0.14	0.11	**0.91**	0.11
Carbon Capture Storage technology (CCS)	0.07	0.02	–0.01	0.16	**0.97**

Source: own calculations.

Notes

Question: Please assess the necessity of the following measures in [reference area] by 2025. Basis: All respondents; $n = 830$. Kaiser–Meyer–Olkin: 0.899. Bartlett Test: $***p < 001$. Number of factors: theoretical derivation. Extraction method: principal component analysis. Rotation method: varimax with Kaiser normalisation. Results after 6 iterations. Total variance explained: 82%.

Items in brackets have been removed due to parallel correlation and/or low consistency values (for details, see Heimann 2017).

- *Ecological* and *technological sinks* are clearly distinguishable, each with high correlations to the fourth and fifth factors.
- Considering the first factor, *consistency* and *efficiency strategies* are barely distinguishable. All items belonging within the discursive tradition of *ecological modernisation* show high correlations with this factor.
- Items from the *sufficiency strategy* correlate (>0.7) with the remaining factors (2) and (3). Thus a distinction can be drawn here between specific transport- and building-related measures (factor 2), and the density of settlements (factor 3).
- However, some variables correlate to more than one factor. Actors draw an association between efficiency measures and aspects of sufficiency ('Energy-optimised buildings' correlates slightly to 'Sufficiency I' (0.40)).[1] The item 'more local public transport' correlates strongly to 'Sufficiency I' while also demonstrating a parallel correlation to 'Ecological modernisation'.[2]

Which classification scheme should now be used for determining climate-mitigation cultures in the cluster analysis, and which measures can be combined into indices?[3] Given the observed results, the category *ecological modernisation* will be used here, and no further differentiation will be made between efficiency and consistency. To be able to distinguish adequately between ecological modernisation and sufficiency strategies, all items with parallel correlations have been removed.

The result of the factor analysis raises the question whether the two sufficiency strategies need to be differentiated. As the reliability remains strong when both are summarised, only one sufficiency scale is used here.[4]

To summarise, a *four-dimensional relational cultural space* forms the basis for the following cluster analysis, examining which actors share similarities in terms of preferences for the four above-mentioned core strategies. Thus four indices are now formed, each representing the preferences of actors along the dimensions '*ecological modernisation*', '*sufficiency*', '*ecological sinks*', and '*technological sinks*'.

10.2.2 Climate-mitigation cultures according to the relational concept of space: universalists, classics, moderates, and minimalists

With the aid of the hierarchical cluster analysis and Ward's procedure, four climate-mitigation cultures can be determined (Figure 10.3). These follow the relational understanding of cultural space and distinguish themselves according to shared preferences for the climate-mitigation practices mentioned previously (for more on the relational understanding of cultural space, see Chapter 2). The climate-mitigation cultures thus differentiate themselves according to preferences for 'ecological modernisation', 'sufficiency', 'ecological sinks', and 'technological sinks'.

The groups can be divided into universalists, classic climate mitigators, moderates, and minimalists:

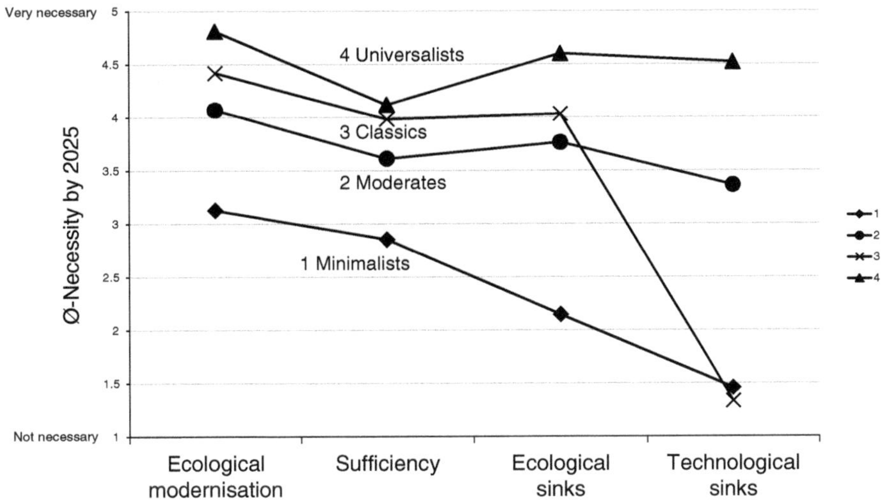

Figure 10.3 Climate-mitigation cultures: description of clusters.

Source: own calculations.

Notes

Question: Please assess the necessity of the following measures in [reference area] by 2025: (MS1 positions 2–5); n = 652 (DE: 385, NL: 88, DK: 86, PL: 93; administration: 362, politics: 190, economy: 23, NGO: 77).

To be able to investigate specific climate-mitigation cultures, actors who generally rejected measures reducing climate mitigation as well as sink strategies (each position 1) were encoded as rejecting individual measures (position 1). Hierarchical cluster analysis, proximity measure: squared Euclidean distance; grouping procedure: Ward.

- The first group of actors shows high preference values along all four indices. This culture is characterised by openness to ecological modernisation, sufficiency, and technological and ecological sinks (in each case Ø necessity >4). Sufficiency receives slightly lower preference than the remaining measures, but nevertheless continues to enjoy high acceptance. This group is labelled *universalists* (Group 4).
- One other group shares all preference values with the exception of technological sinks (Group 3). As this corresponds to the early IPCC climate-mitigation discourse from the 1990s, when CCS technologies had barely been considered, this culture is labelled *classic climate mitigationists*.
- Another group is characterised by its moderate preference for all the measures. Members of this group are therefore named *moderate climate mitigationists*.
- Finally, actors in the last group demonstrate comparatively low preferences for all four climate-mitigation strategies. It must be noted, however, that this is only in comparison to the other groups, as some preferences still obtain medium-strength values. Therefore we cannot speak of complete

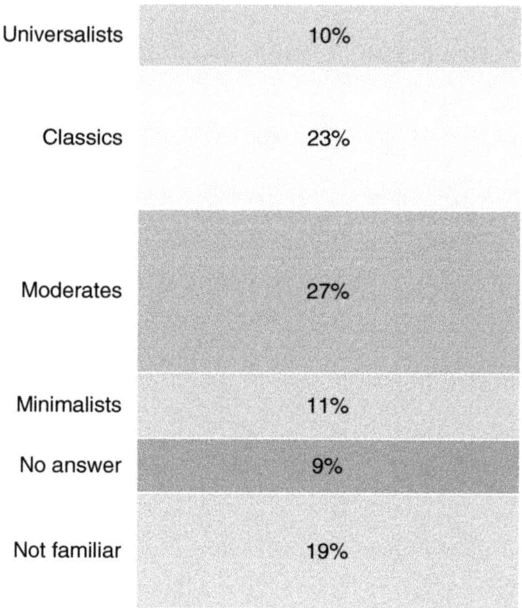

Figure 10.4 Climate-mitigation cultures: distribution of clusters – total.

Source: own calculations.

Note
Basis: All respondents; $n = 830$.

rejection, and in the case of technological modernisation the rejection is lowest (Ø necessity: 3.2). The group is thus termed the *minimalists*.

Considering the distribution of the groups in the sample (Figure 10.4), a third of respondents (33%) are among the two most highly motivated groups, the classic climate mitigationists and the universalists. Of these, the classic mitigationists form the larger share (23%), while only one in 10 (10%) is a universalist who additionally supports CCS technologies. Overall, the moderate climate mitigationists are the most represented (27%). Minimalists make up a tenth (11%). In total, actors who favour climate mitigation predominate, albeit moderately. In addition, around one-fifth (19%) of respondents indicate that they are not familiar with climate-mitigation measures in spatial development, while another tenth (9%) provided no answer to the question.

10.2.3 Explanation of differences

How can the differences between the groups be explained? In the following, the question of the role played by different forms of background knowledge is examined, as well as the extent to which other social factors, such as field affiliation and national background, can explain differences.

Background knowledge

On the basis of a multinomial logistic regression, I first examine the extent to which the forms of background knowledge discussed can explain different affiliations with the four climate-mitigation cultures (Table 10.3).

The group of minimalists will serve as the reference group, as it provides the advantage that differences are always considered in comparison to the group with the lowest climate-mitigation acceptance values.[5]

First, the core hypotheses set out in Chapter 5 will be applied to the groups identified. Again, it should be noted that the groups do not vary greatly between strategies such as sufficiency, efficiency, and consistency. Only the highly motivated groups differ in that universalists advocate CCS technology while classic climate mitigationists do not. The other groups differ primarily in their levels of overall acceptance of climate mitigation. This raises the general question as to whether climate-mitigation measures and all their consequences are accepted or rejected on a wholesale basis. Applying the hypotheses to the given groups as follows:

- The first values-based hypothesis stated that actors with a higher *universal-altruism orientation* also prefer ecologically oriented measures. This should lead to an increasing universal-altruism orientation, raising the likelihood of belonging to one of the more highly motivated groups.
- By contrast, an increased *human-altruism orientation* leads to measures being preferred that benefit other humans. As climate-mitigation strategies are meant to benefit human communities in the long term, this should also lead to a clear separation with the minimalists in favour of one of the more highly motivated groups.
- *Self-enhancement orientation* could lead to measures associated with higher costs being avoided. This could thus increase the probability of affiliation to the minimalists.
- Greater *openness to change* should lead to new, innovative measures being preferred. This pertains to climate mitigation in general, so that in comparison with minimalists, there should be an increased probability of an affiliation to the other groups.
- Additionally, various forms of conservatism should lead to comparatively new forms of climate-mitigation strategies being rejected. *Security* and *conformity/traditionalist orientations* should therefore shift the probability of group affiliation towards the minimalists.
- Apart from the value orientations, internalisation of the *New Ecological Paradigm* should lead to ecologically oriented measures being preferred, including sufficiency strategies. Measures associated with a threat to nature and the environment might be preferred less. This is why NEP orientation

Table 10.3 Multinomial regression models (1–3): climate-mitigation cultures

	Model 1: Values			*Model 2: Values + NEP + RI*			*Model 3: Values + NEP + RI + KV*		
	$\beta^{Sig.}$ *[odds ratio]*			$\beta^{Sig.}$ *[odds ratio]*			$\beta^{Sig.}$ *[odds ratio]*		
Reference group: Minimalists	*Universalists* n = *82*	*Classics* n = *181*	*Moderates* n = *211*	*Universalists* n = *82*	*Classics* n = *181*	*Moderates* n = *211*	*Universalists* n = *79*	*Classics* n = *179*	*Moderates* n = *202*
	Minimalists n = *86*			*Minimalists* n = *86*			*Minimalists* n = *81*		
Background knowledge									
Values									
Self-transcendence I: Universalism	0.57** [1.77]	0.88*** [2.42]	0.24* [1.28]	0.39* [1.47]	0.73*** [2.08]	0.18	0.39* [1.47]	0.76*** [2.13]	0.21
Self-transcendence II: Human altruism	0.19	–0.06	0.07	0.14	–0.09	0.03	0.13	–0.1	0.03
Self-enhancement	0.04	0.04	0.09	0.02	0.03	0.07	0.07	0.02	0.04
Openness to change	0.26	–0.24	–0.08	0.3	–0.2	–0.06	0.31	–0.18	–0.04
Conservation I: Security	–0.28* [0.75]	–0.36** [0.70]	–0.29* [0.75]	–0.29* [0.75]	–0.36** [0.7]	–0.29* [0.75]	–0.25* [0.78]	–0.34** [0.71]	–0.26* [0.77]
Conservation II: Tradition/ Conformity	0.17	–0.05	0.23* [1.26]	0.22	–0.02	0.23* [1.26]	0.25	0.01	0.27* [1.31]
Beliefs									
New Ecological Paradigm (NEP)				1.00*** [2.73]	0.86*** [2.35]	0.39* [1.48]	0.88** [2.41]	0.81*** [2.25]	0.41* [1.51]
Identity construction									
Local identification				0.16	0.14	0.19	0.12	0.08	0.19
Control variables (CV)									
Age							0.02	0	–0.02
Gender (0 = female, 1 = male)							–0.18	–0.12	–0.13
Constants	–4.57***	–1.24	–0.36	–7.88***	–4.14**	–1.98	–8.54***	–4.00**	–1.42
Model accuracy									
Pseudo R^2 (Nagelkerke)	0.136			0.171			0.183		
LR test	75.39***			96.42***			100.47***		
N	560			560			541		

Source: own calculations.

Notes

β ***$p<0.001$; **$p<0.01$; *$p<0.1$. Values in brackets [odds ratio].

should increase the probability of affiliation with the motivated groups, while HEP orientation should have the opposite effect.

- It was postulated that *local identification* could increase the acceptance of climate-mitigation measures if the possibility of development in the place/region can be positively ascribed to them. In this case, increased local identification would raise the probability of belonging to one of the more highly motivated groups.

Observing the results once again, on the basis of the Likelihood Ratio test (LR) one can reject the hypothesis for all three models that they do not contribute to group differentiation with a high level of significance ($p < 0.001$). The explained variance proportions are very moderate, lying between 13% (Model 1: Pseudo $R^2_{Nagelkerke}$) and 18% (Model 3).

Comparing all three models, the explanatory patterns appear consistently:

- A significant explanatory contribution is made by *universal-altruism orientation* (Self-transcendence I: *Universalism*) as well as both conservative orientations (Conservation I: *Security*, Conservation II: *Tradition/conformity*).
- The dissemination of the *New Ecological Paradigm* may provide additional explanation for the separation of climate-mitigation cultures.
- On the other hand, no significant explanatory power can be ascribed to local identification, human-altruism orientation, self-enhancement, or openness to change.

Considering universalism orientation: as already indicated in many other studies which were discussed in Chapter 5, these respondents demonstrate a high separation force for climate-mitigation acceptance, significantly increasing the probability of belonging to one of the groups with an affirmative attitude to climate mitigation in comparison to the group of minimalists (Model 1: universalists $\beta = 0.57$**; classics: $\beta = 0.88$***, moderates: 0.24*). The odds ratio values (in brackets) also reveal a clear logic to the strength of the influence: the more nature-oriented the group, the greater the explanatory power of universalism orientation. The explanatory power is highest for classic mitigationists (odds ratio: 2.42), followed by the universalists (odds ratio: 1.77), who advocate technological, in addition to natural, sinks. In last place are the moderates (odds ratio: 1.28). This basic schema is revealed in all three models.

Unsurprisingly, there is a clear pattern regarding the two conservative items. With an increased security orientation, the probability of belonging among the minimalists increases (Model 1: universalists: $\beta = -0.28$*; classics: $\beta = -0.36$**; moderates: $\beta = -0.29$*). The moderates are characterised not only by their forming the largest group, but also by their moderate, more consensus-oriented bearing towards climate-mitigation. It can be seen that with increasing conformity orientation, the probability of belonging among the group of moderates increases (Model 1: $\beta = 0.23$*).

In the second model, the explanatory power of NEP orientation and place identification are considered as well. As expected, an increase in NEP orientation also increases the probability of belonging to the groups with higher climate-mitigation preferences, in comparison to the minimalists (Model 2: universalists: $\beta = 1.00^{***}$; classics: $\beta = 0.86^{***}$; moderates: $\beta = 0.39^{*}$). It is also noticeable that partial mediator effects again appear between NEP and universalism orientations, as their beta values decrease significantly once NEP orientation is taken into account (see also Urban and Mayerl 2011: 304 and Section 9.1.3). Conversely, local identification appears to have no significant explanatory power for group affiliation. This is also the case for age and gender (Model 3).

An additional detail deserves our attention: universalists, who are highly motivated climate-mitigationists, differ from classic climate mitigators in their support for CCS technology. At the time of publication, CCS technology is still undergoing trials. In this context, beta coefficients for 'openness to change' indicate higher probabilities of belonging among the universalists, in contrast to the other groups (Model 1: universalists: $\beta = 0.26$, classics: $\beta = -0.24$, moderates: $\beta = -0.08$). These values do not, however, appear significant.

In conclusion, it is obvious that value orientations, above all universalism and conservatism, together with beliefs, especially the New Ecological Paradigm, make significant contributions for explaining affiliation with the respective climate-mitigation cultures.

- High universalism orientation increases the probability of belonging among an affirmative climate-mitigation culture. In addition, the more nature-oriented the group, the higher the explanatory force of universalism orientation.
- In comparison with the minimalists, affiliation among the moderates can be explained by a higher conformity orientation.
- With an increasing NEP orientation, the probability of belonging to the affirmative climate-mitigation cultures also increases.

Field affiliation and national background

To what extent does national background separate the climate-mitigation cultures discussed here? Does field affiliation also explain differences?

Turning again to the frequency distributions of climate-mitigation cultures by field and country (Figure 10.5), we see very little difference between the fields. Among environmental NGOs the share of actors motivated towards climate mitigation is the highest, since universalists, classics, and moderates together compose more than two-thirds (70%) of these respondents. This should not come as a surprise, since the efforts of these NGOs are usually focused on placing environmental issues on the social agenda. Nevertheless, actors from politics follow closely (68%), and supporters of climate mitigation in each of these three groups also form the majority in the economy (58%) and

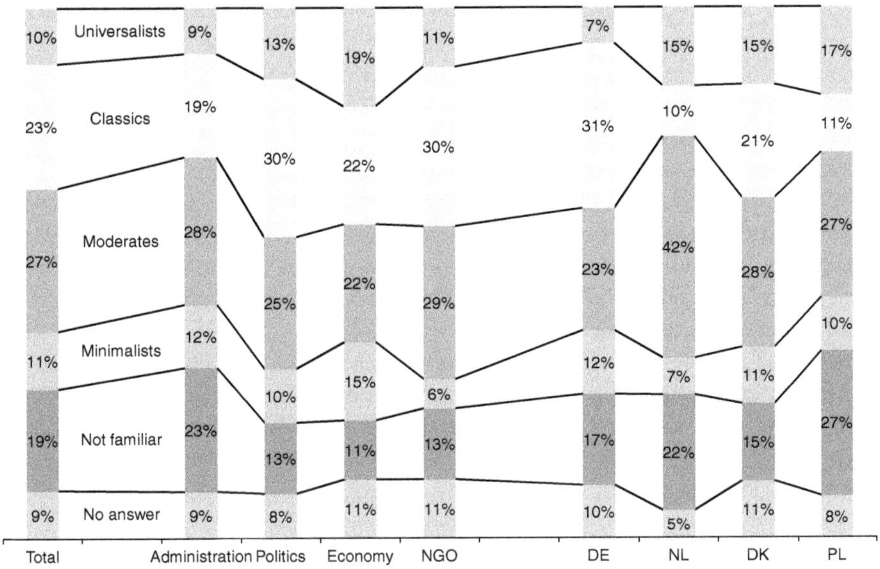

Figure 10.5 Climate-mitigation cultures: distribution of clusters – countries and fields.

Source: own calculations.

Notes

Question: Please assess the necessity of the following measures in [reference area] by 2025.
Basis: All respondents; $n = 830$.

administration (56%). On the fundamental issue of the need for climate mitigation, the fields are thus hardly polarised. Considering the two most highly motivated groups, the universalists and the classics, a further difference is apparent. Universalists, who are also open to CCS technology, are most often found in the economy. One in five economic actors belong to this group (19%), whereas for NGOs (11%) and administrative actors (9%), only about one in 10 are universalists. On the whole, however, it should again be emphasised that only a few, insignificant differences exist between the fields.

Considering national background, the share of the three climate-affirmative groups is lowest in Poland, although it still accounts for 55% of respondents. Additionally, the proportion of actors unfamiliar with climate-change measures is higher in Poland than in other countries, accounting for more than one-quarter (27%). Accordingly, although 'other priorities' prevail in Poland (Raabe and Härtel 2007), there can be no suggestion of a fundamental rejection of climate mitigation.[6] Differences also occur between the three affirmative groups. In Germany, nearly one third of the respondents are classic climate mitigators (31%). Technological sinks, on the other hand, are rarely advocated in Germany (universalists: 7%). In the other countries, universalists are represented somewhat more frequently (the Netherlands: 15%; Denmark: 15%;

Poland: 17%). On the other hand, classical climate mitigationists are comparatively rare in the Netherlands (10%) and Poland (11%). In the Netherlands, the group of moderates is the largest (42%). To summarise initially, the frequency distributions thus demonstrate stronger differences in their composition according to national background.

Taking a further look at regression Models 4 to 6 (Table 10.4), calculated using a procedure analogous to the one used for the vulnerability constructions (Section 9.1.3), the LR test shows that these models also differ significantly from the null model, and thus have a significant explanatory power for climate-mitigation cultures. Model 4, which takes account of field affiliation, national background, age, and gender, still provides a moderate explanatory power to the total variance being explained (Model 4: Pseudo $R^2_{\text{Nagelkerke}}$: 15.3%). It can thus be assumed that field affiliation and national background are of relevance to distinct climate-mitigation cultures, but only make up a small part of the explanation of the total variance. If background knowledge is additionally taken into account, the explanation of variance increases further (Model 5: Pseudo $R^2_{\text{Nagelkerke}}$: 19.5%). Finally, the overall variance can be assumed to have a sufficient explanatory level in the final model (Model 6: Pseudo $R^2_{\text{Nagelkerke}}$: 30.8%).

As already observed for frequency distribution, under mutual control of the variables field affiliation and national background, Model 5 displays markedly larger and more frequent effects resulting from the latter. This indicates the greater relevance of national background for the separation of climate-mitigation cultures compared to field affiliation.

For the fields, administration is again chosen as the reference group. A significant effect appears here. Compared with the minimalists in administration, NGO affiliation increases the probability of belonging to the group of classic climate mitigators (Model 4: NGO to classics: $\beta = 1.1^{*}$). Thus field affiliation might be able at least to explain affiliation to the classic climate mitigators. But what are the underlying reasons for this? Sequential regression analysis is again able to provide an answer. As previously observed in the vulnerability constructions, the beta value in Model 5 drops once again to an insignificant level (Model 5: NGO to classics: $\beta = 0.46$), confirming that mediator effects are present, in this case because background knowledge systematically varies among NGO actors compared to those in other fields. Since in comparison to Model 3 all the significant values fall slightly, it is not possible to determine to what extent exactly the effect can be explained indirectly, such as by higher NEP or universalism orientations, or lower conservatism values. Considering the aggregate data in Table 9.3, it becomes clear that NGOs have increased NEP and universalism orientations and, at the same time, the lowest conservatism orientation. All these knowledge constructions therefore contribute explanations to the separation of the groups.

Considering the separation force of national background, Model 4 shows that membership to the Dutch, Danish, and Polish groups significantly increases the probability of belonging among the universalists (NL: $\beta = 1.79^{***}$; DK: $\beta = 1.14^{*}$; PL: $\beta = 1.46^{***}$).[7] This result is again confirmed under control by the

Table 10.4 Multinomial regression models (4–6): climate-mitigation cultures

	Model 4: Field + Country + CV			*Model 5: Field + Knowledge + CV*			*Model 6: Field + Country + Knowledge + KV*		
	$\beta^{Sig.}$ *[odds ratio]*			$\beta^{Sig.}$ *[odds ratio]*			$\beta^{Sig.}$ *[odds ratio]*		
Reference group: Minimalists	*Universalists* n = *84*	*Classics* n = *189*	*Moderates* n = *212*	*Universalists* n = *79*	*Classics* n = *179*	*Moderates* n = *202*	*Universalists* n = *79*	*Classics* n = *179*	*Moderates* n = *202*
	Minimalists n = *82*			*Minimalists* n = *81*			*Minimalists* n = *81*		
Background knowledge									
Self-transcendence I: Universalism				0.39* [1.48]	0.75*** [2.11]	0.17	0.46* [1.59]	0.78*** [2.18]	0.19
Self-transcendence II: Human altruism				0.13	–0.12	0.04	0.17	–0.13	0.09
Self-enhancement				0.08	0.02	0.06	0.42* [1.52]	0.04	0.31
Openness to change				0.3	–0.19	–0.05	–0.14	–0.19	–0.31* [0.73]
Conservation I: Security				–0.25* [0.78]	–0.33** [0.72]	–0.26* [0.77]	–0.43** [0.65]	–0.32* [0.73]	–0.36** [0.7]
Conservation II: Tradition/ Conformity				0.26	0.02	0.28* [1.32]	0.24	0.01	0.33* [1.39]
New Ecological Paradigm (NEP)				0.88** [2.42]	0.79*** [2.21]	0.39* [1.48]	1.24*** [3.44]	0.81*** [2.24]	0.61* [1.84]
Local identification				0.11	0.05	0.19	0.24	0.01	0.25
Field affiliation									
Politics	0.49	0.52	0.12	0.09	0.42	0.04	0.08	0.42	–0.11
Economy	0.52	–0.13	–0.56	0.53	0.06	–0.43	0.51	–0.09	–0.53
NGO	0.54	1.1* [3.00]	0.64	0.27	0.46	0.68	–0.22	0.52	0.28
(*Reference group*: Administration)									

continued

Table 10.4 Continued

	Model 4: Field + Country + CV			*Model 5: Field + Knowledge + CV*			*Model 6: Field + Country + Knowledge + KV*		
	$\beta^{Sig.}$ *[odds ratio]*			$\beta^{Sig.}$ *[odds ratio]*			$\beta^{Sig.}$ *[odds ratio]*		
Reference group: Minimalists	*Universalists* n = *84*	*Classics* n = *189*	*Moderates* n = *212*	*Universalists* n = *79*	*Classics* n = *179*	*Moderates* n = *202*	*Universalists* n = *79*	*Classics* n = *179*	*Moderates* n = *202*
	Minimalists n = *82*			*Minimalists* n = *81*			*Minimalists* n = *81*		
National background									
NL	1.54** [4.66]	–0.76	1.09* [2.96]				2.44*** [11.50]	–0.35	1.99*** [7.35]
DK	0.92* [2.51]	–0.37	0.2				1.25* [3.48]	–0.5	0.7
PL	1.59*** [4.90]	–0.56	0.42				2.40*** [11.05]	0.12	1.06* [2.88]
(*Reference group*: DE)									
Control variables (CV)									
Age	0.03* [1.03]	0	–0.01	0.02	0	–0.02	0.04* [1.04]	0	0
Gender (0 = female, 1 = male)	–0.47	–0.27	–0.28	–0.19	–0.13	–0.11	–0.15	–0.1	–0.15
Constants	–2.17*	0.83	1.21*	–8.61***	–3.68*	–1.28	–11.45***	–3.68*	–3.4*
Model accuracy									
Pseudo R^2 (Nagelkerke)	0.153			0.195			0.308		
LR test	86.42***			107.62***			181.24***		
N	567			541			541		

Source: own calculations.

Notes
β ***p < 0.001; **p < 0.01; *p < 0.1. Values in brackets [odds ratio].

observations from Figure 10.5, where in comparison to Germany, comparably larger groups of universalists were to be found in the Netherlands, Denmark, and Poland (DE: 7%; NL: 15%; DK: 15%; PL: 17%). The large proportion of moderates in the Netherlands (42%) also emerges under control conditions, as Dutch origin significantly raises the probability of affiliation to this group (Model 4: 1.09*). In summary, there appear to be significant differences between Germany and the other countries regarding affiliation to the universalists (greater in the Netherlands, Denmark, and Poland) and moderates (greater in the Netherlands). Affiliation to the group of classic climate mitigators, on the other hand, cannot be predicted on the basis of national background.

To what extent can variations in field and country distributions be explained by differences in background knowledge? To answer this question, the micro and macro levels must be viewed together. When considering the models individually, the effects of universal-altruism and NEP orientations on the micro level are continuously observable. Both increase the individual probability in all models of belonging to the more motivated groups. It is clear in the aggregate comparison (Table 9.3) that the ecocentric NEP worldview occurs somewhat more frequently in Germany and Denmark than in the Netherlands, and less frequently in Poland (Germany: 83%, Denmark: 81%, the Netherlands: 69%, Poland: 60%). There are also differences in the prevalence of universal-altruism orientation, which occurs most often in Denmark (Denmark: 92%, the Netherlands: 85%, Germany: 82%, Poland: 77%). As effects are visible here both at the micro and macro levels, they quite reasonably explain why universalists and classic climate mitigators together form a larger group, especially in Germany and Denmark, and are less frequently represented in the Netherlands and Poland.

The relevance of national background in the separation of the groups is even clearer in Model 6, where a significant influence continues to be exerted on the climate-mitigation cultures discussed for universalists and moderates. In fact, the effect strengths of national background all increase markedly, and it is clear that the effect of national background on separating climate cultures cannot be explained by the observed variables alone.

In summary it can be stated that, at least in regard to German actors, national background continues to manifest a separating force on the majority of climate-culture orientations. A 'disappearance of national borders' can only be spoken of to a limited extent with regard to climate-mitigation cultures as these cultures appear transnationally throughout the reference area, while significant differences in their regional frequency are clearly observable. By contrast, field affiliation is less likely to contribute to the separation of climate-mitigation cultures, with the important exception of environmental protection organisations, which are clearly distinguished from the other fields by higher proportions of highly motivated groups.

10.3 Adaptation practices against sea flooding

10.3.1 Categorical framework: protection, retreat, and integration

The following investigates which adaptation practices against sea flooding actors consider necessary. I shall first examine whether the division of measures into three strategic categories, namely 'protection', 'retreat', and 'integration', as discussed in planning discourses (Chapter 4), can be empirically reproduced. To this end, consideration will now be made of the respondents' evaluations of the need for flood-related measures in their own reference area up to 2025.

In the principal component analysis, the categorical schema is reproduced virtually without deviation (Table 10.5). Thus the item 'improvement of dykes and protective constructions' has a very high correlation (0.94) along the first factor. This can therefore be interpreted as the protection strategy. The second factor has a high correlation to retreat-strategy measures, such as 'deconstruction of dykes' (0.87) and 'displacement of dykes inland' (0.81). The third factor correlates to measures that are a component of integration strategies, namely 'shelters' (0.74) and 'water-adapted settlement structures' (0.84). A parallel is noticeable, however. The measure 'deconstruction of settlement areas near the sea' loads on both the factors retreat (0.52) and integration (0.65). From a planning perspective, the parallel correlation is quite justified, as this measure can serve both integration of and retreat from water.

Thus for the following cluster analysis, a three-dimensional concept of cultural space is used as a basis for examining the extent to which actors share preferences for the three core strategies. Three indices are accordingly formed, one each for *protection*, *retreat*, and *integration*.[8]

Table 10.5 Factor analysis: strategies against sea flooding

	Protection	*Retreat*	*Integration*
Necessity of measures against sea flooding	1	2	3
Improvement of dykes and protective constructions	**0.94**	0.09	0.19
Displacement of dykes inland	0.31	**0.81**	0.18
Deconstruction of dykes to renaturalise the coastline	–0.04	**0.87**	0.19
(Deconstruction of settlement areas near the sea)	–0.19	**0.52**	**0.65**
Water-adapted settlement structures	0.19	0.09	**0.84**
(Population shelters)	0.22	0.22	**0.74**

Source: own calculations.

Notes

Question: Please assess the necessity of the following measures against sea flooding in [reference area] by 2025. Basis: Respondents who did not completely reject measures against sea flooding in general by 2025 (as per Question MWB1_1a, positions 2–5); $n = 647$. Kaiser–Meyer–Olkin: 0.763. Bartlett Test: $^{***}p < 0.001$. Number of factors: theoretical derivation and scree-plot. Extraction method: principal component analysis. Rotation method: varimax with Kaiser normalisation. Results after 5 iterations. Total variance explained: 77.1%.

Items in brackets have been removed due to parallel correlation and/or low consistency values (for details, see Heimann 2017).

10.3.2 Sea flooding adaptation cultures according to the relational understanding of space: protectionists, coastline retreaters, moderates, integrators, and minimalists

Which socially shared patterns of preference can be observed with regard to adaptation practices against sea flooding? Five cultures can be established through cluster fusioning according to Ward's procedure, each characterised by varying preferences for protection, retreat, and integration measures. These are presented in Figure 10.6.

It is possible to distinguish between pure protectionists, coastline retreaters, moderates, integrators, and minimalists:

- The first group shows a high preference for protection strategies (Ø necessity: protection >4). Retreat and integration measures are rejected. The coastline should therefore be held ('hold the line' strategy, see Chapter 4). Actors in this culture are therefore called *protectionists*.
- The second group, by contrast, is characterised by a high preference for the deconstruction of dykes and their removal inland (Ø necessity: retreat >4). Dykes in inland areas are to be expanded, while integration measures are

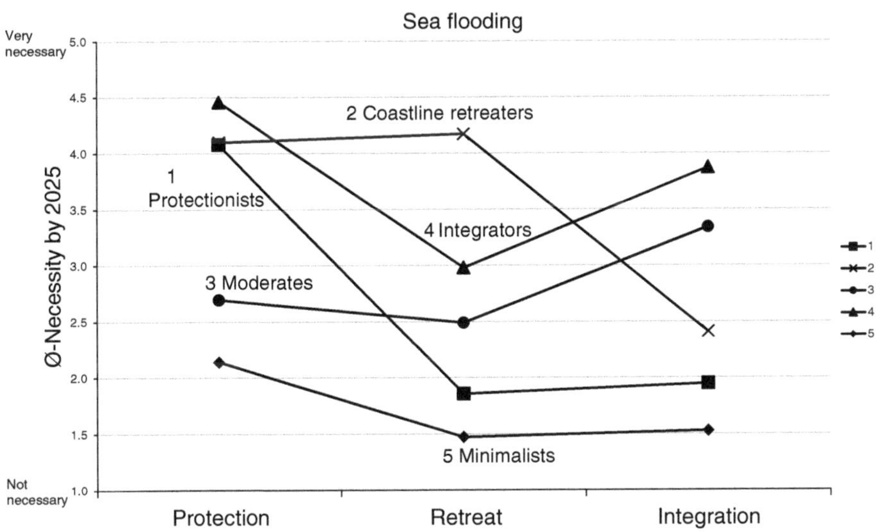

Figure 10.6 Climate-adaptation cultures against sea flooding: description of clusters.

Source: own calculations.

Notes

Question: Please assess the necessity of the following measures against sea flooding in [reference area] by 2025.

Basis: Respondents familiar with measures against sea flooding (Question MW1_SQ001 (positions 2–5), who did not completely reject measures against sea flooding in general by 2025 (as per Question MWB1_1, positions 2–5); $n = 647$. Hierarchical cluster analysis, proximity measure: squared Euclidean distance; grouping procedure: Ward.

rarely preferred (Ø requirement: protection >4, integration <2.5). This culture adopts the 'retreat the line' strategy (ibid.), and the actors are therefore referred to as *coastline retreaters*.

- The third group, meanwhile, has moderate preference levels for all three strategies, though preference for integration seems somewhat higher (Ø necessity between 2.5 and 3.5). This group is therefore referred to as the *moderates*.
- A fourth group demonstrates greater differences in practice preferences, showing greater preference for water-adapted settlement structures (Ø necessity: integration = 3.8) and for the development of dykes and protective constructions (Ø necessity: protection = 4.4). By contrast, the group has only a moderate preference for retreat. Land use should therefore largely continue as before, even if that entails encroachment of water inland. The actors in this group are thus named *integrators*.
- Group 5 finds little need for further adaptation measures up to 2025 (Ø necessity <2.5 and below). This culture is therefore labelled as *minimalists*.

Considering the frequencies of the individual groups in the sample (Figure 10.7), the largest group wishes to maintain coastlines as they are, relying

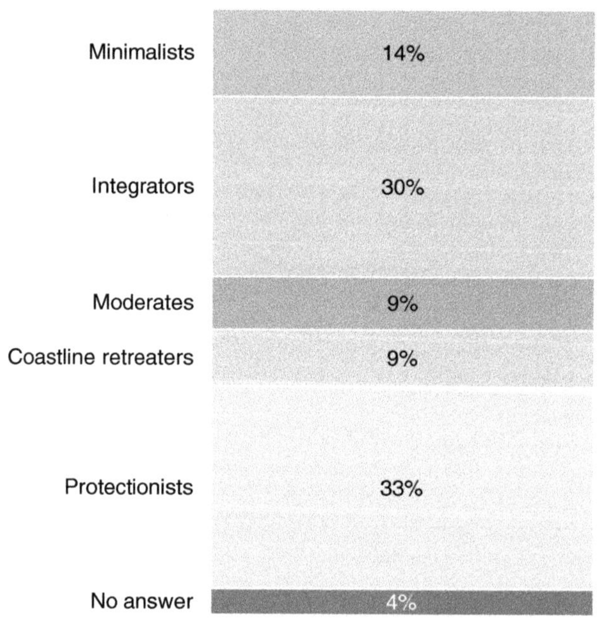

Figure 10.7 Climate-adaptation cultures against sea flooding: distribution of clusters – total.

Source: own calculations.

Notes

Basis: Respondents familiar with measures against sea flooding (as per Question MW1_SQ001, positions 2–5), who did not completely reject measures against sea flooding by 2025 (as per Question MWB1_1, positions 2–5); $n = 647$.

on the development of dykes and protective constructions up to 2025 (protectionists: 33%). Another third also backs integration (integrators: 30%). Actors preferring retreat strategies form a minority, with no more than about one in 10 (9%) who can be described as a coastline retreater. Minimalists (14%) and moderates (9%) also constitute rather smaller groups.

10.3.3 *Explanation of differences*

The question now arises as to which factors explain the different climate cultures for coastal protection.

Background knowledge

Hypotheses were formulated in Chapter 5 that can now be applied to the coast protection cultures discovered.

- As previously discussed, it is assumed that with higher *universalism and NEP orientations*, ecologically oriented measures will also be increasingly preferred. The probability of belonging to the group of coastline retreaters or integrators should thus also increase.
- The higher the *openness to change*, the more actors will be open to newly discussed measures, which in turn would increase the probability of affiliation with the integrators or coastline retreaters.
- Forms of conservatism can also play a role in the choice of measures. It was deduced in Chapter 5 that a higher *traditionalist orientation* can lead to more traditional, protectionist measures being preferred, as these have already conserved the coastline over many centuries. Conservatism should therefore increase the probability of belonging among the group of protectionists. *Security orientation* should also lead to preferences for practices to which protection from, and thus security against the sea can be ascribed. These should thus increase the probability of belonging to the group of protectionists.
- The relevance of *local identification* has also been discussed. Increasing local identification should accordingly manifest a preference in measures that are of high relevance to the identity of the place or region in question. In this case, the probability of belonging among the protectionists should increase.

More detail can be gained by considering Table 10.6 more closely. Looking at the model accuracy in the Likelihood Ratio test, all three models distinguish themselves from the null model, so that a certain explanatory force can be assumed, albeit at under 20% in each of the three cases (Pseudo $R^2_{\text{Nagelkerke}}$: Model 1: 11%; Model 2: 13%; Model 3: 18%). Background knowledge both alone and in combination with the control variables therefore explains only a small share of the total variance among the groups.

Table 10.6 Multinomial regression models (1–3): sea flooding

	Model 1: Values				*Model 2: Values + NEP + RI*				*Model 3: Values + NEP + RI + CV*			
	$\beta^{Sig.}$ *[odds ratio]*				$\beta^{Sig.}$ *[odds ratio]*				$\beta^{Sig.}$ *[odds ratio]*			
Reference group: Protectionists	*Coastline retreaters* n = 58	*Moderates* n = 54	*Integrators* n = 184	*Minimalists* n = 82	*Coastline retreaters* n = 58	*Moderates* n = 54	*Integrators* n = 184	*Minimalists* n = 82	*Coastline retreaters* n = 56	*Moderates* n = 52	*Integrators* n = 182	*Minimalists* n = 79
	Protectionists n = 204				*Protectionists* n = 204				*Protectionists* n = 196			
Background knowledge												
Values												
Self-transcendence I: Universalism	0.26	0.13	0.19	–0.12	0.17	0.09	0.11	–0.09	0.14	0.07	0.09	–0.04
Self-transcendence II: Human altruism	0.17	–0.27	0	0.01	0.15	–0.22	–0.01	0.02	0.07	–0.28	–0.06	0.03
Self-enhancement	0.25	0.34* [1.40]	0.28* [1.32]	0.00	0.26	0.37* [1.44]	0.29* [1.33]	0.01	0.22	0.29	0.25* [1.29]	0.04
Openness to change	0.18	0.3	0.28* [1.33]	0.21	0.21	0.29	0.31* [1.37]	0.20	0.27	0.36* [1.44]	0.34** [1.40]	0.14
Conservation I: Security	0.07	–0.02	–0.08	0.15	0.08	–0.04	–0.08	0.15	0.08	–0.07	–0.08	0.17
Conservation II: Tradition/ Conformity	–0.63*** [0.53]	–0.41* [0.67]	–0.27* [0.77]	–0.35* [0.71]	–0.59*** [0.55]	–0.31* [0.73]	–0.22* [0.80]	–0.35* [0.71]	–0.59*** [0.56]	–0.33* [0.72]	–0.22* [0.81]	–0.35* [0.71]
Beliefs												
New Ecological Paradigm (NEP)					0.42	0.25	0.36* [1.43]	–0.14	0.49* [1.63]	0.4	0.42* [1.53]	–0.24
Identity construction												
Local identification					–0.09	–0.48** [0.62]	–0.17	–0.09	–0.09	–0.41* [0.66]	–0.12	–0.14
Control variables (CV)												
Age									–0.02	–0.03* [0.97]	–0.01	0.02
Gender (0 = female, 1 = male)									–0.46	–0.69* [0.50]	–0.51* [0.60]	0.56
Constants	–2.96*	–1.51	–1.84*	–0.46	–3.90*	–1.01	–2.42*	0.18	–2.56	0.53	–1.25	–1.00
Model accuracy												
Pseudo R^2 (Nagelkerke)	0.107				0.133				0.176			
LR Test	61.96***				78.14***				102.94***			
N	582				582				565			

Source: own calculations.

Notes
β ***$p<0.001$; **$p<0.01$; *$p<0.1$. Values in brackets [odds ratio].

Let us now consider the explanatory force of the variables in more detail. The protectionists serve as the reference group in each case.

- Significant contributions to the explanation of group separation are made on multiple occasions by *traditionalist/conformity orientations*.
- Additionally, *openness to change* and *self-enhancement orientation* are of relevance.
- Also of interest is that *security orientation* does not contribute significant explanations here.
- The internalisation of the *New Ecological Paradigm (NEP)* and
- *local identification* may also explain group affiliation.

Protection generally counts as a traditional coastal protection strategy (see Chapter 4). It is thus noticeable that, as previously postulated, conservatism in the form of tradition and conformity orientations consistently increases the probability of belonging to the group of protectionists (see Models 1, 2, and 3). Considering the odds ratio values in square brackets, the effect regarding coastline retreaters is greatest (Model 1: –0.63*** [0.53]). Moderates ([0.67]) and minimalists ([0.71]) follow with almost the same values, followed lastly by the integrators ([0.77]).

Additionally, the hypothesis that, in comparison to protectionist orientations, ecocentric beliefs (NEP) increase the probability of accepting alternative coastal protection paradigms is here confirmed, albeit with some restrictions. Chapter 4 discussed in detail how protection strategies repeatedly debated in the literature are often combined with notions of a binary separation between humans and nature, as reflected in the frequently used phrase of the 'fight against nature'. Thus, internalisation of the Human Exceptionalism Paradigm (HEP) should increase the probability of belonging to the group of protectionists. In the other direction, increasing internalisation of the ecocentric New Ecological Paradigm (NEP) should increase the probability of alternative group affiliations.

These ideas are mirrored here when considering the polarity of the NEP-scale beta coefficients alone. In both Model 2 and Model 3, increased NEP orientation raises the probability, compared to the protectionists, of belonging among the coastline retreaters, the moderates, and the integrators. Significant explanatory force through the NEP appears initially only in relation to the integrators (Model 2: $\beta = 0.36$*, Model 3 analogous), but under controls for age and gender it also emerges for coastline retreaters (Model 3: $\beta = 0.49$*).

A paradigm shift from protectionist strategies to integration and retreat can therefore be traced back to a general paradigm shift in worldview from the Human Exceptionalism Paradigm (HEP) to the New Ecological Paradigm (NEP). On the other hand, NEP orientation does not indicate any significant separation force between protectionists, minimalists, and moderates.

In addition, affiliation to the group of integrators increases in all three models with increased openness to change and self-orientation.[9] By contrast, the moderates distinguish themselves in comparison with the protectionists through higher self-orientation and lower tradition and conformity orientations. With higher local identification, the probability also increases of belonging to the group of protectionists (Model 2: $\beta = -0.48^{**}$, Model 3 analogous).

It was previously assumed that local identification will usually increase the probability of belonging to groups with protectionist orientations. Here, though, only the differences between moderates and protectionists appear significant. Looking at all the polarity values of the beta coefficients, the results suggest that local identification can universally increase the probability of belonging to the group of protectionists, whom all coefficients appear to favour. However, the values are only significant for the moderates. As this group is comparatively small, it cannot be concluded that this is solely due to differences in group size. At this point, the issue of why local identification only seems significant for moderates must remain open.

Regarding the hypotheses restated at the beginning of this chapter, security orientation has still not been considered. Interestingly, it indicates no significant contribution to explaining the different affiliations to the adaptation cultures in question. This is a surprising result, as the climate-adaptation measures discussed aim to reduce the threat posed by the sea and should therefore be associated with various security beliefs. Apparently, though, the other factors mentioned above – conservatism, openness to change, self-orientation, and ecocentrism, play a much greater role than security aspects in current coastal protection preferences (see also Schmitt 2015: 301; Christmann *et al.* 2012). To summarise the results once again:

- In comparison with the protectionists, coastline retreaters significantly distinguish themselves through lower tradition and conformity orientations, as well as through a higher tendency to reject the Human Exceptionalism Paradigm in favour of the New Ecological Paradigm.
- Moderates distinguish themselves through higher self-enhancement orientation, lower tradition and conformity orientations, and lower local identification.
- Integrators distinguish themselves from protectionists through higher self-enhancement orientation and lower traditionalist values. Additionally, they appear more open to change and more frequently share an ecocentric (NEP) worldview, questioning a categorical separation between humans and nature.
- Minimalists distinguish themselves above all through a decreased tendency towards conservation presented in the form of lower tradition and conformity orientations.[10]

Field affiliation and national background

It is also sometimes possible to attribute differentiating effects to field affiliation and national background when it comes to the evaluation of adaptation practices against sea flooding (Figure 10.8).

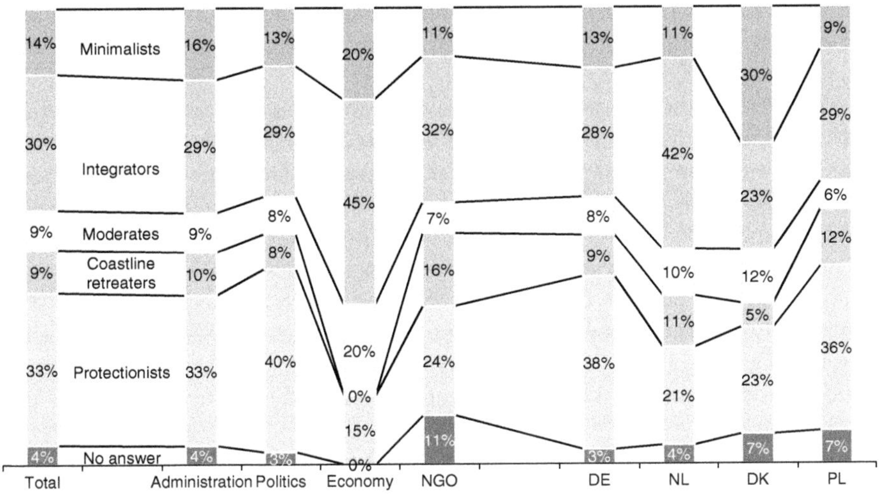

Figure 10.8 Practices against sea flooding: distribution of clusters – countries and fields.

Source: own calculations.

Notes

Question: Please assess the necessity of the following measures in [reference area] by 2025.

Basis: Respondents familiar with measures against sea flooding (as per Question MW1_SQ001, positions 2–5), who did not completely reject measures against sea flooding by 2025 (as per Question MWB1_1, positions 2–5); $n=647$ (DE: 375, NL: 98, DK: 77, PL: 97, administration: 370, politics: 182, economy: 20, NGO: 75).

Regarding field affiliation, major differences appear between the economy and the other groups. Here the proportion of integrators, that is, those who prefer measures that are comparatively new and innovative, is clearly over-represented in comparison to the other groups, comprising nearly half of actors from the field (45%). By contrast, a maximum of one third of actors in the remaining fields are integrators (administration and politics each 29%; NGO: 32%).

Among the group of protectionists there are also visible differences in distribution by field. Traditional coastal protection still has some weight chiefly in politics and administration. While almost 40% of politicians from planning and environmental committees and at least one-third of actors in administration (33%) support classical coastal protection, the group of protectionists themselves are clearly under-represented in the economy, comprising only 15%. Among NGOs too, outright protectionists are comparatively rare, with one in four (24%) being counted among this group.

How does it look for the remaining possible ways of dealing with the threats posed by the sea? While no economic actors support abandoning the land to the sea (economy: 0% coastline retreaters), a small group of NGO actors, namely one in six (16%), could be comfortable with the idea. Coastline retreaters continue to represent a minority in all groups, even in politics (8%) and administration (10%). Minimalists and moderates also remain minorities in all groups, although they form somewhat larger groups in the economy (minimalists and moderates each 20%).

In addition, differences in the frequency of groups appear according to actors' national background. Starting with the integrators, it is hardly surprising to find that these are represented most frequently among the Dutch than in the remaining countries (the Netherlands: 42% compared with Germany 28%, Denmark: 23%, Poland: 29%). Strictly, of course, a comprehensive reception study would be necessary before drawing a conclusion between these results and Dutch discourse in recent decades (see Chapter 6). The results do suggest though, that slogans such as 'Make room for the climate' (VROM *et al.* 2007), 'Building with nature' (DeVriend and Van Koningsveld 2012) or 'Working together with water' (Deltacommissie 2008) have been internalised by many actors (see also Chapter 6).

At the same time, it must be noted that integrators are also represented in the other countries. In fact, integrators and protectionists are present everywhere, making up the largest groups in almost every country, except Denmark. Here, in addition to a small group of integrators (23%), there is a significant group of minimalists (30%). Regarding the distributing of protectionists, it is also noticeable that in Germany (38%) and Poland (36%) traditional coastal protection appears to play a slightly greater role than in Denmark (23%) and the Netherlands (21%).

To summarise, the relative frequencies indicate the distinguishing effects of both field affiliation and national background. It should now be examined whether the patterns described continue to be apparent when the mutual influence of variables at the individual level is controlled for. Table 10.7 shows the results of logistic regression models 4 to 6.[11] It is clear that a great many of the separation forces of field affiliation and national background discussed above are also observable under conditions of mutual control.

The same higher proportions of coastline retreaters among NGOs mentioned above are also present here. Indeed, the probability of belonging with the coastline retreaters is significantly higher compared with protectionists in administration (Model 1: $\beta = 0.77^{*}$). Significant differences also appear between economy and administration (minimalists: $\beta = -18.48!^{**}$[12]; moderates: $\beta = 1.58^{*}$). There are also clear differences regarding minimalists in administration and politics ($\beta = -0.6^{*}[0.55]$). In summary, even when controlling for national background and social demographics, differentiation between climate cultures can be ascribed to field affiliation.

This observation also holds for national background. In the following analysis, Germany will again be placed in central position for the consideration of possible climate-culture boundaries. Regarding the frequency of protectionists, integrators, and coastline retreaters, Germany differs significantly from the Netherlands (NL: coastline retreaters compared to protectionists: $\beta = 0.74^{*}$; integrators: $\beta = 0.91^{**}$). On the other hand, differences with regard to Poland

Table 10.7 Multinomial regression models (4–6): sea flooding

	Model 4: Field + Country + CV				*Model 5: Field + Knowledge + CV*				*Model 6: Field + Country + Knowledge + CV*			
	$\beta^{Sig.}$ *[odds ratio]*				$\beta^{Sig.}$ *[odds ratio]*				$\beta^{Sig.}$ *[odds ratio]*			
Reference group: Protectionists	*Coastline retreaters* n = 59	*Moderates* n = 54	*Integrators* n = 188	*Minimalists* n = 88	*Coastline retreaters* n = 56	*Moderates* n = 52	*Integrators* n = 182	*Minimalists* n = 79	*Coastline retreaters* n = 56	*Moderates* n = 52	*Integrators* n = 182	*Minimalists* n = 79
	Protectionists n = 205				*Protectionists* n = 196				*Protectionists* n = 196			
Background knowledge												
Self-transcendence I: Universalism					0.09	0.09	0.09	–0.03	0.13	0.07	0.1	–0.06
Self-transcendence II: Human altruism					0.1	–0.28	–0.03	0.08	0.13	–0.29	0	0.04
Self-enhancement					0.23	0.29	0.26* [1.29]	0.03	0.32	0.29	0.34* [1.40]	0.02
Openness to change					0.27	0.34* [1.40]	0.33* [1.39]	0.14	0.21	0.32	0.25* [1.29]	0.1
Conservation I: Security					0.08	–0.06	–0.09	0.15	0.08	–0.07	–0.1	0.11
Conservation II: Tradition/ Conformity					–0.57*** [0.56]	–0.32* [0.72]	–0.21* [0.81]	–0.36* [0.70]	–0.60*** [0.55]	–0.27	–0.21* [0.81]	–0.29* [0.75]
New Ecological Paradigm (NEP)					0.47	0.45	0.45* [1.57]	–0.23	0.54* [1.71]	0.44	0.57** [1.77]	–0.28
Local identification					–0.05	–0.47** [0.62]	–0.1	–0.09	–0.09	–0.44* [0.64]	–0.1	–0.01
Field affiliation												
Politics	–0.25	0.02	–0.21	–0.6* [0.55]	–0.24	0.34	–0.27	–0.7* [0.49]	–0.35	0.28	–0.38	–0.71* [0.49]
Economy	–18.48!* [0.00]	1.58* [4.87]	1.07	0.89	–18.24** [0.00]	2.09* [8.08]	1.21* [3.35]	0.64	–18.28!** [0.00]	1.99* [7.32]	1.11	0.57
NGO	0.77* [2.16]	0.07	0.27	–0.53	0.57	0.11	0.18	–0.47	0.45	0.03	–0.01	–0.71
(*Reference group*: Administration)												
National background												
NL	0.74* [2.10]	0.54	0.91** [2.49]	0.52					0.8	0.47	1.02** [2.79]	0.38
DK	–0.51	0.68	0.06	1.3*** [3.66]					–0.94	0.34	–0.14	1.06* [2.90]
PL	0.17	–0.81	–0.15	–0.37					0.26	–0.57	0.17	–0.41
(*Reference group*: DE)												

continued

Table 10.7 Continued

	Model 4: Field + Country + CV				*Model 5: Field + Knowledge + CV*				*Model 6: Field + Country + Knowledge + CV*			
	$\beta^{Sig.}$ *[odds ratio]*				$\beta^{Sig.}$ *[odds ratio]*				$\beta^{Sig.}$ *[odds ratio]*			
Reference group: Protectionists	*Coastline retreaters* n = 59	*Moderates* n = 54	*Integrators* n = 188	*Minimalists* n = 88	*Coastline retreaters* n = 56	*Moderates* n = 52	*Integrators* n = 182	*Minimalists* n = 79	*Coastline retreaters* n = 56	*Moderates* n = 52	*Integrators* n = 182	*Minimalists* n = 79
	Protectionists n = 205				*Protectionists* n = 196				*Protectionists* n = 196			
Control variables (CV)												
Age	–0.01	–0.03* [0.97]	–0.01	0.03* [1.02]	–0.02	–0.03* [0.97]	–0.01	0.03* [1.03]	–0.01	–0.03* [0.97]	–0.01	0.03* [1.03]
Gender (0 = female, 1 = male)	–0.60* [0.55]	–0.64* [0.53]	–0.55* [0.58]	0.13	–0.42	–0.72* [0.48]	–0.5* [0.61]	0.57	–0.38	–0.8* [0.45]	–0.5* [0.61]	0.44
Constants	–0.18	0.6	0.92*	–2.38**	–2.69	0.39	–1.62	–1.67	–3.42*	0.51	–2.5*	–1.58
Model accuracy												
Pseudo R^2 (Nagelkerke)	0.135				0.206				0.250			
LR Test	81.49***				122.71***				152.37***			
N	594				565				565			

Source: own calculations.

Notes

β ***$p<0.001$; **$p<0.01$; *$p<0.1$. Values in brackets [odds ratio].

!* Empty cells: Economic actors who preferred coastline retreat are not considered. Therefore the testing requirements that both possible y-values be present cannot be fulfilled here. Distortion of the associated beta value, which here appears at a significantly increased value, needs to be taken into account (Urban and Mayerl 2011: 349).

and Denmark are not so pronounced. In Denmark, more minimalists are apparent than in Germany, compared with protectionists, even under control conditions ($\beta = 1.3$***). No significant differences are observable between Poland and Germany. This confirms the picture in Figure 10.8, in which both countries share many similarities, such as higher preference for protection and somewhat lower preference for integration, along with smaller groups of minimalists, moderates, and coastline retreaters.

Finally, the influence of background knowledge should be considered under control of actors' field affiliation and national background, as well as possible reasons for the differences in frequency of climate cultures.

The significant differences in affiliation to the groups of coastline retreaters and protectionists between actors from administration and NGOs disappear between Models 4 and 5 under the additional control of background knowledge. This once again indicates a mediator effect. The significant separation force between NGOs and administration can accordingly be ascribed in part to differences in background knowledge. The beta value for traditionalist and conformity orientations falls slightly between Models 3 and 5, and somewhat more for the New Ecological Paradigm. Both variables show significant separation force in differentiating the groups at the individual level. In the aggregated data it is also apparent that the New Ecological Paradigm is markedly more widely distributed (administration: 74%, NGO: 84%) (see Table 9.3). NGO actors are also less frequently oriented towards tradition or conformity (administration: 34%, NGO: 20%). These differences in characteristics can be cited as causes for the lower representation of protectionists and higher frequency of coastline retreaters among NGOs.

What of differences according to national background? Can the higher proportions of integrators in the Netherlands compared with Germany be explained by controlling for background knowledge? At the individual level, it can be seen in all models that with increasing openness to change, self-orientations, and internalisation of the New Ecological Paradigm, the probability of belonging among the integrators rather than the protectionists increases. In addition, in the multivariate analysis, higher traditionalist and conformity orientations shift the probability in favour of belonging to the protectionists. At the aggregate level, there is a significant increase in the proportion of actors with greater openness to change in the Netherlands than in Germany (DE: 40%, NL: 69%), while the traditionalist orientation is somewhat higher in Germany (DE: 34%, NL: 24%). These two orientations can thus partially explain the differences. In contrast to the results at the individual level, though, Germany displays higher NEP orientation (DE: 83%, NL: 69%) and self-enhancement orientation (DE: 36%, NL: 22%). Additionally, the Dutch beta coefficient increases still further between Models 4 and 5 for the integrators (Model 4: $\beta = 0.91$**[2.49]; Model 5: $\beta = 1.02$**[2.79]). This indicates that additional, unaccounted for factors remain to be found in order to explain the higher proportion of integrators in the Netherlands.

On the other hand, the higher proportion of protectionists in Germany and of minimalists in Denmark can be partially explained by traditionalist and conformity orientations. At the individual level, these significantly shift the probability of belonging among the protectionists in all relevant models. The beta coefficient also decreases slightly between Models 4 and 6 in Denmark (minimalists, DK compared with DE: Model 4: $\beta = 1.3^{***}[3.66]$; Model 5: $\beta = 1.06^{*}[2.90]$), while at the same time traditionalist and conformity orientations continue to exert significant influence (Model 6: $\beta = -0.29^{*}[0.75]$). At the aggregate level, Denmark actually displays a somewhat lower proportion of traditionalist and conformity orientations than in Germany (DE: 34%, DK: 27%). Assuming otherwise constant conditions, one could suppose that the higher proportion of traditionalist and conformity orientations in Poland contributes to the higher frequency of protectionists here than in the Netherlands or Denmark (PL: 49%, DE: 34%, NL: 24%, DK: 27%).

To summarise, there are significant differences in the frequencies of climate-adaptation cultures relating to sea flooding according to field affiliation and national background. These appear in different constellations to those observed for vulnerability constructions or climate-mitigation cultures, however. Much like in the case of the other climate-knowledge aspects, significant differences exist between NGOs and the other groups, although variation is also found between the economy, politics, and administration. Regarding national background, the greatest difference in comparison to German actors occurs in the Netherlands, where a markedly higher proportion of integrators and smaller number of traditional protectionists is evident. In Denmark too, traditional protectionists are also comparatively rare, while minimalists are more frequently represented, and integrators less so. Germany and Poland are most similar regarding coastal protection, whose deviation from the other countries can in part be explained by higher traditionalist and conformity orientations. Further potential explanatory factors nonetheless still need to be investigated.

10.4 Adaptation practices against inland flooding

10.4.1 Categorical framework: protection, retreat, and integration

Protection, retreat, and integration have also been proposed as potential core climate-adaptation strategies against inland flooding. Chapter 4 previously presented how possible concrete measures involved in these strategies differ in comparison to coastal protection. A separate set of tools is therefore developed to assess preferences here. I shall first consider the respondents' evaluations of the necessity of measures against inland flooding up to 2025. The three-factor solution to the principal component analysis can be interpreted as follows (see Table 10.8).

- The first factor corresponds to practices that can be applied within a framework of technologically oriented *protection* strategies. Alongside the 'expansion of dykes and protective constructions' (0.86), the item 'installation of watersheds' appears with a relatively high correlation (0.70).

Table 10.8 Factor analysis: strategies against inland flooding

	1 *Protection/ Technological construction*	2 *Retreat*	3 *Integration*
Necessity of measures against inland flooding			
Expansion of dykes and protective constructions against inland flooding	**0.86**	0.01	0.10
Reintroduction of natural watercourses	0.17	**0.85**	0.06
(Deconstruction of settlements in areas of inland flooding)	**0.46**	**0.47**	0.24
(Installation of watersheds)	**0.70**	0.30	0.09
(Desealing surfaces)	0.11	**0.66**	**0.50**
(Greening of roofs and facades)	–0.05	0.39	**0.78**
Water-adapted settlements	0.38	–0.02	**0.80**

Source: own calculations.

Notes
Question: Please assess the necessity of the following measures in [reference area] by 2025. Basis: Respondents familiar with measures against inland flooding (MW1_SQ002 positions 2–5; $n = 687$), who did not completely reject measures against inland flooding in general by 2025 (as per Question MWB1_8, positions 2–5); $n = 598$. Kaiser–Meyer–Olkin: 0.779. Bartlett Test: $^{***}p < 0.001$. Number of factors: theoretical derivation. Extraction method: principal component analysis. Rotation method: varimax with Kaiser normalisation. Results after 5 iterations. Total explained variance: 69%.
Items in brackets have been removed due to parallel correlation and/or low consistency values (for details, see Heimann 2017).

- The second factor includes the items ‘reintroduction of natural watercourses’ (0.85) and ‘desealing surfaces’ (0.66). In addition, there is a medium correlation with the item ‘deconstruction of settlements’ (0.47). These measures are thus grouped together as *retreat* strategies. However, the item ‘deconstruction of settlements’ also indicates a medium correlation to the factor ‘protection’ (0.46).
- Finally, along the third factor are loaded practices for the *integration* of water, such as ‘water-adapted settlements’, ‘greening of roofs and facades’, and ‘desealing surfaces’ (see Section 4.3). Desealing surfaces thus demonstrates a double correlation and can be interpreted both as an integration and as a retreat strategy.

To what extent can measures be summarised as preference indicators? Again, a three-dimensional relational cultural space appears reasonable here. The following cluster analysis therefore differentiates between preferences for *protection* (here: ‘improvement of dykes’), *retreat* (here: ‘reintroduction of natural watercourses’), and *integration* (here: ‘water-adapted settlement structures’). After removing parallel correlations and testing reliability, three indices are calculated to form an image of actors’ preference with regard to inland-flood protection (for further details, see Heimann 2017).

10.4.2 *Inland-flooding adaptation cultures: universalists, protectionists, renaturators, and minimalists*

Four cultures appear for inland-flood protection according to the relational concept of space (Figure 10.9).[13] Two of these groups previously appeared in relation to coastal protection:

- *Protectionists* wish to continue concentrating on traditional constructions such as classic dykes and protective constructions and they reject renaturation and integration measures (Ø necessity: protection = 3.8; renaturation = 1.7, integration = 2.3).
- The group of *minimalists* sees little need for the named strategies (Ø necessity <2 in nearly all cases).

In addition there are two groups specific to the ways of dealing with inland flooding:

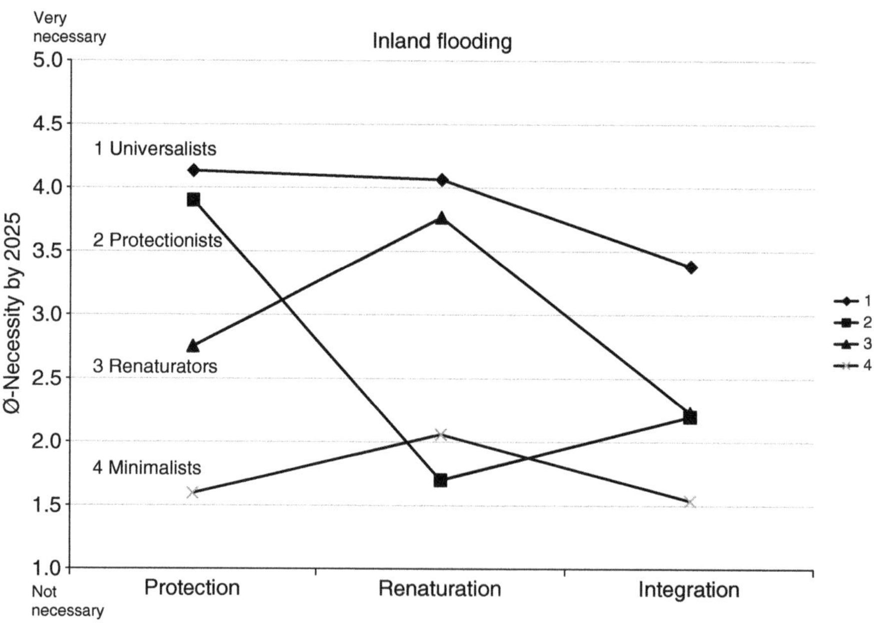

Figure 10.9 Climate-adaptation cultures against inland flooding: description of clusters.

Source: own calculations.

Notes

Question: Please assess the necessity of the following measures in [reference area] by 2025.

Basis: Respondents familiar with measures against inland flooding (MW1_SQ002, positions 2–5), who did not completely reject measures against inland flooding in general by 2025 (as per Question MWB1_8, positions 2–5); $n = 598$. Hierarchical cluster analysis, proximity measure: squared Euclidean distance; grouping procedure: Ward.

- *Renaturators* desire above all to limit intervention and to restore supposed 'natural' watercourses. Technological protective and integration measures have only low levels of acceptance (Ø necessity: renaturation = 3.7; protection = 2.3; integration = 2.2).
- By contrast, the group of *universalists* draws on all registers of flood protection. Where it seems reasonable, dykes and protective constructions should be expanded, but renaturation measures should also be undertaken along with integrated settlements to provide more room for water (Ø necessity protection and renaturation each >4; integration 3.4).

Considering the groups and their frequencies reveals clear differences regarding coastal protection (Figure 10.10). Section 4.3 discussed the changing notions in the inland-flooding discourse of the old EU countries that have been visible since the 1970s, ranging from pure protectionist orientation to an understanding that water should be given 'more room'. Comparing the group frequencies shows that these approaches have also been largely internalised by the actors surveyed. Only 10% of respondents desire that a purely protectionist

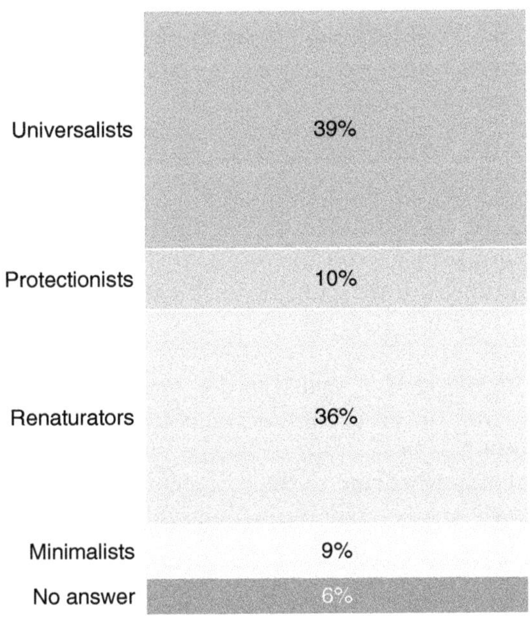

Figure 10.10 Climate-adaptation cultures against inland flooding: distribution of clusters – total.

Source: own calculations.

Notes

Basis: Respondents familiar with measures against inland flooding (MW1_SQ002 positions 2–5), who did not completely reject measures against inland flooding in general by 2025 (as per Question MWB1_8, positions 2–5); $n = 598$.

approach be pursued (protectionists). More than a third of actors back renaturation above all (renaturators: 36%). Still more frequently represented are those actors who wish to draw upon all registers of flooding protection, including the expansion of dykes where it is deemed necessary, alongside renaturation and integration measures (universalists: 39%). Approximately one in ten rejected any further measures up to 2025 (minimalists: 9%).

10.4.3 Explanation of differences

Which factors explain climate-culture affiliation for inland-flood protection?

Background knowledge

Before considering the empirical results, the hypotheses formed in Chapter 5 can be applied to the groups found.

- Accordingly, universalism and NEP orientations should increase the probability of belonging to a group that prefers ecologically oriented measures. Here this applies to renaturators, minimalists, and (with limitations) universalists, in contrast to protectionists.
- *Human-altruism orientation* should decrease affiliation to groups that share preferences for measures associated with restrictions and costs to humans (here: renaturation measures).
- *Self-enhancement orientation*, by contrast, should lead to an increase in preference for measures associated with individual or collective gains or the potential for new development (here: universalists).
- *Openness to change* should increase the probability of belonging to innovation-oriented groups (here: universalists).
- *Security orientation* should increase the probability of belonging to groups associated with low risks or which have already proven their effectiveness (here: protectionists).
- *Conformity and traditionalist orientations* should lead to traditional measures being preferred, and thus shift the probability in favour of belonging to the protectionists.
- *Local identification* should also contribute to the probability of belonging to a group preferring measures that guarantee the continued existence of a personally significant place or region (protectionists) or which can make a positive contribution to its development (universalists).

While the highly represented group of protectionists was selected as the reference category in the previous section on coastal protection, the same group is in the minority for inland-flood protection (Figure 10.10). Therefore the reference group chosen for inland flooding is the renaturators, as these form one of the two largest groups, along with the universalists.[14]

It can be concluded on the basis of the Likelihood Ratio test (Table 10.9) that significant deviations from the null model are observable, and thus a

Table 10.9 Multinomial regression models (1–3): inland flooding

	Model 1: Values			*Model 2: Values + NEP + RI*			*Model 3: Values + NEP + RI + CV*		
	$\beta^{Sig.}$ *[odds ratio]*			$\beta^{Sig.}$ *[odds ratio]*			$\beta^{Sig.}$ *[odds ratio]*		
Reference group: Renaturators	*Universalists* n = *224*	*Protectionists* n = *55*	*Minimalists* n = *48*	*Universalists* n = *224*	*Protectionists* n = *55*	*Minimalists* n = *48*	*Universalists* n = *220*	*Protectionists* n = *53*	*Minimalists* n = *46*
	Renaturators n = *201*			*Renaturators* n = *201*			*Renaturators* n = *191*		
Background knowledge									
Values									
Self-transcendence I: Universalism	–0.18	–0.45** [0.64]	–0.56** [0.57]	–0.2	–0.34* [0.71]	–0.54** [0.58]	–0.24* [0.79]	–0.3	–0.57** [0.56]
Self-transcendence II: Human altruism	0.2* [1.22]	0.2	0.07	0.2* [1.22]	0.23	0.06	0.21* [1.24]	0.34* [1.41]	0.08
Self-enhancement	–0.13	0.05	–0.06	–0.13	0.06	–0.07	–0.17	0.05	–0.01
Openness to change	0.37** [1.45]	0.03	–0.03	0.37** [1.45]	–0.02	–0.04	0.36** [1.43]	–0.06	–0.1
Conservation I: Security	0.04	0.21	0.22	0.03	0.23* [1.27]	0.23	0.04	0.26* [1.30]	0.2
Conservation II: Tradition/ Conformity	0	0.08	–0.15	0.02	0.01	–0.18	0	0.03	–0.19
Beliefs									
New Ecological Paradigm (NEP)				0.08	–0.67* [0.51]	–0.11	0.1	–0.71** [0.49]	–0.1
Identity construction									
Local identification				–0.03	0.08	0.11	0.01	–0.01	0.03
Control variables (CV)									
Age							–0.02* [0.98]	0	0.02
Gender (0 = female, 1 = male)							–0.31	0.47	0.25
Constants	–1.03	–1.29	1.05	–1.2	0.38	1.12	0.14	0.02	0.54
Model accuracy									
Pseudo R^2 (Nagelkerke)	0.078			0.095			0.128		
LR test	38.85**			47.71**			63.10***		
N	528			528			510		

Source: own calculations.

Notes
β ***$p < 0.001$; **$p < 0.01$; *$p < 0.1$. Values in brackets [odds ratio].

certain explanatory force can be assumed (LR: Model 1: 38.85**, Model 2: 47.71**, Model 3: 63.10***). The total variance information is, however, relatively poor (Pseudo $R^2_{Nagelkerke}$: Model 1: 7.8%, Model 2: 9.5%, Model 3: 12.8%). It must therefore be emphasised that in the inland-flooding discourse, many further factors besides the background knowledge observed are of significance to group affiliation.

Let us now consider the explanatory force of the value orientations for climate-culture group affiliation.

- In inland-flood protection, as anticipated, both *universal-* and *human-altruism orientation* demonstrate significant values.
- The same is also true of *openness to change.*
- In contrast to sea flooding, where *security orientation* did not appear to play a role (see Section 10.3), this is of definite significance for inland-flood protection.
- Apart from the values themselves, human and ecocentric orientations (the *New Ecological Paradigm*) contribute to the explanation of the various group affiliations.

Beginning with the New Ecological Paradigm, the assumption that ecocentric worldviews also lead to preferences for ecocentric measures is evident here. With increased ecocentrism, the probability significantly shifts towards affiliation to the group of *renaturators* in comparison to the *protectionists* (Model 2: $\beta = -0.67$*, Model 3 analogous). As already observed for coastal protection, the results give force to the thesis that the Human Exceptionalism Paradigm (HEP) is closely connected to the approach of *protection* that was dominant up into the twentieth century. In contrast, the (comparatively) New Ecological Paradigm (NEP) stands in close connection to the notion that sheer technological protection from rivers is to be understood in many ways as an ‘error’ (Patt and Gonsowski 2011: 159). In the new ecological worldview, renaturation appears a much more appropriate way of dealing with water. This suggests that the transformation of actors’ collectively shared worldviews is ultimately able to bring about systematic changes in (physically) spatial structures, namely when actors begin to develop new strategies for dealing with their physical environment on the basis of internalised knowledge constructions.

A similar pattern occurs in the case of the universalism orientation. Here the initial hypothesis proposed that these should contribute to a preference for ecologically oriented measures. The *renaturators* pursue a balance between humans and nature most consistently, as renaturation can be interpreted to mean that human interest in flood protection needs to be carried out in harmony with other living creatures by ‘making space’ for them (see Section 4.3.1). The results conform to this hypothesis, as the beta coefficients take negative values for all groups in comparison with the renaturators.

‘Creating room for nature’ through renaturation, on the other hand, is simultaneously interpreted as coming at a cost to the possibilities for human land use (see Section 4.3.1). Renaturation therefore brings limitations for humans, and it can be assumed that this would be of significance in regard to human-altruism orientation. As already discussed, in comparison with renaturators, *Universalists* prefer situationally adapted measures in which renaturation, protection, and integration are balanced. Humans, in this case, are again placed more centrally amid the efforts made. It is therefore hardly surprising that human-altruism orientation comes at the expense of renaturators, increasing the probability of belonging to the Universalists (Model 1: $\beta = 0.2^*$, Models 2 and 3 analogous).

Protection also serves to preserve existing human utilisation possibilities. Controlling for age and gender, human altruism can significantly shift the probability of group affiliation in favour of the protectionists rather than the renaturators (Model 3: $\beta = 0.34^*$).

Openness to change is also expected to make a significant contribution to group affiliation when the actions to be taken are particularly novel or innovative. *Universalists* prefer new combinations of different measures, of which some, as in the case of integration, have only just started to undergo a renaissance in recent years. Openness to change thus also produces significant shifts in group affiliation probability at the expense of renaturators and in favour of universalists (Model 1: $\beta = 0.37^{**}$, Models 2 and 3 analogous).

As previously mentioned, security orientation should contribute to a preference for measures associated with low risks, or which have already proven effective. This is particularly true for technological measures. It is possible that the outcomes of renaturation measures appear less controllable. Security orientation shows positive beta coefficients in all models, and the probability of belonging among *renaturators* thus falls with higher security orientation. The value for the *protectionists* appears significant (Model 2: $\beta = 0.23^*$). This is surprising, as it was shown in Chapter 4 that specialist discourses in western Europe since the 1970s have suggested that protection measures can also contribute to an increase in flooding, such as when dyke structures allow for construction in riverbed areas or floodplains (see Section 4.3.1). Evidently, protection measures are still associated with higher security by actors with higher security orientation. To summarise, with reference to renaturators one can differentiate:

- Universalists, who above all demonstrate higher human altruism and greater openness to change.
- Protectionists, who have lower universalism orientation, higher human-altruism orientation, higher security orientation, and a greater tendency to hold a worldview of human-exceptionalism.
- Minimalists, who by contrast are only distinguishable by their lower universalism orientation.

Field affiliation and national background

The relevance of field affiliation and national background as discriminants of climate culture will now be considered with respect to inland-flooding practices.

The distribution according to field and country in Figure 10.11 indicates that for inland-flood protection there are substantial differences between countries and fields.

For field affiliation by and large there are two groups facing one another, that is, the renaturators and the universalists. In politics and administration these nearly balance out, with each making up slightly more than a third (universalists: administration: 39%, politics: 38%; renaturators: administration: 36%, politics: 35%). For the economy and NGOs, on the other hand, there are deviations. The group of renaturators is markedly smaller in the economy (25%), whereby more protectionists are present (economy: 25%, administration: 10%, politics: 12%, NGO: 0%). For NGOs, renaturators are more frequently represented (38%), and universalists comprise the largest proportion (45%). The results thus indicate a substantial degree of rejection of protectionist flood protection among NGOs. In the economy, by contrast, a noticeably larger group of actors advocates protectionism.

Greater differences can be found in the frequency distributions between actors of different national background. Denmark and Germany seem to resemble each other in their overall profiles. Focusing on renaturation seems to be an especially important topic in Germany and Denmark (renaturators: DE and DK each 43%). Significantly lower shares of renaturators are to be found in

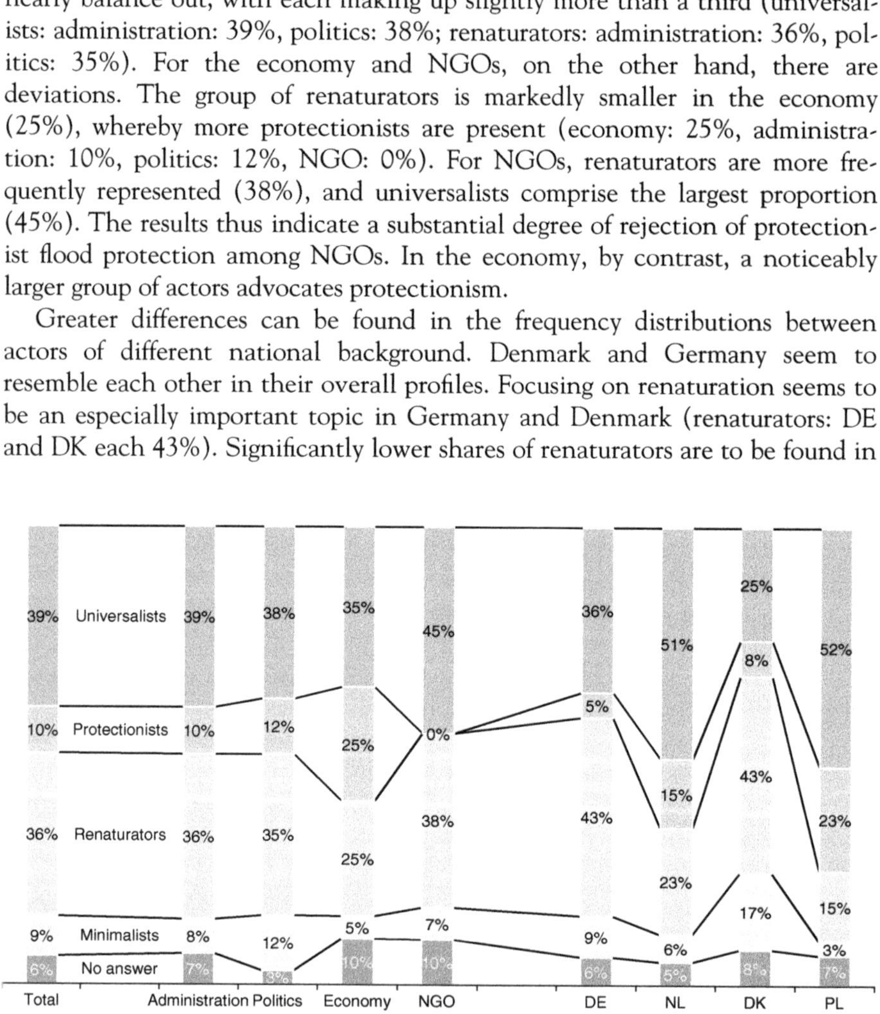

Figure 10.11 Climate-adaptation cultures against inland flooding: distribution of clusters – countries and fields.

Source: own calculations.

Notes

Basis: Respondents familiar with measures against inland flooding (MW1_SQ002 positions 2–5), who did not completely reject measures against inland flooding in general by 2025 (as per Question MWB1_8, positions 2–5); *n* = 598 (DE: 330, NL: 99, DK: 77, PL: 92, administration: 342, politics: 165, economy: 20, NGO: 71).

the Netherlands (23%) and Poland (15%). By contrast, protectionists in Poland achieve notable levels, with nearly one quarter belonging to this group (23%). In the Netherlands protectionists are less frequent but they appear more often (15%) that in Denmark (8%) and Germany (5%), where they are markedly more of a minority.

Universalists, i.e. actors who wish to see a balance between protection, integration, and renaturation measures, compose around half of actors in the Netherlands (51%) and Poland (52%). In Germany almost one in three (36%) belong among the universalists, while in Denmark they make up only a quarter (25%).

Do these results hold true if the variables are mutually controlled for at the individual level? Table 10.10 shows that there are no significant differences to be found between administration and politics. On the other hand, clear differences exist when comparing administration with the economy and NGOs. The national differences discussed above are also observable, with significant differences between Germany and the Netherlands, as well as Germany and Poland with respect to the German reference group. Regarding Denmark, significant differences only appear in the probability of belonging to the minimalists. Thus the differentiating effects of field affiliation and national background observed through the relative group frequencies also appear at the micro level in the regression analysis.

To what extent can varying background knowledge explain the differences found in the distribution of climate-culture orientations according to field affiliation and national background?

It is difficult to state unambiguously the degree to which background knowledge can explain the relatively high frequency of protectionists and low proportion of renaturators in the economy in comparison with NGOs, which have a high frequency of renaturators and low frequency of protectionists. In the sequential comparison of Models 4 and 5, the beta coefficient falls slightly for the economy once background knowledge is taken into account (economy: $\beta_{Model4} = 1.22^*$; $\beta_{Model5} = 1.18^*$). It seems possible that a mediator effect is traceable back to high security orientation ($\beta_{Model5} = 0.29^*$) and/or low internalisation of the New Ecological Paradigm ($\beta_{Model5} = -0.65^*$). At the macro level though, it is observable that actors with high security orientation occur in the same proportions in the economy and NGOs (each 33%). The New Ecological Paradigm also does not appear noticeably less often in the economy (81% in comparison to 84% for NGOs).

At this point, an important central question can be answered aspectually for flood protection. With respect to the influence of background knowledge, there are clear national background effects, so that in this regard one can speak of the relevance of container-space oriented national backgrounds in inland-flood protection. This can also be shown in the example of the New Ecological Paradigm. The significant effect disappears when national context is controlled for. Therefore the previously postulated shift from the Human Exceptionalism Paradigm to the New Ecological Paradigm as the background for a change of preference from protectionism to restoration only has validity in specific

Table 10.10 Multinomial regression models (4–6): inland flooding

	Mod. 4: Field + Reg. + CV			Mod. 5: Field + Knowledge + CV			Mod. 6: Field + Reg. + Knowledge + CV		
	$\beta^{Sig.}$ [odds ratio]			$\beta^{Sig.}$ [odds ratio]			$\beta^{Sig.}$ [odds ratio]		
Reference group: Renaturators	*Universalists* n = 230	*Protectionists* n = 55	*Minimalists* n = 50	*Universalists* n = 220	*Protectionists* n = 53	*Minimalists* n = 46	*Universalists* n = 220	*Protectionists* n = 53	*Minimalists* n = 46
	Renaturators n = 198			*Renaturators* n = 191			*Renaturators* n = 191		
Background knowledge									
Self-transcendence I: Universalism				–0.26* [0.77]	–0.23	–0.59** [0.55]	–0.21	–0.22	–0.64*** [0.53]
Self-transcendence II: Human altruism				0.21* [1.24]	0.3	0.06	0.28* [1.32]	0.35* [1.43]	0.05
Self-enhancement				–0.16	0.01	–0.02	–0.01	0.42* [1.52]	0.04
Openness to change				0.36** [1.43]	–0.06	–0.1	0.21	–0.42* [0.66]	–0.21
Conservation I: Security				0.04	0.29* [1.34]	0.21	0	0.17	0.16
Conservation II: Tradition/ Conformity				0.01	0.02	–0.18	–0.08	–0.08	–0.18
New Ecological Paradigm (NEP)				0.1	–0.65* [0.52]	–0.1	0.31	–0.2	–0.11
Local identification				0.02	–0.1	0	0.03	0.07	0.11
Field affiliation									
Politics	0.08	0.52	0.29	–0.02	0.51	0.41	0	0.58	0.52
Economy	0.07	1.22* [3.37]	–0.16	–0.08	1.18* [3.25]	–0.34	0.03	01.02.17	–0.34
NGO	0.23	-20.99!**	–0.45	0.28	-20.59!**	0.14	0.21	-20.65!**	0.09
(*Reference group*: Administration)									

National background									
NL	1.02*** [2.78]	1.88*** [6.53]	0.33				1.06** [2.89]	2.25*** [9.48]	0.34
DK	–0.42	0.76	0.85* [2.34]				–0.64* [0.53]	1.04* [2.83]	0.97* [2.64]
PL	1.39*** [4.02]	2.8*** [16.39]	0.32				1.69*** [5.43]	3.26*** [26.09]	0.51
(Reference group: DE)									
Control variables (CV)									
Age	–0.01	0.01	0.02	–0.02* [0.98]	0	0.01	–0.01	0.01	0.02
Gender (0 = female, 1 = male)	–0.19	0.61	0.15	–0.3	0.38	0.25	–0.23	0.38	0.17
Constants	0.38	–3.35***	–2.97***	0.17	0	0.84	–1.46	–3.72*	0.67
Model accuracy									
Pseudo R^2 (Nagelkerke)	195			0.161			0.286		
LR test	103.91***			80.24***			152.96***		
N	533			510			510		

Source: own calculations.

Notes
β ***$p<0.001$; **$p<0.01$; *$p<0.1$. Values in brackets [odds ratio].
!**Empty cells: There are no NGO actors who prefer protection. Therefore the test requirement that both possible y-values should be present at this point cannot be met, and distortions of the associated beta value are to be expected. The value at this point therefore appears significantly increased (see also Urban and Mayerl 2011: 349).

contexts, and probably does not apply to all the countries considered. In addition, the beta coefficients for national background increase further in Model 6, indicating further undisclosed country-specific factors that differentiate the groups. The relevance of container-space oriented national-background logics can thus not be excluded in the case of inland flooding. Although once again we observe a transnational climate-culture orientation, it is not yet possible to speak of convergence or homogeneity when considering its distribution.

10.5 Adaptation practices against heat

10.5.1 Categorical framework: building optimisation, land restoration, surface design

Regarding adaptation measures against heat, Chapter 4 differentiated between building design, settlement structure, and surface design. In the factor analysis (Table 10.11) this three-dimensional structure is clearly reproduced.

- The item 'energy-optimised buildings' loads onto the first factor (0.95). This factor is thus labelled 'energy optimisation of buildings'.
- The items 'creation of fresh- and cold-air corridors' (0.84) and 'reduction in building density' (0.88) indicate high loading along the second factor, which is thus interpreted as 'deconstruction of settlements'.
- The items 'desealing surfaces' (0.84) and 'greening of roofs and facades' (0.81) are considered together as 'surface design' measures along the third factor.

Table 10.11 Factor analysis: strategies against heat

	1	*2*	*3*
	Energy-optimised buildings	*Settlements: Deconstruction*	*Surface: Desealing/greening*
Necessity of measures against heat			
Creation of fresh- and cold-air corridors	0.17	**0.84**	0.19
Reduction in building density	–0.05	**0.88**	0.15
Desealing surfaces to reduce heat stress	0.10	0.23	**0.84**
Energy-optimised buildings to reduce heat stress	**0.95**	0.07	0.25
Greening of roofs and facades to reduce heat stress	0.27	0.12	**0.81**

Source: own calculations.

Notes

Question: Please assess the necessity of the following measures in [reference area] by 2025. Basis: Respondents familiar with measures against heat (MR1_SQ001 positions 2–5; $n = 536$) who do not completely reject measures against heat in general by the year 2025 (as per Question MR2_1a, positions 2–5); $n = 426$. Kaiser–Meyer–Olkin: 0.707. Bartlett Test: $***p < 0.001$. Number of factors: theoretical derivation. Extraction method: principal component analysis. Rotation method: varimax with Kaiser normalisation. Results after 5 iterations. Total explained variance: 81%.

As the factor analysis clearly allows for an unambiguous division into three central categories, the following cluster analysis will be built upon a three-dimensional cultural space concept, with an analogous differentiation between the central categories of 'energy optimisation of buildings' (item: energy-optimised buildings), 'deconstruction' (items: creation of fresh- and cold-air corridors, reduction in building density) and 'green surface design' items: desealing surfaces, greening of roofs and facades).

10.5.2 Heat-adaptation cultures: universalists, building and open-area designers, building optimisers

The cluster analysis results in three adaptation cultures against heat that can be differentiated according to their preferences for building optimisation, deconstruction, and green surface design (Figure 10.12).[15] These groups are the universalists, building and open-space designers, and building optimisers:

- First there is a group that stands relatively open to all three strategies (Ø necessity: building optimisation >4, deconstruction and green surfaces each >3.5). This group is thus labelled as *universalists*.

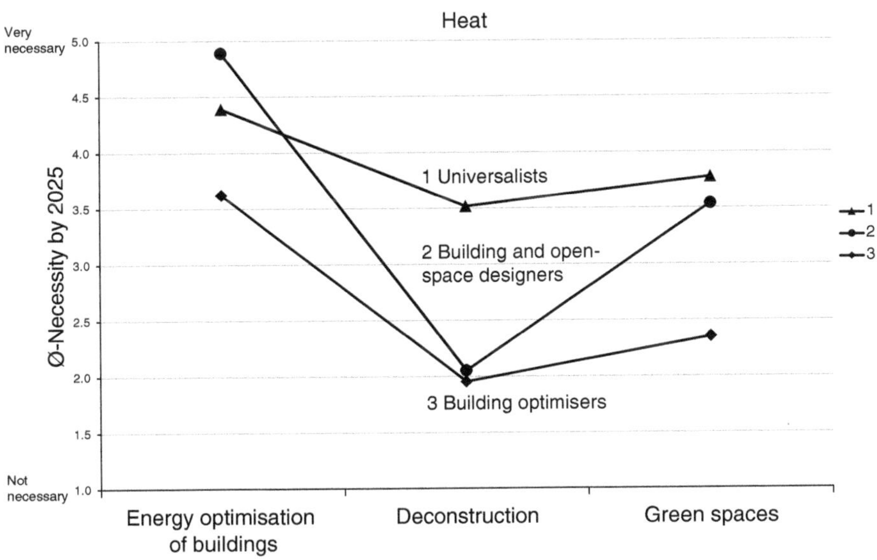

Figure 10.12 Climate-adaptation cultures against heat: description of clusters.

Source: own calculations.

Notes

Question: Please assess the necessity of the following measures in [reference area] by 2025.

Basis: Respondents familiar with measures against heat (MR1_SQ001, positions 2–5), who did not completely reject measures against heat in general by 2025 (as per Question MR2_1a, positions 2–5); $n = 426$. Hierarchical cluster analysis, proximity measure: squared Euclidean distance; grouping procedure: Ward.

The two remaining groups differentiate themselves through their preferences for deconstruction and green surfaces:

- *Building and open-space designers* are positioned critically towards the creation of fresh- and cold-air corridors or a reduction in building density (Ø necessity: deconstruction = 2.1). They share a comparatively strong preference for building optimisation and green surface design (Ø necessity: building optimisation = 4.9; green surfaces = 3.6).
- The last group is termed minimalistic *building optimisers*. Here, building optimisation alone is preferred, although this preference is itself only moderate (Ø necessity: 3.6). Surface greening and deconstruction measures are rejected by the group (Ø necessity: deconstruction = 1.9; green surfaces = 2.4).

Here it is important to remember that the point in question is not whether adaptation to heat is necessary in general, but much more which specific measures actors prefer (of those who do not reject these measures in general). As already discussed in Chapter 3, heat has been more a topic of discussion in

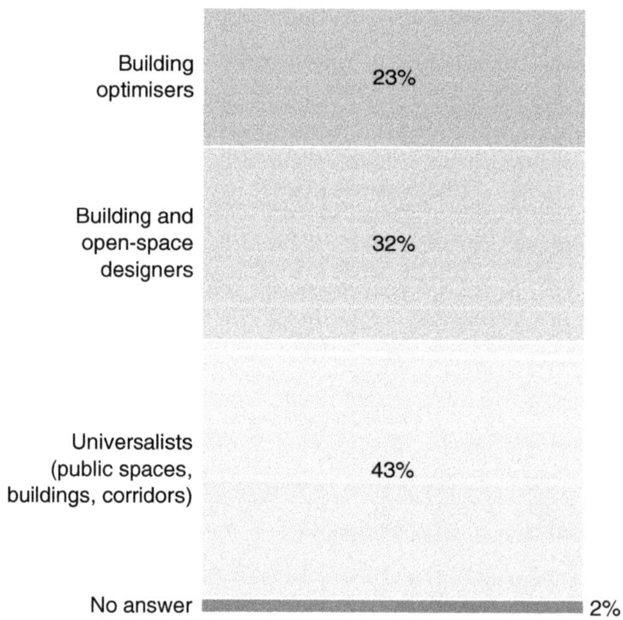

Figure 10.13 Climate-adaptation cultures against heat: distribution of clusters – total.

Source: own calculations.

Notes

Basis: Respondents familiar with measures against heat (MR1_SQ001, positions 2–5), who did not completely reject measures against heat in general by 2025 (as per Question MR2_1a, positions 2–5); n = 426.

inland regions, and in the present sample too, only half of respondents viewed measures against heat as a relevant option for the development of their reference area up to 2025 ($n = 426$). Of these actors, the largest group would like to see all registers of spatial development drawn upon. Thus universalists make up 43% of respondents (Figure 10.13). A further third, though, reject the deconstruction measures required for the creation of fresh-air corridors and back surface design and building optimisation above all (building and open-space designers: 32%). Near to a quarter would prefer to limit measures to the energy optimisation of buildings (building optimisers: 23%).

10.5.3 *Explanation of differences*

Let us now look at the degree to which existing background knowledge, field affiliation, and national background contribute to the separation of climate cultures against heat.

Background knowledge

Applying the hypotheses derived in Chapter 5 to the group affiliations found here:

- The greater the *universalism* and *NEP orientations*, the higher the probability of belonging to groups that prefer ecocentric measures. The probability should therefore increase of being affiliated to the universalists, or the building and open-space designers, as greening is explicitly incorporated here.
- The greater the *openness to change*, the higher should be the probability of belonging to a group that prefers measures associated with more substantial changes (here: building and open-space designers, being universalists in comparison to those who are solely building optimisers).
- Additionally, the conservatism orientation exerts several kinds of possible influence:
  - *Security orientation* could contribute to obtaining the greatest possible security against heat, and thus to the acceptance of the highest possible number of measures to counter it. On the other hand, security orientation can also lead to 'playing it safe' initially, and to reject all changes, in which case the opposite direction of influence will be observed, and minimalist-oriented measures should be preferred.
  - *Conformity and traditionalist orientations* could lead to a desire to implement long-standing measures, or to maintain the current conditions for as long as possible. These should lead to an increase in the probability of belonging to the group of minimalistic building optimisers. Local identification would, in addition, lead to a preference for measures that promote the development of a place or region or which maintain its current condition. The clear consequences of these orientations cannot be derived here a priori.

The total explanation of variance provided by the models remains at a low level, namely between 9.1% (Pseudo $R^2_{\text{Nagelkerke}}$: Model 1) and 13.1% (Model 3) (Table 10.12). On the basis of the Likelihood Ratio tests, though, significant explanatory differences can be assumed compared to the null model. The observed background knowledge is thus of relevance to the separation of groups, but many further undetermined factors apparently also play a role.

The building optimisers serve in the following as the reference group, as these prefer to follow a minimalistic programme and thus tend to reject further measures such as surface greening and open-space design, or even changes to building structures in order to create fresh- and cold-air corridors.

- *Universalism orientation*, *openness to change*, and the *New Ecological Paradigm (NEP)* have significant explanatory force for differentiation of this group from adaptation cultures.
- By contrast, *human-altruism*, *self-enhancement*, *security*, and *conformity and traditionalist orientations* provide no significant explanatory contribution. This is also true for *local identification*.

Universalism orientation significantly increases the probability of preference for ecologically oriented practices. Greater universal-altruism orientation increases the probability of belonging to groups that not only want to optimise buildings, but also wish to implement greening or create fresh- and cold-air corridors (building and open-space designers: Model 1: $\beta = 0.59^{***}$, Models 2 and 3 analogous; universalists: $\beta = 0.49^{***}$, Models 2 and 3 analogous).

In addition, openness to change has significant explanatory force for various group affiliations. As expected, with higher openness comes an increase in the probability of being more than a 'mere' building optimiser, but instead a *building and open-space designer* (Model 1: $\beta = 0.34^{*}$, Models 2 and 3 analogous) or *universalist* ($\beta = 0.31^{*}$, Models 2 and 3 analogous) to endorse still greater changes.

Universalists go the furthest, in that they are also prepared to implement comprehensive fresh- and cold-air corridors within the municipal structure, dissolving both cognitively and physically the binary notion of the boundary between humans and nature that still dominates modern western thinking. As anticipated, those with a disinclination to accept the Human Exceptionalism Paradigm have a significantly higher probability of belonging to the group of universalists (Model 2: $\beta = 0.46^{*}$, Model 3 analogous). To summarise:

- The building and open-space designers can be distinguished from the building optimisers by their higher universalism orientation and greater openness to change.
- This is also true for universalists, who in addition indicate higher levels of internalisation of the New Ecological Paradigm.

Table 10.12 Multinomial regression models (1–3): heat

	Model 1: Values		*Model 2: Values + NEP + RI*		*Model 3: Values + NEP + RI + CV*	
	$\beta^{Sig.}$ *[odds ratio]*		$\beta^{Sig.}$ *[odds ratio]*		$\beta^{Sig.}$ *[odds ratio]*	
Reference group: *Minimalists – building optimisers*	*Universalists: buildings and open spaces, deconstruction* n = *174*	*Building and open-space designers* n = *131*	*Universalists: buildings and open spaces, deconstruction* n = *174*	*Building and open-space designers* n = *131*	*Universalists: buildings and open spaces, deconstruction* n = *167*	*Building and open-space designers* n = *123*
	Minimalists n = 90		*Minimalists* n = 90		*Minimalists* n = 85	
Background knowledge						
Values						
Self-transcendence I: Universalism	0.49*** [1.64]	0.59*** [1.81]	0.43** [1.54]	0.54*** [1.72]	0.37* [1.44]	0.51** [1.67]
Self-transcendence II: Human altruism	0.03	0.10	0.04	0.09	0.01	0.05
Self-enhancement	–0.07	–0.12	–0.07	–0.13	–0.06	–0.15
Openness to change	0.34* [1.40]	0.31* [1.36]	0.34* [1.41]	0.32* [1.38]	0.43* [1.54]	0.4* [1.50]
Conservation I: Security	–0.1	–0.11	–0.12	–0.12	–0.09	–0.1
Conservation II: Tradition/Conformity	0.1	0.03	0.15	0.05	0.11	0.06
Beliefs						
New Ecological Paradigm (NEP)			0.46* [1.58]	0.36	0.48* [1.61]	0.34
Identity construction						
Local identification			–0.17	–0.01	–0.12	–0.01
Control variables (CV)						
Age					0	0.02
Gender (0 = female, 1 = male)					–0.81* [0.45]	–0.38
Constants	–2.93**	–3.49**	–3.85**	–4.56**	–3.45*	–5.17**
Model accuracy						
Pseudo R^2 (Nagelkerke)	0.091		0.105		0.131	
LR Test	33.12***		38.30***		46.79***	
N	395		395		381	

Source: own calculations.

Notes
β ***$p < 0.001$; **$p < 0.01$; *$p < 0.1$. Values in brackets [odds ratio].

Field affiliation and national background

A few differences in the frequency distributions are evident. For field affiliation, the distributions between administration, politics, and the economy indicate virtually no great deviations (Figure 10.14). Building and open-space designers appear somewhat more frequently in politics (38%, compared to administration: 30%, economy: 29%). Universalists are somewhat more frequently represented in administration (44%, compared to politics: 34% and economy: 36%). Clearer differences appear in comparison to NGOs, though. Predominantly universalists, 61% prefer a comprehensive programme of building optimisation, open-space design, and deconstruction of existing buildings. Those who favour building optimisation alone are comparatively under-represented among NGOs (11%, compared to administration: 25%; politics: 24%, economy: 21%). Accordingly, markedly larger groups are present who endorse the most comprehensive programme of city and community development.

The results by country at first glance appear surprising. First of all, building optimisers continue to be distributed very similarly, from 19% in Poland to 26% in Denmark. In addition, Denmark has a relatively high proportion of building and open-space designers (44%), followed by Germany (35%) and the Netherlands (25%). The proportion is lowest in Poland (16%), together with those of the other more minimalist-oriented groups focused on mere building optimisation (19%).

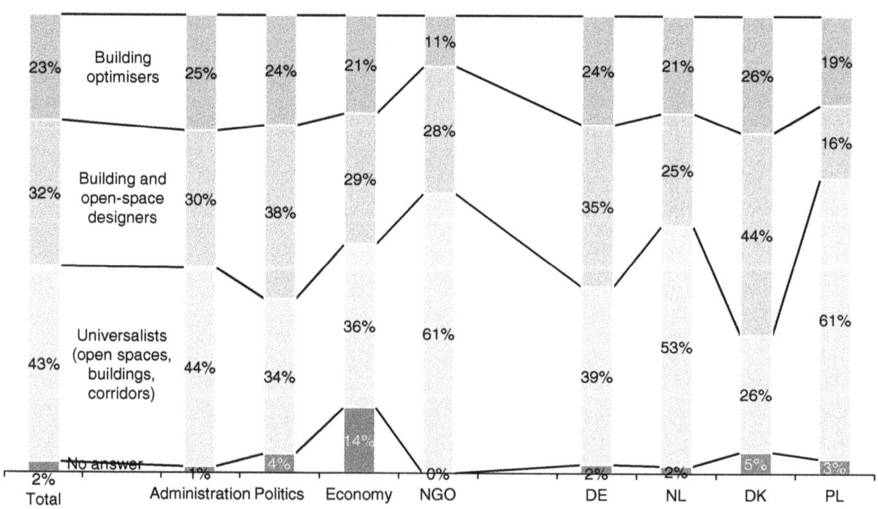

Figure 10.14 Climate-adaptation cultures against heat: distribution of clusters – countries and fields.

Source: own calculations.

Notes

Basis: Respondents familiar with measures against heat (MR1_SQ001, positions 2–5; $n=536$), who did not reject measures against heat in general by 2025 (as per Question MR2_1a, positions 2–5); $n=426$ (DE: 260, NL: 61, DK43, PL: 62, administration: 237, politics: 121, economy: 14, NGO: 54).

As with inland flooding, it appears that the Netherlands and Poland resemble one another in their frequency distributions. Especially among the universalists, and in contrast to Germany and Denmark, majorities are to be found here (Netherlands: 53%, Poland: 61%). Regarding the heat measures discussed, the proportion of actors who consider high levels of activism to be necessary is comparatively high in Poland, at about the level of environmental NGOs (61%).

This result is initially surprising. It was seen in the previous section that the probability of belonging among the universalists rose with increasing universal-altruism and NEP orientations, as well as greater openness to change. The aggregate comparison (see Table 9.3) shows that in Poland universal-altruism and NEP orientations are relatively low, and that openness to change is moderate in comparison with the other countries.[16] It was also evident, however, that the models considered together with background knowledge could only explain a comparatively small amount of the total variance. It is therefore conceivable that other unaccounted factors could explain the high level of activism in Poland. A possibility would be the strong focus of the Polish adaptation strategy on the overall development of the country (discussed in Section 6.3.5). In addition, perception of the quality of the existing infrastructure should be examined more closely. Universalists focus strongly on deconstruction and redevelopment of open space and buildings, all of which can contribute to stock renewal. The issue cannot be definitively explained here, however, and offers scope for further research in Poland.

At the same time, the observed differences need to be put in perspective by comparing the results when variables are mutually controlled for at the individual level (Table 10.13).[17] With administration and Germany selected as reference groups, no further significant differences appear according to national background. While the beta coefficients continue to suggest that, compared with building optimisers, the probability of belonging among the universalists increases with Polish or Dutch origin (PL: $\beta = 0.57$; NL: $\beta = 0.32$), the values are not significant compared to Germany. Thus, the differences under mutual control of the variables do not appear as strongly pronounced here as they may initially appear in merely comparing the frequencies.

By contrast, the climate-culture separation force of NGOs is clear. As already observed with the relative frequencies, the probability increases for NGO actors, in comparison to building optimisers in administration, of belonging among the universalists (Model 4: $\beta = 1.06^{*}$). The difference to the NGOs can be partially explained here by disparities in actors' background knowledge. In the sequential comparison of Models 4 and 5 it becomes clear that there is a mediator effect here. After taking into account background knowledge, the NGO beta coefficient drops to an insignificant level. Since universalism orientation, openness to change, and the New Ecological Paradigm remain significant at this point, and also occur much more frequently among NGOs (see Table 9.3), it can be assumed with little probability of error that these three variables contribute to explaining the more frequent occurrence of universalists among NGO members compared to the actors from the other fields.

Table 10.13 Multinomial regression models (4–6): heat

	Model 4: Field + Country + CV		*Model 5: Field + Knowledge + CV*		*Model 6: Field + Country + Knowledge + CV*	
	$\beta^{Sig.}$ *[odds ratio]*		$\beta^{Sig.}$ *[odds ratio]*		$\beta^{Sig.}$ *[odds ratio]*	
Reference group: *Minimalists – Building optimisers*	*Universalists: buildings and open spaces, deconstruction* n = *173*	*Building and open-space designers* n = *133*	*Universalists: buildings and open spaces, deconstruction* n = *167*	*Building and open-space designers* n = *129*	*Universalists: buildings and open spaces, deconstruction* n = *167*	*Building and open-space designers* n = *129*
	Minimalists n = *92*		*Minimalists* n = *85*		*Minimalists* n = *85*	
Background knowledge						
Self-transcendence I: Universalism			0.35* [1.41]	0.53** [1.70]	0.39* [1.48]	0.52** [1.68]
Self-transcendence II: Human altruism			0.03	0.07	0.03	0.07
Self-enhancement			–0.03	–0.14	0.07	–0.18
Openness to change			0.43* [1.54]	0.41* [1.50]	0.39* [1.48]	0.46* [1.58]
Conservation I: Security			–0.11	–0.12	–0.11	–0.1
Conservation II: Tradition/Conformity			0.14	0.05	0.09	0.08
New Ecological Paradigm (NEP)			0.5* [1.64]	0.34	0.58* [1.78]	0.32
Local identification			–0.08	0	–0.11	–0.02
Field affiliation						
Politics	–0.1	0.14	–0.44	–0.19	–0.44	–0.23
Economy	–0.02	0.11	0.24	0.31	0.16	0.29
NGO	1.06* [2.90]	0.39	0.55	–0.22	0.5	–0.25
(*Reference group*: Administration)						

National background						
NL	0.32	–0.11			0.41	–0.2
DK	–0.46	0.33			–0.84	–0.13
PL	0.57	–0.36			0.47	–0.53
(*Reference group*: DE)						
Control variables (CV)						
Age	0.01	0.03* [1.03]	0	0.02	0.01	0.02
Gender (0 = female, 1 = male)	–0.61* [0.54]	–0.29	–0.76* [0.47]	–0.38	–0.69* [0.50]	–0.39
Constants	0.42	–0.81	–3.93*	–5.37**	–4.58**	–5.22**
Model accuracy						
Pseudo R^2 (Nagelkerke)	0.101		0.152		0.187	
LR Test	36.96**		54.86***		68.45***	
N	398		381		381	

Source: own calculations.

Notes
β ***$p < 0.001$; **$p < 0.01$; *$p < 0.1$. Values in brackets [odds ratio].

In summary, with respect to strategies against heat, national background is barely a discriminatory factor in climate-culture affiliation. Actors' field affiliation, however, does appear significant. Differences between environmental NGOs and the other fields are also evident.

10.6 Adaptation practices against storms

10.6.1 Categorical framework: technological building adaptation, shelters

As already discussed, few measures have been debated for responding to possible changes in wind conditions (see Chapter 3). As only a few measures have appeared to date in the discourse, they are collected together here under three items (Table 10.14). In the two-factor solution, one item is loaded along the second factor, namely 'technical adaptation of buildings' (0.96).[18] Loaded along the first factor are the items 'installation of population shelters' (0.75) and 'changes to wind load regulations' (0.88). If we follow the substance of the regional development discourse on storm measures, which was briefly presented in Chapter 4, then we should expect that technical adaptations to buildings and wind load regulations would appear as a common factor, since both items refer to possible technical adjustments to buildings. Here, however, another aspect appears, which seems difficult to interpret.

Let us take a closer look here at the results of the reliability analysis (see Heimann 2017). The inter-item correlation between 'changes to wind load regulations' and 'installation of shelters' lies just below the middle of the scale (0.449). Cronbach's alpha shows a moderate value of 0.619. The use of either item in a Likert scale would therefore still need to be considered.

Table 10.14 Factor analysis: strategies against storms

	1	*2*
	Factor difficult to interpret	*Technological adaption of buildings*
Necessity of measures against storms		
Technical adaptation of buildings against storms	0.23	**0.96**
Installation of population shelters against storms	**0.75**	0.33
Changes to wind load regulations for buildings	**0.88**	0.13

Source: own calculations.

Notes

Question: Please assess the necessity of the following measures in [reference area] by 2025. Basis: Respondents who did not reject measures against storms in general by 2025 (as per Question MR2_7a; positions 2–5) $n = 537$. Kaiser–Meyer–Olkin: 0.666. Bartlett Test: $^{***}p < 0.001$. Number of factors: theoretical derivation. Extraction method: principal component analysis. Rotation method: varimax with Kaiser normalisation. Results after 3 iterations. Total explained variance: 82%.

Ultimately, a decision was taken against the Likert scale in order to do justice to the content of the discourse's discussed strategies (Chapter 4), and hence the items 'installation of shelters' and 'technical adaptation of buildings' alone were approved for the cluster analysis.

10.6.2 Storm-adaptation cultures: building optimisers, building and shelter protectionists, minimalists

The cluster analysis reveals three storm-adaptation cultures according to the relational understanding of cultural space, distinguishable according to their preference for 'technical adaptation of buildings' and 'installation of shelters'. They are labelled building and shelter protectionists, building optimisers, and minimalists (Figure 10.15):

- The group of *building and shelter protectionists* tends towards both technical building optimisation and the building of shelters. It must be noted though, that their preferences are moderate on the whole (Ø necessity: technical building adaptation = 3.8; installation of shelters = 3.4), which somewhat relativises their attributed urgency.

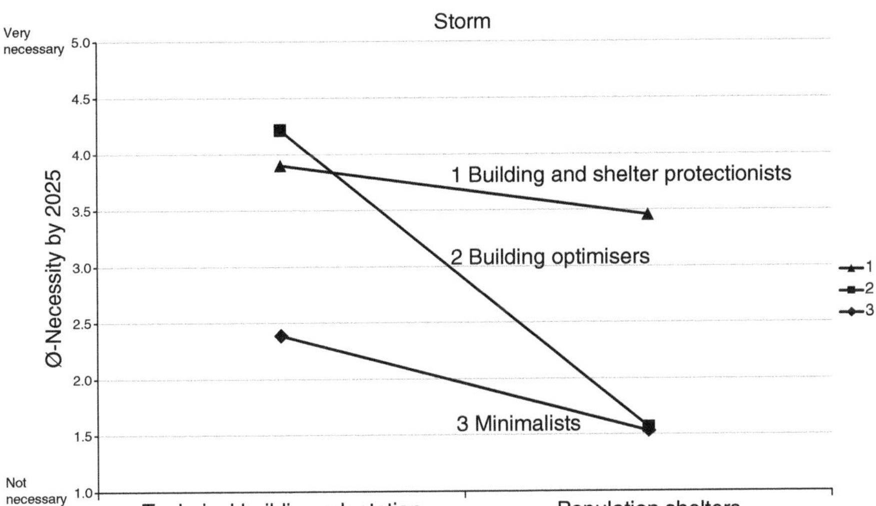

Figure 10.15 Climate-adaptation cultures against storms: description of clusters.

Source: own calculations.

Notes

Question: Please assess the necessity of the following measures in [reference area] by 2025.

Basis: Respondents familiar with measures against storms (MR1_SQ002, positions 2–5; *n* = 598), who did not entirely reject measures against storms in general by 2025 (as per Question MR2_7a, positions 2–5); *n* = 537. Hierarchical cluster analysis, proximity measure: squared Euclidean distance; grouping procedure: Ward.

- The second culture of *building optimisers* would like to restrict action until 2025 to the technical adaptation of buildings. Actors in this group reject the installation of shelters (Ø necessity: technical building adaptation >4; installation of shelters <2).
- Finally, there is a culture of *minimalists*. These actors see only a limited need for technical building adaptations (Ø necessity <2.5). As with building optimisers, the group rejects the installation of shelters (Ø necessity <2).

Considering the distributions in Figure 10.16, two groups of approximately equal size appear. On the one hand are actors who barely view any spatial development measures as a priority (minimalists: 37%), while on the other is a larger group of building and shelter protectionists (39%), who would like both to implement technical reinforcement of buildings as well as install shelters. There is also a somewhat smaller group of building optimisers (21%).

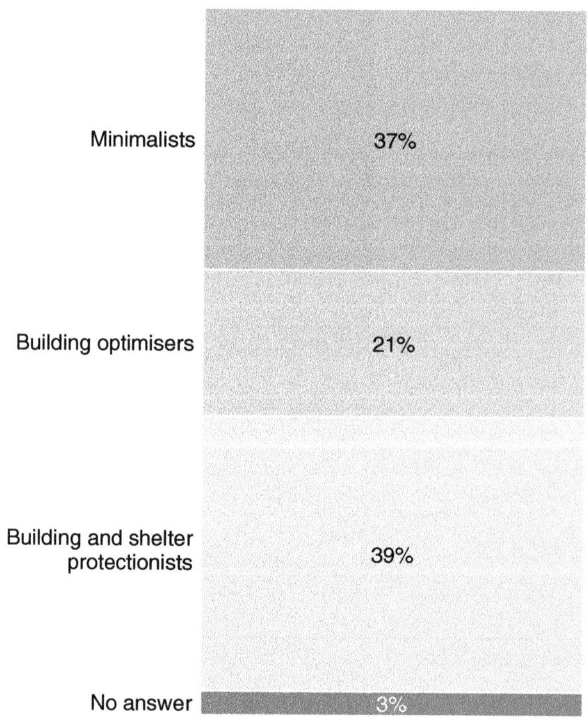

Figure 10.16 Climate-adaptation cultures against storms: distribution of clusters – total.

Source: own calculations.

Notes

Basis: Respondents familiar with measures against storms (MR1_SQ002, positions 2–5), who did not entirely reject measures against storms in general by 2025 (as per Question MR2_7a, positions 2–5); $n = 537$.

10.6.3 Explanation of differences

What explanatory force do background knowledge, field affiliation, and national background have for the separation of climate adaptation cultures against storms?

Background knowledge

The hypotheses formulated in Chapter 5 can be applied to the given storm-adaptation cultures as follows:

- At this point it is not possible to define exactly which measures appear ecologically oriented regarding protection from storm damage. Thus the hypothesis that with increasing *universal-altruism* and *NEP orientations* comes an increase in preferences for ecologically oriented measures cannot be properly applied here.
- The next hypothesis was that higher *human-altruism orientation* results in a higher probability that measures which benefit humans will be preferred. They should therefore result in a shift, relative to minimalists, to affiliation with both the building optimisers and the building and shelter protectionists, as these involve investing in enhanced security of human communities.
- In addition, the greater the *openness to change*, the greater must be the probability of belonging to groups that prefer measures associated with more substantial changes, i.e. building and shelter protectionists and building optimisers.
- *Security orientation* should increase preferences in favour of both building optimisers and building and shelter protectionists.
- *Traditionalist and conformity oriented* actors should prefer measures that have proven themselves in the past and/or which cause the least change. Shelters against storms are still a novelty along European coasts (see also Chapter 4). Transfers of group affiliation will therefore probably be in favour of minimalists and building optimisers.
- Finally, there remains the question of the relevance of *local identification*. The hypothesis states that the greater the local identification, the likelier it is that climate-adaptation measures will be preferred to which positive opportunities for local development can be ascribed. Accordingly, relative to minimalists, increasing local identification should be accompanied by an increase in the probability of belonging among the building optimisers as well as building and shelter protectionists.

The Likelihood Ratio test (LR) indicates a certain explanatory force for the models compared to the null model, although at only a 0.1 significance level in the case of Models 1 and 2. The overall variance explanation lies only between 5.5% (Pseudo $R^2_{\text{Nagelkerke}}$) in Model 1 and 11.6% in Model 3. The results therefore indicate that factors as yet unaccounted for might still further explain group affiliation. Considering the results in Table 10.15, it is noticeable that:

Table 10.15 Multinomial regression models (1–3): storms

	Model 1: Values		*Model 2: Values + NEP + RI*		*Model 3: Values + NEP + RI + CV*	
	$\beta^{Sig.}$ *[odds ratio]*		$\beta^{Sig.}$ *[odds ratio]*		$\beta^{Sig.}$ *[odds ratio]*	
Reference group: Minimalists	*Building and shelter protectionists* n = *200*	*Building optimisers* n = *105*	*Building and shelter protectionists* n = *200*	*Building optimisers* n = *105*	*Building and shelter protectionists* n = *195*	*Building optimisers* n = *102*
	Minimalists: n = *185*		*Minimalists:* n = *185*		*Minimalists:* n = *178*	
Background knowledge						
Values						
Self-transcendence I: Universalism	–0.15	0.05	–0.18	0.00	–0.22* [0.80]	0
Self-transcendence II: Human altruism	0.29* [1.33]	0.38* [1.46]	0.30* [1.35]	0.36* [1.43]	0.31* [1.36]	0.37* [1.45]
Self-enhancement	–0.01	0.22	0	0.21	–0.04	0.15
Openness to change	0.24* [1.27]	0.11	0.25* [1.29]	0.12	0.26* [1.30]	0.11
Conservation I: Security	–0.07	0.01	–0.08	0.01	–0.11	0
Conservation II: Tradition/Conformity	0.12	–0.05	0.16	–0.04	0.18	–0.03
Beliefs						
New Ecological Paradigm (NEP)			0.2	0.23	0.3	0.3
Identity construction						
Local identification			–0.15	0.06	–0.06	0.13
Control variables (CV)						
Age					–0.03*** [0.97]	–0.02* [0.98]
Gender (0 = female, 1 = male)					–0.68* [0.51]	0.11
Constants	–1.57*	–3.58***	–1.83*	–4.34***	0.05	–3.52**
Model accuracy						
Pseudo R^2 (Nagelkerke)	0.055		0.064		0.116	
LR Test	24.29*		28.49*		51.32***	
N	490		490		475	

Source: own calculations.

Notes
β ***$p < 0.001$; **$p < 0.01$; *$p < 0.1$. Values in brackets [odds ratio].

- Significant contributions to the separation of group are observable in the cases of *human-altruism orientation* and *openness to change.*
- This is also true for *universal-altruism orientation* controlling for age and gender. The other variables provide no significant contributions.

As expected, the probability of endorsing storm measures increases with higher human-altruism orientation, and thus with affiliation to the groups of building and shelter protectionists or building optimisers (building and shelter protectionists: $\beta_{Model1} = 0.29^*$, Models 2 and 3 analogous; building optimisers $\beta_{Model1} = 0.38^*$, Models 2 and 3 analogous).

Along European coasts, storm shelters are particular likely to be associated with major changes because, as discussed in Chapter 4, this practice is still rare and is more likely to be used in areas where hurricanes, typhoons, or tornadoes occur more frequently. Thus with greater openness to change, the probability of belonging to the group of building and shelter protectionists increases in comparison to minimalists (Model 1: $\beta = 0.24^*$).

In addition, controlling for age and gender, a significant influence can be detected from universal-altruism orientation. although no clear direction of influence could be derived. In the present case, however, it appears that actors with increased universal-altruism orientation tend to belong more among the minimalists than the protectionists (Model 3: $\beta = -0.22^*$). In summary, it can be formulated that:

- Building and shelter protectionists, in comparison with minimalists, are distinguished by more pronounced human-altruism orientation and openness to change, and (under control for age and gender) lower universal-altruism orientation.
- Differences between building optimisers and minimalists can be explained by more pronounced human-altruism orientation among the building optimisers.

Field affiliation and national background

The distributions according to field affiliation and national background indicate the differentiating effects of both (Figure 10.17). The distributions according to national background are much more unbalanced and the differentiation effects thus slightly larger.

There are also striking differences in relation to field affiliation, however. While the proportions for building and shelter protectionists are similar among administration, politics and NGOs (administration: 40%, politics: 37% NGO: 41%), they fall somewhat among the economy (29%). Minimalists form a larger proportion among environmental NGOs (46%), followed by administration (39%), politics (30%), and the economy (29%). Differences are also visible for the pure building optimisers, who are more frequently represented in the economy (36%) and markedly rarer among NGOs (9%). In between are administration (19%) and politics (29%).

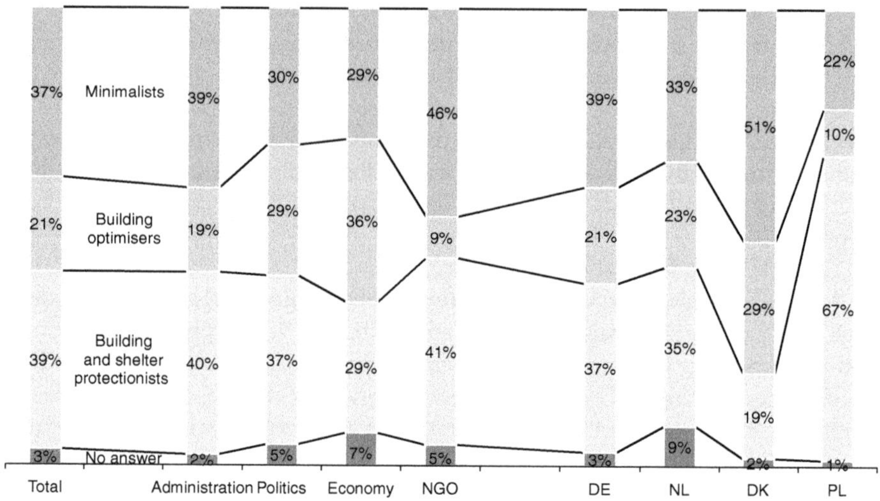

Figure 10.17 Climate-adaptation cultures against storms: distribution of clusters – countries and fields.

Source: own calculations.

Notes
Basis: Respondents familiar with measures against storms (MR1_SQ002, positions 2–5; $n=598$), who did not entirely reject measures against storms in general by 2025 (as per Question MR2_7a, positions 2–5); $n=537$ (DE: 331, NL: 69, DK: 59, PL: 78, administration: 303, politics: 154, economy: 14, NGO: 66).

Considering distribution by national background, there are some differences, but not between all groups. For storm cultures, the frequencies between Germany and the Netherlands appear relatively similarly distributed. Significant differences occur, however, in Poland, as well as in Denmark in some cases. In the latter example, building optimisers occur at a similar frequency (29%) as in the Netherlands (23%) and Germany (21%), but there are clear deviations in the case of minimalists, who narrowly make up the majority in Denmark (51%), as well as the building and shelter protectionists, who occur much less often (19%). Thus storm measures appear to be of less relevance in Denmark than in Germany, the Netherlands, or Poland. In Poland in particular, actors who do not completely reject measures against storms have predominantly activist orientations. The proportion of minimalists is comparatively low here (22%). The majority (67%) even consider the comprehensive programme of building optimisation and shelter installation as necessary by 2025.

The significant effects of field affiliation and national background continue to be visible even under the mutual control of variables (Table 10.16). The total variance explanation is once again only moderately successful, leading one to conclude that other factors not considered here are of influence.[19] Compared with minimalist administrative actors, significant differences between

Table 10.16 Multinomial regression models (4–6): storms

Reference group: Minimalists	Model 4: Field + Reg. + CV		Model 5: Field + Knowledge + CV		Model 6: Field + Reg. + Knowledge + CV	
	$\beta^{Sig.}$ [odds ratio]		$\beta^{Sig.}$ [odds ratio]		$\beta^{Sig.}$ [odds ratio]	
	Building and shelter protectionists n = *204*	*Building optimisers* n = *106*	*Building and shelter protectionists* n = *195*	*Building optimisers* n = *102*	*Building and shelter protectionists* n = *195*	*Building optimisers* n = *102*
	Minimalists: n = *187*		*Minimalists:* n = *178*		*Minimalists:* n = *178*	
Background knowledge						
Self-transcendence I: Universalism			–0.23* [0.79]	0.03	–0.19	0.01
Self-transcendence II: Human altruism			0.29* [1.34]	0.35* [1.42]	0.29* [1.34]	0.36* [1.44]
Self-enhancement			–0.05	0.1	0.08	0.16
Openness to change			0.25* [1.28]	0.1	0.2	0.05
Conservation I: Security			–0.09	0.01	–0.1	–0.01
Conservation II: Tradition/Conformity			0.18	–0.07	0.08	–0.06
New Ecological Paradigm (NEP)			0.29	0.32	0.44* [1.55]	0.4
Local identification			–0.09	0.03	–0.11	0.08
Field affiliation						
Politics	0.66* [1.94]	0.91** [2.48]	0.4	0.71* [2.03]	0.52* [1.69]	0.68* [1.97]
Economy	–0.07	1.07	0.1	1.08	–0.05	1.07
NGO	0.42	–0.69	0.26	–0.67	0.41	–0.76
(*Reference group:* Administration)						
National background						
NL	–0.07	0.36			0.07	0.55
DK	–0.9* [0.41]	0.36			–1.04* [0.35]	0.29
PL	1.16*** [3.19]	0.10			1.3*** [3.66]	0.26
(*Reference group:* DE)						
Control variables (CV)						
Age	–0.03*** [0.97]	–0.02* [0.98]	–0.04*** [0.96]	–0.03* [0.97]	–0.03** [0.97]	–0.02* [0.97]
Gender (0 = female, 1 = male)	–0.46* [0.63]	0.02	–0.7** [0.50]	0.07	–0.58* [0.56]	0.02
Constants	1.91***	0.31	0.39	-3.01*	–0.53	-3.66*
Model accuracy						
Pseudo R^2 (Nagelkerke)	0.155		0.145		0.208	
LR Test	72.71***		64.76***		96.26***	
N	497		475		475	

Source: own calculations.

Notes
β ***$p<0.001$; **$p<0.01$; *$p<0.1$. Values in brackets [odds ratio].

administration and politics are observable. On the other hand, the differences discussed for NGOs which should exist on the basis of the given polarities remain insignificant. This outcome may also lie with the choice of administrative minimalists as the reference group, since these fall in the mid-range (minimalists: administration 39%). The difference between administration and NGOs (46%) thus does not carry too much weight (Model 1: building and shelter protectionists: $\beta = 0.42$, building optimisation alone: $\beta = 0.69$). Considering national background, and taking Germany once again as the reference group, the previously mentioned differences continue to exist even under mutual controls. Thus in comparison to German minimalists, as already discussed, the probability of affiliation to the building and shelter protectionists falls for Denmark (Model 4: $\beta = -0.9^{*}$, Model 6 analogous), but, as expected, increases for Poland (Model 4: $\beta = 1.16^{***}$, Model 6 analogous).

This again raises the question to what degree the observed differences between field affiliation and national background are attributable to the forms of background knowledge being considered. Politics clearly differentiates itself from administration here. Is it possible to explain why fewer minimalists occur in politics, and thus a greater proportion are affiliated to action-oriented groups? Considering the micro level first, the comparison between Models 4 and 5 shows that, controlling for background knowledge, the significant separation between administration and politics subsides (politics: $\beta_{Model4} = 0.66^{*}$; $\beta_{Model5} = 0.4$), while universal-altruism, human-altruism, and openness to change continue to remain significant. Mediator effects are thus indicated here. Universal-altruism should shift the probability of individual affiliation towards the minimalists. When we take the macro level into account, the proportion of universal-altruists in politics (86%) is still somewhat larger than in administration (80%). Observation of both micro and macro levels together thus barely provides any explanatory contribution. Looking then at the other values: with increasing human-altruism orientation and openness to change, the probability of belonging among the minimalists falls. At the same time, at the macro level, politics has more human-altruists than administration (administration: 75%, politics: 87%), as well as a larger proportion of actors with a high degree of openness to change (administration: 45%, politics: 54%). The smaller proportion of minimalists in politics can thus be attributed to slight differences in values. With the additional control of national background it is revealed that human-altruism orientation in politics can consistently be interpreted transnationally as an explanation for the smaller proportion of minimalists.

At the same time, however, this indicates national context effects. Under the additional control of national background in Model 6, not only is there a decline in beta coefficients for universal-altruism orientation, but the influence of the New Ecological Paradigm also significantly contributes to the separation of groups ($\beta_{Model5} = 0.44^{*}$). Thus, it can be assumed that NEP orientation in particular has different effect strengths according to country. A multi-level analysis might cast further light on the more precise correlations of moderator effects, but is not feasible where only four countries are being investigated.

Additionally, the national background beta coefficients in Model 6 regain strength, suggesting that other factors thus far unaccounted for might further explain group affiliation within the different countries.

In conclusion, it can be stated that for storm-adaptation cultures, transnational groups with shared preferences are detectable. It is also evident though that significant field-specific differences occur in the distributions while simultaneously national peculiarities remain observable. These cannot be explained easily using the currently investigated differences in background knowledge. National context effects are also apparent. Thus, even in the case of storm-adaptation cultures, it seems true that new cultural constellations appear. At the same time, there can only be limited talk here of deterritorialisation or a 'disappearance of national (container-)space'.

10.7 Interim conclusion

This chapter has investigated which commonalities and differences are observable with regard to resilience constructions among spatial development actors in the research area. In addition, the question was posed as to which factors contribute to the differentiation of various climate-culture orientations.

Subsequently, the question was pursued as to which climate cultures are observable according to the relational understanding of cultural space (Chapter 2). The following groups were spread throughout the entire research area, each with shared patterns of preference:

- *Climate-mitigation*: universalists, classics, moderates, and minimalists.
- Adaptation to threats posed by *sea flooding*: protectionists, integrators, coastline retreaters, moderates, and minimalists.
- Adaptation to threats posed by *inland flooding*: protectionists, renaturators, universalists, minimalists.
- Adaptation to *heat*: universalists, building and open-space designers, building optimisers.
- Adaptation to *storms*: building optimisers, building and shelter protectionists, minimalists.

Beyond the issue of how patterns of socially shared measure preferences arise, the question also arose as to what relationships exist between the observed patterns of preference and general knowledge stocks. It was thus necessary to investigate which contributions to explanation of climate-culture orientations were provided by different forms of background knowledge, in particular: values, ecological beliefs, and local identification. It emerged that a majority of the variables developed in Chapter 5 for the VBI model were regularly able to explain differences in socially shared vulnerability and resilience constructions.

Field affiliation and national background were also responsible for differentiating climate cultures. These can appear in different spatial formations, however, depending on which aspect the researcher focuses on. Of the different fields, members of environmental protection organisations most often differed from the other fields, although not fundamentally so. Differences between actors of different national backgrounds were also visible. In these cases, climate-culture formations were founded upon various aspects. In the case of climate-mitigation cultures, national peculiarities between actors from Germany, Poland, and the other countries were evident, although for coastal protection, there were greater similarities between Germany and Poland.

Possible consequences for future social-science research arising from these results will be discussed in more detail in the next chapter.

Notes

1 Their similarity of substance seems plausible when one realises that eco-efficiency stands not only for 'uncoupled growth' (Huber 2011: 141), but also for the 'rationalisation of consumption' (ibid.: 166).

2 This also appears plausible when one considers that the development of local public transport can also be interpreted as a measure for modernisation instead of personal restriction.

3 As already discussed in Chapter 3, the German version of this book contains additional multi-dimensional scaling and reliability analyses for detailed discussion of the creation of the indices (see Heimann 2017). An English version of the Appendix is available from the author on request.

4 The Likert scale for recording sufficiency preferences shows an acceptable Cronbach's alpha coefficient (0.790), which does not increase further if the item 'Increase of building density' is excluded (Cronbach's alpha if the item is deleted: 0.790). I have therefore opted to use a common Likert scale for sufficiency preferences (see Heimann 2017).

5 One must note critically that the group sizes sometimes vary. In particular, 'moderate climate mitigationists' are much more frequently represented than other groups. As a consequence of the higher group size, possible significances can appear here more often than with the other groups.

6 Owing to low response rates, it is however possible that the actual proportion of advocates is slightly lower, as actors more open to the issue of climate change might be more prepared to complete the survey than those who have a more negative attitude towards climate issues.

7 The reference group here is the German minimalists.

8 In order to enhance reliability. the item 'deconstruction of settlement areas near the sea' has been excluded due to its parallel correlation, as has the item 'population shelters' (see Heimann 2017).

9 At this point it must be noted that the calculation of significance depends on the group size, and that this varies in the following cases. It might thus be possible that values for the integrators show more significance than for other groups.

10 Owing to the varying group sizes, it should be noted that other factors might possibly have assumed significance for larger group sizes, especially among moderates and minimalists. A degree of uncertainty must be accepted here.

11 The LR test shows that all models display explanatory force for the total variance, although these appear relatively low to moderate (Pseudo $R^2_{\text{Nagelkerke}}$: Model 1: 13.5%; Model 2: 20.6%; Model 3: 25%).

12 Here it can be seen once again that no coastline retreaters are to be found in the economy, and that this group is comparatively small. The beta coefficients are therefore rather large and the calculated probability of error assumes the value zero.

13 Summing of the least squares reveals more strongly pronounced leaps for inland-flood protection in the case of three- and six-cluster solutions (see Heimann 2017: Appendix). The decision was thus made to use a four-cluster solution as these were more consistent and produced clearly interpretable groups. Thus the criterion of a logically compelling, content-related interpretation is here prioritised above that of summing the least squares.

14 It should also be noted that the groups are of very different sizes, which might tend to give the universalists' calculated p-values higher significance earlier on. Hence in some instances the non-significant values are scrutinised more closely.

15 For the procedure used, see Section 9.1.3. Observing the course of the least squares method, it was noticeable that five-, four-, and three-cluster solutions appeared possible. The three-cluster solution was chosen, as very similar groups would otherwise have had to be distinguished (see Heimann 2017: Appendix).

16 Universal-altruism: DE: 82%, NL:85%, DK: 92%, PL: 77%; New Ecological Paradigm: DE: 83%, NL: 69%, DK: 81%, PL: 60%; Self-enhancement: DE: 36%, NL: 22%, DK: 26% PL: 31% (top 3 boxes).

17 The maximum Likelihood Ratio test (LR) shows that all models provide significant explanatory contributions. The explained total variance of Models 4 to 6 indicates a low to moderate level (Pseudo $R^2_{\text{Nagelkerke}}$: Model 4: 10.1%; Model 5: 15.2%; Model 6: 18.7%).

18 According to the Kaiser–Meyer–Olkin criterion, the present correlation matrix is still just sufficient for a factor analysis (KMO = 0.666). Additionally, according to the Bartlett test ($p < 0.001$), the null hypothesis of an existing unit matrix for the population can be rejected and a factor analysis can be allowed.

19 The separation from the null model is significant ($p < 0.001$) for all three models. The explained variance lies, as can be read in Table 10.16, between 14.5% (Pseudo $R^2_{\text{Nagelkerke}}$: Model 2) and 20.8% (Model 3).

References

Christmann, Gabriela B., Heimann, T., Mahlkow, N., and Balgar, K. (2012) 'Klimawandel als soziale Konstruktion?' *Zeitschrift für Zukunftsforschung 1*. (urn:nbn:de:0009–32–34126) Available at: www.zeitschrift-zukunftsforschung.de/ausgaben/2012/1/3412/Christmann_et_al (accessed 28 October 2013).

Deltacommissie (2008) *Working Together with Water. A Living Land Builds for its Future. Findings of the Deltacommissie 2008 [Samen werken met water. Een land dat leeft, bouwt aan zijn toekomst. Bevindingen van de Deltacommissie 2008]*. Alkmaar: Hollandia Printing. Available at: www.deltacommissie.com/doc/deltareport_full.pdf (accessed 15 May 2012).

DeVriend, H.J. and Van Koningsveld, M. (2012) *Building with Nature: Thinking, Acting and Interacting Differently*. Dordrecht: Alphen a/d Rijn.

Heimann, Thorsten (2013) 'Konsens oder Dissens. Herausforderungen des Klimawandels für die deutschen Küstenstädte und –gemeinden aus Expertensicht'. In: Heimann, Thorsten and Christmann, Gabriela B. (2013): *Klimawandel in den deutschen Küstenstädten und –gemeinden. Befunde und Handlungsempfehlungen für Praktiker*, pp. 22–72.

Heimann, Thorsten (2017) *Klimakulturen und Raum. Umgangsweisen mit Klimawandel an europäischen Küsten*. Wiesbaden: Springer VS.

Huber, Joseph (2011) *Allgemeine Umweltsoziologie*. Wiesbaden: Springer VS.

Patt, Heinz and Gonsowski, Peter (2011) *Wasserbau. Grundlagen, Gestaltung von wasserbaulichen Bauwerken und Anlagen*. Heidelberg/Dordrecht/London/New York: Springer.

Raabe, Stephan and Härtel, Janina (2007) *Andere Prioritäten. Der Diskurs über den Klimawandel in Polen*. Warszawa: Konrad-Adenauer-Stiftung in Polen. Available at: www.kas.de/polen/de/publications/11117/ (accessed 15 October 2010).

Reckwitz, Andreas (2000) *Die Transformation der Kulturtheorien*. Weilerswist: Velbrück.

Schmitt, Lea (2015) *Klima, Raum und Zeit im Wandel. Eine ethnographische Untersuchung von Adaptationen und Konflikten auf der westfriesischen Insel Ameland*. München: Oekom.

Urban, Dieter and Mayerl, Jochen (2011) *Regressionsanalyse: Theorie, Technik und Anwendung*. Wiesbaden: VS Verlag für Sozialwissenschaften/Springer.

VROM Ministry of Housing, Spatial Planning and the Environment/V&W Ministry of Transport, Public Works and Water Management/LNV Ministry of Agriculture, Nature and Food Quality/EZ Ministry of Economic Affairs/IPO Association of Provincial Authorities/VNG Association of Netherlands Municipalities/UvW Association of Water Boards (ed.) (2007) *Make Room for the Climate! Initial Steps Towards a National Adaptation Strategy*. Memorandum for policy Discussion. Den Haag: VROM.

11 Conclusion

What climate cultures can be observed along European coasts? What factors separate these cultures, and what explanations can be found for differences in climate culture?

On the one hand, the central research questions of this work set out the challenge in relation to the *description of climate cultures* of responding to the social-science debates about cultural globalisation present in spatial theory in recent decades. These debates have increasingly questioned the relationship between place/region and culture, resulting in a need for new theoretical and methodical approaches to investigating cultural spatialisation. Furthermore, the present work delivers an additional contribution to the discourse around *explanations for environmental and climate-change behaviour* (Scannell and Gifford 2013: 66; Dunlap and van Liere 1977: 202). The findings are summarised below and the consequences for current debates are discussed, along with some questions that remain unresolved.

11.1 Summary of results

Authors with differing positions appear in the discourse surrounding cultural globalisation, ranging from proponents of the emergence of a 'world society' and its associated cultural homogenisation, to those who argue for continuing regional differentiation (see Chapter 2). In order to precisely understand both theoretically and empirically the constellation of climate cultures and the possible differentiating factors between 'zones of cultural friction' (Crate 2011: 188), the present work proposes an understanding of culture that pursues a relational conception of space. The central idea of the *relational approach to cultural space* ('relationaler Kulturraum', 'culture as relational space') thus developed is founded on the idea that actors who share certain kinds of knowledge constructions will be positioned either more or less distant to one another as a consequence. From this perspective, cultures are viewed as arranging themselves according to certain aspects. Consequently, climate cultures are defined as the relational positions of actors who share similar knowledge related to climate change. The study distinguishes between shared knowledge of vulnerability, such as perceptions of problems and opportunities related to climate change, and shared knowledge of resilience, such as a preference for measures to adapt to or mitigate climate change.

The *theoretical approach* to climate cultures enables their *empirical observation*. Eight hundred and thirty actors in the field of spatial development along the coasts of Germany, the Netherlands, Denmark, and Poland were surveyed online. The actors assessed possible opportunities and problems of climate change as well as the necessity of climate-mitigation and adaptation measures. In addition, various values, beliefs, and identities were investigated.

Authors of the younger generation of communicative constructivists, such as Keller, Knoblauch, or Christmann, have emphasised the high *significance of communication* for the genesis of shared knowledge (Keller *et al.* 2013). Before dedicating ourselves to the specific systems of climate knowledge investigated, I therefore investigated to what extent actors communicate about climate change, and what media they use to do so. Throughout the study, respondents were found whose communication partners were dispersed across Europe and worldwide. Globally authored media, such as the IPCC reports or academic publications, were also read throughout the entire field. At the same time, significant differences could be found. Particularly noticeable were actors from non-governmental organisations, who displayed a significantly greater trans-European and global scope than other actors. In addition, interviewees of Polish origin showed a much lower frequency of climate-change communication than actors from other countries.

The answers to the first central question of this study can now be summarised: *What climate cultures can be observed on European coasts?* The question can be further specified: What views of opportunities and problems (knowledge constructions of vulnerability) are shared by actors, and what measures are preferred (constructions of resilience)? Table 11.1 gives an overview of the shared knowledge constructions of vulnerability.

Cluster analysis was able to reveal four groups among the shared vulnerability constructions, namely:

- *Glocally convinced*: These are characterised by a strong belief in the occurrence and human cause of climate change. In addition, they view climate change as a major local and global problem.
- *Globally convinced*: These also believe in a currently occurring, human-made climate change that poses a major worldwide problem. Local problems are rarely considered.

Table 11.1 Summary of knowledge constructions of vulnerability

Climate-change beliefs/Constructions of vulnerability	*Frequency*	*Climate change is occurring*	*Climate change is human-made*	*Global problem*	*Local problem*
Glocally convinced	45%	yes	yes	yes	yes
Globally convinced	18%	yes	yes	yes	unsure
Anthropogenic sceptics	15%	yes	unsure	yes	unsure
Climate sceptics	15%	unsure	unsure	unsure	unsure
No response	7%	–	–	–	–

- *Anthropogenic sceptics*: These are convinced that climate change is occurring and poses a global problem. However, they doubt human influence and hardly see any major problems in their own local setting.
- *Climate sceptics*: This group is characterised by high uncertainty. They are neither convinced that climate change is taking place, nor that it is human-made. In addition, they are not able to recognise any great problem for the world or in their local setting.

The convinced groups formed the majority, but even so one third of actors shared a sceptical position. Thus one can hardly speak of uniformity in the shared climate-change knowledge of vulnerability along European coastal areas.

Furthermore, the study observed four climate-mitigation cultures. These characterise themselves around differences in their preferences for strategies concerning ecological modernisation, sufficiency, and ecological and technological countersinking (Table 11.2).

- *Universalists and classic mitigators*: Two highly motivated groups emerged who see a high necessity for all partial strategies, albeit with one difference: classic climate mitigators are sceptical about technological sink methods such as CCS technology.
- *Moderate mitigators*: In addition, there was a group who saw some necessity but each measure, but who overall were only slightly motivated.
- *Minimalists:* Finally, a *minimalistically* oriented group could be observed.

All of the climate-mitigation cultures share moderate positions, at least in regard to the need for ecological modernisation and sufficiency practices in spatial development. While climate mitigation does not manifest itself enthusiastically among the survey's participants, there is a greater consensus with regard to ecological modernisation, efficiency, and consistency strategies, as well as sufficiency practices.

Finally, I was able to discern different adaptation cultures to possible challenges posed by sea and inland flooding, heat, and storms (Table 11.3). A significant difference emerged between the water-related measures, depending on whether the decisions dealt with sea or inland flooding:

Table 11.2 Resilience constructions I: summary of climate-mitigation cultures

Climate mitigation		*Ecological modernisation*	*Sufficiency*	*Ecological sinks*	*Technological sinks*
Universalists	10%	pro	pro	pro	pro
Classic mitigators	23%	pro	pro	pro	against
Moderate mitigators	27%	moderate	moderate	moderate	moderate
Minimalists	11%	moderate	moderate	rather against	rather against
Unfamiliar	19%	–	–	–	–
No response	9%	–	–	–	–

Table 11.3 Resilience constructions II: summary of climate-adaptation cultures

Climate adaptation against …				
… sea flooding *Measures are familiar and generally necessary**	78%			
	of which:	*Protection*	*Retreat*	*Integration*
Protectionists	33%	pro	against	against
Integrators	30%	pro	rather pro	pro
Coastline retreaters	9%	pro	pro	rather against
Moderates	9%	moderate	moderate	moderate
Minimalists	14%	rather against	rather against	rather against
No response	4%	–	–	–
… inland flooding *Measures are familiar and generally necessary**	72%			
	of which:	*Protection*	*Retreat*	*Integration*
Universalists	39%	pro	pro	pro
Renaturators	36%	moderate	pro	rather against
Protectionists	10%	pro	against	rather against
Minimalists	9%	against	rather against	against
No response	6%	–	–	–

... heat				
*Measures are familiar and generally necessary**	51%			
	of which:	*Building optimisation*	*Retreat*	*Open-area design*
Universalists	43%	pro	pro	pro
Building and open-area optimisers	32%	pro	against	rather pro
Building optimisers	23%	rather pro	against	against
No response	2%	–	–	–
... storms				
*Measures are familiar and generally necessary**	65%			
	of which:	*Building optimisation*	*Shelter*	
Building and shelter protectionists	39%	pro	rather pro	
Building optimisers	21%	pro	against	
Minimalists	37%	rather against	against	
No response	3%	–	–	–

Notes

* In establishing climate adaptation cultures, only those actors were considered who were familiar with measures and who in general see the need for action against possible concrete challenges (sea flooding, inland flooding, heat, storms). In this way, the disturbance effect of perceived differences in local landscape or physical exposure should be minimised. Percentages add up to ± 100% because of rounding.

- While the largest group on the coast is formed of *protectionists* (actors likely to combat possible threats through the building of dykes and other technological constructions alone), only a minority of pure protectionists emerged in the field of inland flood control.
- A contrasting picture appears for actors advocating retreat strategies. While inland the *renaturators* formed the second-largest group, *coastline retreaters* composed only a small group on the coast.
- In both fields there were actors who tended towards almost all conceivable registers. These (on the coast: *integrators*, inland: *Universalists*) want to create more room for water and, where necessary, also build new dykes. For coastal protection these constitute the second-largest group being nearly a third of the interviewees. Indeed, inland, they were in fact the largest group (39%).

(Re-)forming settlement structures was a central issue for heat- and storm-related adaptation. Measures against heat in general were rejected by half of the actors, while a greater necessity was observed for storm-related measures. In contrast to the water-related practices, measures against storms and heat were considered less urgent.

- Both for heat and storm measures, there are actors who wish to concentrate on technological building alterations (*building optimisers*). These comprise a minority, however.
- Among supporters of heat measures is a group of *universalists*, who want to draw on all registers from building optimisation, and surface-greening to the removal of buildings to create fresh- and cold-air corridors. This represents the largest group, although at the same time it must be noted that these actors stand in opposition to the majority, who reject measures against heat.
- Among supporters of storm measures is a group of actors who take storm shelters into consideration alongside building optimisation: *building and shelter protectionists*. These are the most commonly represented, but they are closely followed by *minimalists*.

In the first analytical stage, the revealing of climate-culture orientations according to a relational understanding of space allowed for an analytical distinction between culture and locality. This approach takes recent methodological debates into account, in which the danger has been discussed that new social or cultural formations could be overlooked by premature adherence to traditional units of comparison. If initially cultures are analysed in isolation, then one can avoid the phenomenon of 'Europe as a blind spot in sociology' ('Europablindheit der Soziologie', Beck 2005: 3).

Subsequently the way is paved for analysis of conceivable factors to explain cultural differences. Consequently, the second central question of this study was: *How can differences in climate culture be explained, and what are the factors that differentiate between climate cultures?* An integrated explanatory model was developed that allows for the identification of the culture-differentiating factors.

Next comes the question whether *interactions with other forms than climate knowledge* are observable. As already mentioned, this connects the present study to the environmental-sociological discourse on the explanation of different forms of *environmental* and *climate change behaviour*. Values, beliefs, and identities were debated as explanatory factors. Many studies have explored only one or other of these, and then only occasionally synthesised them into integrated explanatory models (such as values and beliefs in Stern 2000; Stern *et al.* 1999; Stern and Dietz 1994). Hence I have used these in an integrated explanatory approach: the *VBI model* (*V*alues, *B*eliefs, and *I*dentities). Table 11.4 summarises those variables that contribute at least once to the explanation of differences.

The findings show that climate-culture orientations can often be explained through the forms of background knowledge taken into account here. Namely, *ecological beliefs* can contribute to the explanation of climate-culture differences, lying between the poles of Human Exceptionalism vs. the New Ecological Paradigm. Consequently, there are at least 'two cultures' of background knowledge having an impact.

The influence of both ecological worldviews has, for example, relevance in regard to coastal protection, when the 'struggle against water' by means of technological protection is ended in favour of integration strategies, which are connected to attempts to bring humans into harmony with water/nature. For inland-flood protection, the New Ecological Paradigm explains, in contrast to the Human Exceptionalism Paradigm, why the protection paradigm is discontinued in favour of renaturation measures. In the design of building structures in towns and communities against heat, new ecological beliefs explain preferences for renaturation in favour of newer, 'greener' open spaces or cold-air corridors.[1]

For resilience constructions, the *openness to change* of an actor in virtually all cases corresponded to shared preferences. This study shows how *altruistic and self-orientation values* influence climate-change-related behaviour. Section 5.2.1 discussed how in earlier Anglo-American environmental sociology, values related to environmental protection were initially seen as congruent with altruistic ideals. The distinction was made between altruism in favour of all species ('*land ethic*') and in favour of other humans ('*golden rule*'). With the arrival of the debates concerning ecological modernisation came a discussion about self-orientation values as a motivating element for environmental protection. It often remained unclear what significance these constructions had for problem perceptions and climate-mitigation and adaptation preferences.

In this study a much higher significance was found for *land ethic* as opposed to the *golden rule* for differentiating climate cultures. *Universal-altruism orientation* delivers significant contributions to the explanation of almost all climate-culture differences, with the exception of sea flooding measures. The tendency for higher *universal-altruism* value orientations to result in more ecologically oriented measure preferences could often be observed. *Human-altruism orientation* could also explain some differences, especially when particular measures could

Table 11.4 Explanatory factors for climate-culture differences

	Vulnerability constructions	*Climate mitigation*	*Climate adaptation*			
			Sea flooding	*Inland flooding*	*Heat*	*Storms*
Values						
Self-transcendence I: Universalism	+++	+++		++	+++	+
Self-transcendence II: Human altruism	+			+		+
Self-enhancement	+	(+)	+	(+)		
Openness to change		(+)	++	++	+	+ (–)
Preservation I: Security		++		+ (–)		
Preservation II: Tradition/Conformity	+++	+	+++			
Beliefs						
New Ecological Paradigm (NEP)	+++	+++	+	++ (–)	+	(+)
Identification						
Identification with reference area			++			
Field affiliation	++	+	+++	+++	+	++
National affiliation	+++	+++	+++	+++		+++

Source: own calculations.

Notes

Effect on the level of significance: +++$p<0.001$; ++$p<0.01$; +$p<0.1$. Values in brackets (+++) indicate context effect or moderation: effect first appears or disappears in logistic regression under the control of field affiliation or national affiliation.

come at the heavy expense of other humans. In the case of inland water protection, for example, human-altruism was able to explain why renaturation was focused upon by some groups, in contrast to those who instead promote dykes for the protection of people. It could also explain why actors advocated more comprehensive safeguards against storms.

From an economic point of view, CCS technology can be useful if, for example, the costs of currently more expensive forms of alternative energy production can be saved. It appears that *self-interest orientations* can explain preferences for technological sinks. For coastal protection too, these values can explain differences between groups, differentiating protectionists from integrationists. The current data do not allow me to determine whether integration is here connected to a higher quality of life by actors, or whether some other reason exists.

Forms of *conservatism* also contribute to climate-culture differences. *Conservative orientations* (here: tradition/conformity) increase the probability for climate scepticism. In addition, they explain very well the orientation of actors who endeavour to continue only with traditional technological-protectionist measures against floods.

Surprisingly, constructions were also found that barely contributed to the differentiation of the groups. One would have expected *security orientation* to regularly explain variances in adaptation preferences, as these would create a general level of security against natural hazards. However, it is clear that to date these orientations have hardly played a role. Similar results can also be found in other studies, such as those of Schmitt (2015: 301). Future studies should investigate these phenomena more closely.

Place identification, regularly discussed in IPCC reports, also had hardly any explanatory significance, although, as postulated by the IPCC, it did show itself to be of relevance for protectionist preferences in coastal protection. It appeared to give no further explanatory contribution. It could be possible that this finding relates to the actors' positions as experts, as it is completely conceivable that it makes a difference whether an actor identifies with the region they have professional responsibility for, or with their place of residence. Furthermore, various possible effects are named in the literature (e.g. the halo effect). There is an opportunity to investigate the possible causal framework for this in future studies.

Finally, the question arises, *to what extent can social context factors explain differences*, and in particular, *do cultural formations conform to traditional regional boundaries or are alternative formations appearing?* Here I have focused particularly on national affiliations and field affiliations.

In the sociological debate around cultural globalisation, different positions regarding the present constellation of cultures have been represented. Exponents of homogenisation or convergence theory have argued with recourse to neo-institutionalism that transnational and global cultures would emerge, due to the global formation of the problem of (anthropogenic) climate change, as well as through the creation of joint institutions and coping strategies.

In contrast, other authors have proposed differentiation theories which take the approach that primarily local (climate) cultures would continue to exist (see Chapter 1 and Sections 2.2 and 5.3).

In Chapter 6 of this study, the development of political and administrative institutional frameworks for the research area was considered more closely. With the UN Framework Convention on Climate Change (UNFCCC), important global institutions for the handling of climate change came into being. Likewise in the EU, various institutionalisation processes have been detectable since the implementation of the first European Climate Change Programme (ECCP). These currently assist in enabling the realisation of local adaptation measures, for example through the 20–20–20 policy goal or through fiscal endeavours. At the same time it has become apparent that, beyond these common efforts, national differences persist, such as in the time taken for implementation of climate-mitigation or adaptation strategies, the prioritisation of different fields of action, and the discussion of further steps. On the other hand, I discussed that observed frictions caused by the implementation of measures is not inherently determined by national boundaries. Thus, for example, countries such as Germany or Denmark have revealed similar patterns in the creation of climate-mitigation measures since the 1990s. It was also seen that in all countries, actors from other societal fields such as the economy/business or civil society, were repeatedly involved in the implementation of measures.

Table 11.5 summarises all the cultural formations and sheds light on whether there is empirical evidence for significant differentiation between climate-culture orientations according to national affiliation or field affiliation. These are marked (*sig.*). Empty cells show where no significant differences were apparent, and also to what extent similarities exist in each case in respect to respondents from Germany and the field of administration.

To summarise the key findings. First, national affiliation differentiates climate-culture orientations somewhat more frequently and with higher levels of significance than field affiliation. Frictions nevertheless also show themselves between the societal fields. Most differences appear between actors from NGOs and the other groups, since vulnerability constructions for NGOs are more highly sensitised regarding human-made climate change, and their resilience constructions are frequently more open to ecologically oriented measures than those of other actors. These differences can often be explained by the background knowledge under consideration (in particular the greater dissemination of the ecocentric NEP worldview and universal-altruism values), by reduced traditionalist orientation, and by greater openness to change.

By contrast, climate-culture boundaries appeared less regularly between actors from different nations. Between Germans and Poles there was a greater difference in knowledge concerning climate-change vulnerability constructions, as well as climate-mitigation cultures, and action preferences against floods from inland water. Taken together, these tended to be 'greener' in Germany than in Poland. Greater similarities existed in the vulnerability constructions of Germany, the Netherlands, and Denmark, where actors appeared much more

Table 11.5 Separability of climate-culture orientations by field affiliation and national background

	Administration *Significant deviation from …*			*DE* *Significant deviation from …*		
	Politics	*Business*	*NGO*	*NL*	*DK*	*PL*
Vulnerability constructions						
(*Reference group*: Glocally convinced)						
Globally convinced			sig. [sig.]	[sig.]		
Anthropogenic sceptics			sig. [ ]			sig.
Climate sceptics			sig. [ ]			sig. [sig.]
Climate mitigation						
(*Reference group*: Minimalists)						
Universalists				sig. [sig.]	sig.	sig.
Classics			sig. [ ]			
Moderates				sig. [sig.]		[sig.]
Sea						
(*Reference group*: Protectionists)						
Integrators				sig.		
Coastline retreaters		sig.	sig. [ ]	sig. [ ]		
Moderates		sig.				
Minimalists	sig.				sig. [sig.]	
Inland flooding						
(*Reference group*: Renaturators)						
Universalists				sig. [sig.]	[sig.]	sig.
Protectionists		sig. [ ]	sig.	sig.	[sig.]	sig.
Minimalists					sig.	

continued

Table 11.5 Continued

	Administration *Significant deviation from …*			*DE* *Significant deviation from …*		
	Politics	*Business*	*NGO*	*NL*	*DK*	*PL*
Heat						
(*Reference group*: Building optimisers)						
Universalists			sig. [ ]			
Building and open-area designers						
Storms						
(*Reference group*: Minimalists)						
Building and safe-haven protectionists	sig.				sig.	sig.
Building optimisers	sig. [sig.]					

Notes

Values in brackets [sig.] arise when variations in significance in the sequential multi-variable analysis between model 4 (countries + fields) and model 6 (countries + fields + background knowledge) appear. Empty brackets [ ] indicate that significance completely disappears under the control of background knowledge.

sensitised to problems. The bulk of these observations can in part be explained by the lower dissemination of universal-altruism and NEP orientations in Poland.

In the handling of sea flooding or heat, however, *zones of cultural friction* constellate completely differently. In their adaptation cultures related to threats from the sea, Germans and Poles were, for example, comparatively similar. Both place great weight on protectionist measures. Here the Germans and Poles differentiated themselves more strongly from the Dutch, who placed a much higher emphasis on integration and retreat strategies. In addition, there were differences with the Danes, who exhibited a greater proportion of actors with minimalistically oriented coast protection preferences. This could, moreover, be explained by traditionalist and conformity orientations being more marked. With regard to the Netherlands it was above all clear that further, unobserved explanatory factors should be brought into the picture. These are possibly deducible from the country's history or physical geography.

In comparison, taking together the handling of heat across the survey area, it is clear that this is not yet of great importance at all around European coastal areas, and that hardly any significant differences in the climate cultures were observable between German actors and those from the other countries considered.

It was apparent in a few places that effect sizes also varied sometimes according to social context. Thus, for example, background knowledge first demonstrated itself to have a significant influence on climate-culture affiliation when actors' national affiliation was controlled for. This was the case, for instance, in regard to measures against the consequences of storms, where NEP orientation could only explain affiliations with respect to building and safe-haven protectionism in a few countries. The explanatory force of the self-enhancement orientation for affiliation with inland water protectionism was not applicable to all nations either.

11.2 Consequences and prospects for future research

The proposed idea of culture 'as relational space' allows a broader and more differentiated view on cultural constellations, both worldwide and locally. Through the example of climate cultures I have been able to demonstrate how this can be achieved. The question arising from cultural globalisation discourses of whether cultures become globally homogenous, regionally differentiated, or appear in alternative social constellations can now be answered more precisely. The development of this theoretical approach of observing cultures as relational-space formations of socially shared knowledge enables their empirical description and the unveiling of their differentiating factors.

On the basis of the empirical results, it can be ascertained that both of the discourse's extreme positions can be justified. Thus, on the one hand, within all European coastal areas under observation there are actors who share similar forms of knowledge about climate change; consequently, empirical evidence was provided that climate cultures span right across the selected coastal regions of Europe. This finding is in agreement with the proponents of cultural *homogenisation or deterritorialisation theories*. On the other hand, among the factors that separate different *zones of cultural friction*, local factors such as national affiliation also contribute. Thus the proponents of differentiation theories are also right. Furthermore, *there is also evidence for alternative cultural formations*, as can be seen in the constant difference between actors from NGOs in comparison with other groups such as administration, politics, and the economy.

It must finally be stressed that *zones of cultural friction* appear according to the *aspectual or thematic focus of the observer* in various spatial and social formations. With regard to shared knowledge concerning climate-related problems and climate mitigation, the friction emerged between Polish actors and those from the other three nations. In coastal protection, however, climate-culture boundaries ran between Germany, the Netherlands, and Denmark, while Polish and German actors were markedly more similar. As the *topos* for investigation is selected by the scientific observer, one should regularly reflect that the chosen objects of investigation only ever represent a sample of the complex web of social realities, and this selection is of great relevance with regard to which cultural frictions can ultimately be observed. This point has been willingly overlooked in the sociological debate around cultural globalisation and may be an important reason, alongside the implicit differing conceptions of space, as to why authors have come to different conclusions about the constellations of cultural spaces worldwide.

It must be emphasised that I would not go so far as to say that 'there is no such thing as a boundary' (Thrift 2006). Here, cultural boundaries appear to be 'fluid' with regard to *different aspects of knowledge* that might be shared. Nevertheless, as we have seen in this study, for any *specific aspect*, such as climate mitigation, cultural boundaries emerge and these frictions have real consequences for action.

This study also contributes to the environmental-sociological *climate-behaviour debate*. The VBI model offers an explanatory approach that provides future potential for understanding, or even predicting, different climate-related behaviours. Many of the different forms of shared knowledge discussed in environment-related research can explain why various forms of *climate change behaviour* appear. Even more exciting here is the observation that these exhibit not only climate-problem constructions or protection preferences; they can also explain preferences for adaptation measures, whose original justifications are in fact less aligned with environmental protection than with adaptation to changed environmental conditions.

Finally, I shall consider the remaining questions as well as the limits of the present methodology. First, it is important to highlight the discrepancy – regularly discussed in environmental sociology – between environment- and climate-related knowledge and practice (Diekmann and Preisendörfer 2001: 114ff.). As discussed in Chapter 2, the empirical section of this study focuses on the intentions of actions. Thus future research could illuminate in more detail the relationship between these intentions and the outwardly visible actions.

In the application of the present methodology there also remain a few unanswered questions. The approach of logistic regression, for example, has facilitated comprehensive statements about the differentiating factors of climate cultures. In the present study's reference group formations, only a sample of the possible climate-culture boundaries have been closely observed. In addition, the procedure reached its methodological boundaries when greater differences in group size were visible. Contextual effects could only be explored to a limited extent, as in each case only four different groups of national and field affiliation were available. There is still more potential here for future study: multi-level analyses, for example, would provide more precise results.

The analytical limits of specific objects of this study should be mentioned. This relates above all to the process of the communicative genesis of (climate) knowledge. It was noticeable, for example, that interaction about climate change was not only significantly higher among NGOs and significantly lower among Polish interviewees, but also that these trends and the use of information media corresponded with the awareness of climate change as a problem. (Here one might infer, for example, the hypothesis that the more frequent the communication about climate change, the higher the problem awareness.) In Chapter 9 it was discussed how this study has attempted to investigate climate communication descriptively. Deeper questions about the precise causal interactions between (climate) communication and knowledge would have required still more complex theoretical and empirical elaboration at the intersection of the sociology of knowledge, social psychology, and media and communications studies, which would have far exceeded the capacity of this study. This causal relationship remains open for discussion as a classic 'chicken and egg' problem: does communication about climate change first increase before a high awareness of the problem emerges, or does it only increase once this awareness has already developed? This question offers scope for further study.

The relevance of other contextual factors could also be investigated further. The present study takes no account of economic background or other forms of material safety; topics present since the 1970s in postmaterialistic discourse. To be able to determine the scope of the existing explanatory models, it would also make sense to expand studies of the present kind beyond the borders of Europe. Thus, for example, the Global South could be included in order to determine to what extent the explanatory models detected in the 'Global North' also exist there.

Considered from this perspective, sociological climate research still offers a great deal of scope for comprehensive study.

Note

1 This is reminiscent of the transformation from the (container-space oriented) structure of towns in the Middle Ages, whose walls demarcated a clear separation between 'inside' and 'outside', and the transition to modern cities, where this binary opposition is increasingly fuzzy. Here too, the distinction between 'human' settlement and supposedly 'external nature' is being increasingly erased.

References

Beck, Ulrich (2005) 'Europäisierung. Soziologie für das 21. Jahrhundert'. *Aus Politik und Zeitgeschichte* (APuZ) 34–35: 3–11.

Crate, Susan A. (2011) 'Climate and Culture: Anthropology in the Era of Contemporary Climate Change'. *The Annual Review of Anthropology 2011*. 40: 175–194. DOI: 0.1146/annurev.anthro.012809.104925 (accessed 29 October 2013).

Diekmann, Andreas and Preisendörfer, Peter (2001) *Umweltsoziologie. Eine Einführung*. Reinbek bei Hamburg: Rowohlt.

Dunlap, Riley E. and Van Liere, Kent D. (1977) 'Land Ethic or Golden Rule: Comment on "Land Ethic Realized" by Thomas A. Heberlein'. *Journal of Social Issues* 33 (3): 200–2007.

Keller, Reiner, Knoblauch, Hubert, and Reichertz, Jo (ed.) (2013) *Kommunikativer Konstruktivismus. Theoretische und empirische Arbeiten zu einem neuen wissenssoziologischen Ansatz*. Wiesbaden: Springer VS.

Scannell, Leila and Gifford, Robert (2013) 'Personally Relevant Climate Change: The Role of Place Attachment and Local Versus Global Message Framing in Engagement'. In: *Environment and Behavior* 45 (1): 60–85.

Schmitt, Lea (2015) *Klima, Raum und Zeit im Wandel. Eine ethnographische Untersuchung von Adaptationen und Konflikten auf der westfriesischen Insel Ameland*. München: Oekom.

Stern, Paul (2000) 'Toward a Coherent Theory of Environmentally Significant Behaviour'. *Journal of Social Issues* 56 (3): 407–424.

Stern, Paul C. and Dietz, Thomas (1994) 'The Value Basis of Environmental Concern'. *Journal of Social Issues* 50 (3): 65–84.

Stern, Paul, Dietz, Thomas, Abel, Troy, Guagnano, Gregory A., and Kalof, Linda (1999) 'A Value-Belief-Norm Theory of Support for Social Movements: The Case of Environmentalism'. *Human Ecology Review* 6 (2): 81–97.

Thrift, Nigel (2006) 'Space'. *Theory, Culture & Society* 23 (2–3): 139–155.

Index

Page numbers in **bold** denote tables, those in *italics* denote figures.

www.ingramcontent.com/pod-product-compliance
Ingram Content Group UK Ltd.
Pitfield, Milton Keynes, MK11 3LW, UK
UKHW022012180926
475701UK00003B/324
* 9 7 8 0 3 6 7 5 8 2 8 1 4 *